MAPPING OF THE SOIL

Mapping of the Soil

Jean-Paul Legros
Institute National de la Recherche Agronomique (INRA)
Montpellier, France

Translated from French by
V.A.K. Sarma

Science Publishers
Enfield (NH)   Jersey   Plymouth

SCIENCE PUBLISHERS
An Imprint of Edenbridge Ltd., British Channel Islands.
Post Office Box 699
Enfield, New Hampshire 03748
United States of America

Internet site: *http://www.scipub.net*

sales@scipub.net (marketing department)
editor@scipub.net (editorial department)
info@scipub.net (for all other enquiries)

© 2006 Copyright reserved

ISBN 1-57808-363-X

```
        Library of Congress Cataloging-in-Publication Data
Legros, Jean-Paul.
  [Cartographies des sols. English]
  Mapping of the soil/Jean-Paul Legros; translated from the
    French by V.A.K. Sarma.
       p. cm.
  Rev. translation of: Cartographies des sols. 1996.
  Includes bibliographical references and index.
  ISBN 1-57808-363-X
  1. Soil mapping. I. Title.

S592.147.L44  2005
631.47--dc22
                                                        2005044115
```

Published by arrangement with Presses polytechniques et universitaires romandes, Lausanne, Switzerland.

Translation of: **Cartographies des sols,** *De l'analyse spatiale à la gestion des territoires,* Presses polytechniques et universitaires romandes, Lausanne, Switzerland, 1996. **The author modified and updated the text for the English edition in 2004.**

French edition: © Presses polytechniques et universitaires romandes, Lausanne, 1996.
ISBN 2-88074-298-6

Published by Science Publishers, NH, USA
An Imprint of Edenbridge Ltd.
Printed in India.

FOREWORD TO THE FRENCH EDITION

> *I have read! What did I read? Oh, the great
> big book! The Bible? No! The Earth!*
> [Victor Hugo]

Soil mapping is the product of the work of those who read the land. The British agronomist Arthur Young visited France from 1787 to 1789 and published a map of the agricultural domains of the country containing an evaluation of soils and vegetation. Around 1856, the French scientist Eugène Risler drew a preliminary soil map of Geneva canton in Switzerland. In 1900, at the World Fair in Paris, Dokuchaev of Russia and his students presented the first soil map of western Russia. Thus there have long been in all of Europe a tradition and examples of cartographic inventories, but three wars have cost much time.

Fortunately, Albert Demolon, Inspector-General of Agronomic Research of France had trained remarkable students since 1934. Drouineau, Philippe Duchaufour and Georges Aubert in turn produced students. Some of these: Marcel Jamagne, Alain Ruellan, Emmanuel Servat and myself have in turn taught students, and our students have in their turn been scientifically very productive.

Among these, Jean-Paul Legros has the courage and the credit of describing the tasks accomplished, analysing the principles that guide them and detailing the methods followed and the techniques used. With the vast field experience that years of cartographic surveys have given him, the many achievements of his colleagues in the Soil Science Laboratory at Montpellier and his contacts mostly in French-speaking countries and in particular those attached to the team in Lausanne, he has been able to direct this book very well.

This treatise on soil cartography shows, among other things, particularly rare qualities: it gives with the precision of the experienced practitioner the most important details of the pedologist's work. It deals with the most modern methods and techniques, use of computers and application of statistics. It covers all that is needed for objective interpretation of the results obtained and for finding out the most economical itinerary to attain that purpose.

The vast knowledge of Jean-Paul Legros, who knows to explore other domains of the mind and to draw thence sources to refresh his thinking, has enabled him to conceive, begin and complete this book.

It is said that in the great Canadian North explorers used to find in a cabin the corpse of their predecessor who had gone in advance on the trail holding in the hand a card on which his pencil had written his final message: 'Homage to those who seek adventure'. One of those who had mapped many square kilometres in the nineteen fifties and nineteen sixties ('I have known this life and I forever lament it!') may be permitted here to salute a successor, to wish great success to a brother of the land, to thank a friend for having progressed in intellectual adventure, going further, always further...

Paris
6 January 1996

Professor Jean Boulaine
Member, French Academy of Agriculture

CONTENTS

Foreword to the French Edition   v

General Introduction   ix

1. **Definitions, Objectives and Concepts**   1
   1.1 Definition of the Soil Map and its Components   1
   1.2 Objectives and Applications of Mapping   8
   1.3 The Russian School of Soil Science   17
   1.4 The Issues   23

2. **Various Kinds of Approaches**   36
   2.1 Various Mapping Procedures   36
   2.2 Graphical Treatment of Pedological Objects   57
   2.3 Types of Legend   63
   2.4 Scales   67

3. **Preparation of a Field Work Plan**   71
   3.1 Conditions of Contract   71
   3.2 Estimation of Times and Costs   75
   3.3 Ordering, Preparing and Utilizing Aerial Photographs   78
   3.4 Photogrammetric Documents   85
   3.5 Documents and Equipment   88
   3.6 Problems of Practical Organization   95

4. **Description of Soils in the Field**   100
   4.1 Generalities   100
   4.2 Description of Certain Principal Properties   105
   4.3 Design of a Description System for Computer Processing   124
   4.4 Technique of Describing Soils   130
   4.5 Quality Checks Related to Description of Soils   142
   4.6 Organization of a Profile-Study Programme   147

5. **Map Preparation and Quality Checks**   160
   5.1 Methodological Basis of Zoning   160
   5.2 Preparation of the Map   172
   5.3 Quality Checks Related to Graphic Information   181

6.	**Computer Processing of Data**	**200**
	6.1 Generalities	200
	6.2 Treatment of Graphic Data	201
	6.3 Processing Non-graphic Data Pertaining to the Map	214
	6.4 Processing of Profile Data in France	222
	6.5 Some International Efforts	229
7.	**Modelling and Automation**	**235**
	7.1 Soil Prediction from Aerial Photographs	236
	7.2 Trial of Segmentation of Space	241
	7.3 Generalization of Local Observations	244
	7.4 Measurement of the Similarity between Soils	258
	7.5 Simplification of Boundaries	268
8.	**Principles of Thematic Mapping**	**273**
	8.1 Procedures of Thematic Mapping	274
	8.2 Diffuse Models and Complete Aggregation	281
	8.3 Diffuse Models and Systems of Downgrading	288
	8.4 Diffuse Models and Partial Aggregation	290
	8.5 Diffuse Models and Comparative Reasoning	296
	8.6 Sequence of Spatial Models	297
	8.7 Critical Analysis	304
9.	**Soil Mapping and Multidisciplinary Approach**	**310**
	9.1 Soil Mapping and Agronomy	310
	9.2 Soil Mapping and Civil Engineering	315
	9.3 Soil Mapping and Ecology	320
	9.4 Soil Maps and Climatic Data	325
10.	**Soil Maps of the World and French-speaking Countries**	**332**
	10.1 International Mapping Efforts	332
	10.2 Soil Mapping in Canada	339
	10.3 Soil Mapping in USA	342
	10.4 Soil Mapping in Belgium	343
	10.5 Soil Mapping in Great Britain	347
	10.6 Soil Mapping in France	348
	10.7 Soil Mapping in Switzerland	356
	10.8 Soil Mapping in Africa	357
	10.9 Soil Mapping in Some Other Countries	359

Appendices — 363
A1 Planning a Field Programme — 363
A2 Soils and Landscapes Zoning (Flaine Region, Haute-Savoie) — 368
A3 Planning Land Use — 373

Bibliography — 377

Index — 407

GENERAL INTRODUCTION

Mapping is the art and method of making maps—scientific or artistic documents pertaining to all flat representations, sections or three-dimensional reconstructions of the Earth or other celestial bodies at all scales (adapted from Meynen, 1973).

'... He discovered the very special exhilaration of the mapper. I advised him to try. The pleasure is quite exquisite. When you report on a map your discoveries of the day, your heart throbs more strongly, a kind of pride seizes you, you are the absolute master of the streets drawn by you, of the places, the banks of the stream, the genius of space is yours.... This passion for cartography has still not left me. The more I age, the more I see in it that fellow feeling of method that is the quality of a living thing, a very exalted science: a real physiology of places.' (Found by P. Bonfils in Orsena, 1988, *L'exposition Coloniale*, Éditions du Seuil, Paris, 555 pp.).

1. Aims and Structure of the Book

This book presents the methods of cartography pertaining to the soil mantle. There are many methods of investigation and this will come out clearly in the course of the text; it is not necessary here to dwell on it further. There also exist very different soils that should be demarcated for bringing out these differences.

Cartographic work has three principal objectives; they will be covered in ten chapters:
- The first objective is to **understand** the spatial organization of soils in the natural environment, thus to take part in their study by means of an approach that is at the same time specific and open to other disciplines.
- The second objective is to **summarize** in map form the experience gained. The map is constructed by following step by step the experimental procedure of Claude Bernard: observations, hypotheses, laws, validation and then application. We shall return to it later. But let it be said rightaway that to make a map it is necessary to have in mind a model of the spatial organization of soils (Soil Landscape Hierarchical Model). Just as a good student can do an exercise after doing the

course, a good naturalist who has well understood the structure of the milieu can translate it into a map.
- Lastly, this action aimed at understanding and modelling nature is socially and economically justified only if the corresponding reports can be *applied*. Special effort has been taken in this book to identify what use there is of soil cartography. In the final third of the text will be found: a chapter explaining the principles governing the concept of the applications, a chapter showing how soil cartography can be associated with other disciplines to broaden the range of its possible uses and, finally, a chapter synthesizing the principal achievements in mapping in some countries of the world.

It would be better to read the chapters in the order they are presented. However, to facilitate another way of consulting the book, cross-references are frequently made. They indicate the number of the chapter (Chap. x), and a section (§ x.y) or subsection (§ x.y.z) number. An index is also provided at the end of the book.

2. Significance of the Book

Presentation of new or perfected techniques
Computer methods, very useful in cartography, have evolved considerably in a few years. It is useful to recall them. Figure 1 applies more precisely to the French case.

Scientists in 1976 used only impact printers that were commanded to print several letters of the alphabet so that the area was blackened. Areas with greater or less degree of darkness were thus obtained. By 1978, the University of Montpellier had a high-quality pen plotter. But the user had to develop his own drawing program, generally with stepped boundaries. In 1983, the well-equipped French laboratories obtained a marvellous printer that directly printed small black rectangles. This had followed the American scientists at Harvard, though after a gap of several years (McHaffie, 2000). In 1989, perhaps a year or two earlier, Geographic Information Systems (GIS) that originated in America were installed in several cartographic centres. Quality was not necessarily excellent, but the boundaries of the soil bodies were at last drawn with curved lines. Colour was introduced. In 1990, soils data were crossed with other information, in this case topography. The fact that soils have three dimensions was shown. By 1995 it was understood that improvement in quality and range of treatment possibilities would perhaps never cease!

In other countries, development might have been more rapid or slower. Progress everywhere has been considerable over the past 20 years (GWDSM, 2004), and not in information technology alone. Remote sensing is no longer

GENERAL INTRODUCTION

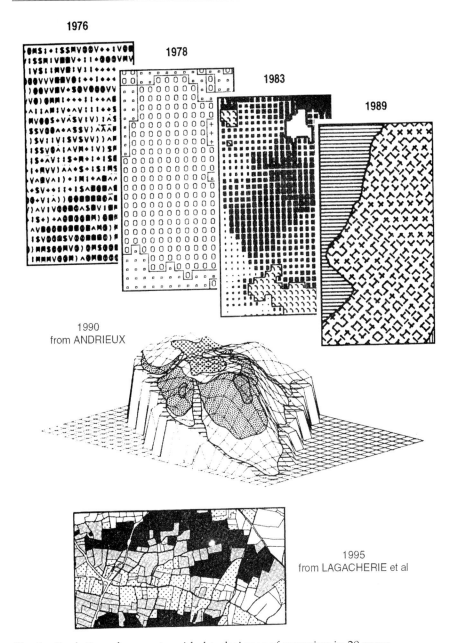

Fig. 1 Evolution of computer-aided techniques of mapping in 20 years.

what it was. Digital Elevation Models (Chap. 8, § 8.6.1) are now commonly available. Geographical Information Systems (GIS) were quickly perfected (Chap. 6, § 6.2). Relational Database Management Systems (RDBMS) are available for all who have stored and managed data (Chap. 6, §§ 6.3-6.5). The Global Positioning System (GPS) has become a common thing (Chap. 3, § 3.5.4). Development has been so much that it will not be possible to present here all the tools and all possibilities. Although cartography, remote sensing and GIS interact (Fischer and Lindenberg, 1989), we are primarily concerned with the first.

Summarizing recent achievements
When things are considered day to day, soil cartography seems somewhat disorganized in Europe and the world. Few countries have completed cartographic coverage of the soils of their territory and few others are still working with dedication in this field. On the other hand, things have greatly changed over the past 15 years. Information technology has everywhere helped in storing and organizing data collected earlier. Making them available to users is the new challenge.

Providing practical tools
Check-lists can be found in this book for calculating the cost of a study (Table 1 in Appendix), for assembling the equipment necessary for a field project (Chap. 3, Tab. 3.3) and for describing soil units (Chap. 5, Tab. 5.1).

3. Readership Targeted

This book is primarily addressed to students of colleges, universities and doctoral courses having soil cartography in their curriculum. For their use, three exercises have been planned in the Appendix, illustrating and rounding off the text. Students of closely related disciplines (ecology, geomorphology, geology, etc.) can find out in what way our procedures converge with or diverge from theirs. Needless to say, professors are not prohibited from reading the book. Their comments would be appreciated in the event of a third edition.

The text certainly is also addressed to the many practising soil scientist-mappers. They accomplish indispensable work the significance of which will continue. This synthesis will show them that they and their maps are remembered, for synthesizing their experiences and at the same time for communicating them to their far or near colleagues, some of whom might be working in connected disciplines. In particular, although remote sensing as a tool does not take part in our field of study, the methods of the specialists in this discipline greatly concern us since they can be applied for quality control of soil maps (Chap. 5, § 5.3).

Lastly, this book is addressed to those who direct the preparation of soil maps. It is evident that understanding of the methods and limits of soil cartography can facilitate discussion with the cartographers who are assigned the work.

4. Directive Principles

First of all we may be permitted not to share the reductionist view that was in fashion in the nineteen-nineties and led to belief in mythical 'changes of scale' to understand the future of our planet based on a few experiments done in the laboratory. All extreme generalization leads to forgetting the new, yet essential, elements at the required scale. Things must not be imagined: knowledge of the land, its protection and its utilization in the service of man should be approached at the appropriate scale, which involves study of the spatial dimensions of soils. Our first objective then is to *bring together the methods of characterization* necessary for the purpose.

To a certain degree, that is, within our means, we should also *restore credibility in soil mapping*. Actually the problem exists. The natural sciences have known their hours of glory, then have been neglected. They seemed to yield place to empirical observation and interpretation. Matters were bad. Actually one could argue strictly about data that are not very reliable. Statistics as a discipline is a good proof of this. Multisearch analysis (Chap. 8, § 8.4) feeds this same ambition. Our second aim then is to present cartographic procedures as being rational, codified and scientific as well, because the quality of the product, the map, can be verified (falsified, according to Karl Popper).

Lastly, we would wish to convince the reader on one point. Besides delicate studies, quite indispensable for scientific progress, it is necessary to learn to look at things globally. *'For more than a century, science has been dominated by the concept of the microscope. We have sought more and more detail about everything, and integration has suffered through neglect. Now with remote sensing and GIS we have, for the first time in history, a rudimentary macroscope'* (Dobson, 1993). Twenty-five years ago such a statement would not have been understood. In our discipline it was only accepted by some leaders, promoters of spatial analysis and by those who had a holistic view of soils. At present the idea is easier to defend. For example, if Geology has made fundamental advances in the preceding decades, it is by learning to consider things globally. Thus earthquakes, volcanism, continental drift, uplift of mountains and palaeomagnetism have been linked to one theory of plate tectonics. Throughout the body of the book, because synthesis is as important as analysis, the reader is invited to place himself at a distance to see things globally. This we shall call *'manipulating the macroscope'* referring to Jerome Dobson and a few others. Besides, did not Lamarck say in 1889 (in *Zoological Philosophy*): *'The right method of getting to understand an object, even in its finest details, is to begin to view it in its entirety.'* In this context, soil cartography is the preferred tool for studying soils as natural bodies of large size.

5. And the Internet?

Growth of the Internet completely changes the significance and content of a written document. By thoroughly searching for a few months, a good part of the material constituting this book will be found on the Web! In such a situation pedagogic objectives change. The printed book is no longer the sole means for acquiring knowledge, but is still a useful tool in which the data provided have been validated, structured and freed of many accessory or useless things. Thus it allows immediate access to the basics and is the best introduction to a discipline. It is also a gateway to the Internet in which it can provide good addresses that allow downloading specialized documents of interest.

However, precise addresses of Web sites pertaining to soil cartography have not been given. Experience has shown that the addresses keep changing. Furthermore, they are not convenient to type on a keyboard and the chances of error are great. We prefer to give in small italic capital letters the keywords that will enable very easy identification of those sites on the Internet using a search engine. For example, for Canada: [search for site: *SOIL MAPS CANADA CANSIS NSDB*] will be given. The last two keywords refer to the acronym of the department responsible for the *National Soil Data Base*.

6. Acknowledgements

Here let me thank all those who have helped me. First of all, those without whom the first French edition of this book would not have appeared:
- Professor Jean-Claude Védy, who gave me the opportunity to give lectures on mapping for years at the École Polytechnique Fédérale de Lausanne (Switzerland) and thus refine my thinking on the subject;
- his collaborators Pauline Dupasquier, Claire Guénat, Catherine Keller and Philippe de Pury; their assistance was efficient in the organization of the courses and practical exercises in cartography;
- the draftsman François Mazzella, the computer virtuoso Pierre Falipou and also the scientists of the team to which I belong; nothing would have been possible without the decisive advances they made in many of the fields mentioned in the book; this applies in particular to Michel Bornand, Jean-Claude Favrot and Jean-Marc Robbez-Masson;
- lastly, the entire management of the Presses Polytechniques et Universitaires Romandes, who work with the authors of their books in a spirit of frank and friendly collaboration.

I express my thanks to Mr Raju Primlani, who gave me the opportunity to prepare a revised and enlarged edition in English of this book.

My sincere gratitude goes to Dr V.A.K. Sarma, who translated with accuracy and reliability all the pages of the text. One often thinks that computer technology keeps people apart by erecting a screen between them. This is quite wrong! Thanks to electronic mail I was able to know and appreciate Dr Sarma. We worked together enthusiastically, although separated by 11,000 kilometres of sea, desert and ocean. This has been a rewarding experience for me and I wish that it happen again with the same fine translator.

To all of them I express my deep, sincere, everlasting gratitude.

ABBREVIATIONS

The following abbreviations are used in the bibliography for journal titles: *EGS* for 'Étude et Gestion des Sols', *PERS* for 'Photogrammetric Engineering and Remote Sensing'; *SdS* for 'Science du Sol' and *SSSAJ* for 'Soil Science Society of America Journal'.

CHAPTER 1

DEFINITIONS, OBJECTIVES AND CONCEPTS

In this first chapter, to begin with, a standard soil map will be defined and its structure and components presented. Then the principal applications of soil mapping will be quickly reviewed so that the utility of the subject is brought out. Thirdly, historical aspects and the development of concepts and methods over time will be overviewed. Later will be considered the principal concepts currently serving as the basis for practical soil mapping.

1.1 DEFINITION OF THE SOIL MAP AND ITS COMPONENTS

1.1.1 Nature of the soil map

A standard soil map is a two-dimensional document presented on paper or on various other kinds of support, and gives a greatly reduced, simplified picture of the spatial organization of soils in the natural environment. The map made available to the viewer is the product of the interaction of four factors (Fig. 1.1).
- First, it is the transposition of a *natural entity* related to the geography of soils. This imposes on the map the essence of its appearance.
- But the map is not a true representation of reality, for otherwise it would have to be as complex as the soil cover and presented at a scale of 1:1! The map is, in fact, a reduced model or, if preferred, a *model* in the strict sense of the term. The reality is simplified according to objectives.

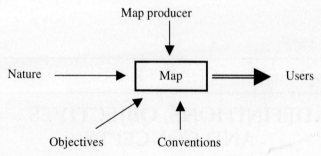

Fig. 1.1 The map is the product of the interaction of four factors.

- The mapper, like a painter, has a style, that is, his own qualities and faults. His drafting style and experience, more or less asserted in the field, lead him to produce documents not quite similar to the same products of his colleagues under the same conditions. Whether one likes it or not, there is a certain *role of subjectivity* linked to the operators.
- In each discipline, and in soil science in particular, there are *conventions* for cartographic representation. For example, the 'clayed' soils are not shown in yellow or red but rather in blue (blue is a cold colour and clayey soils in Europe are often moist). The map is thus part and parcel of a certain scientific, cultural, aesthetic and even emotional context.

The map thus created has *'the impact and power of the picture, and a momentary glimpse of it works as a revelation.... It communicates information, transmits a message, expresses similarities, differences, trends, directions.... It is a tool for work, research, decision-making'* (Meunier and Hardy, 1986).

1.1.2 Definitions

Before proceeding further, the definition of the principal objects represented on a map (Fig. 1.2) in the most classic case must be provided. The others will be examined later.

Elementary soil areal or ***elementary soil delineation*** corresponds, on the map, to an elementary area with closed boundary and defined content represented uniformly by a number or a fixed colour. A map representing such mapping zones is a *chloropleth map* or *categorical map*. But there are maps that do not resort to the concept of mapping areas and show only isovalue lines. These maps are *isolinear maps* or *isopleth maps*. For example, rainfall maps, on which isohyets are drawn, are such maps.

Map unit (MU) is an assemblage of areas of the same nature that appear on the map, for example all the bright red areas. Each MU is described in the

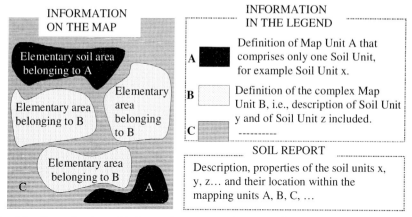

Fig. 1.2 Organization of information on a soil map and associated documents.

map legend. It may happen (but very rarely) that an MU might be represented by one single area.

Soil unit (SU). The map unit can include several soil units whose boundaries are not separated. It will be seen later why one proceeds in this manner (§ 1.4.3). However, when the MU contains only one SU, it is obvious that the boundary of the latter will be the same as that of the former. We shall see in Chapter 6, § 6.3.3 how this is managed on a computer map.

Box. A box is a small rectangle (or square) appearing by the side of the legend to confirm the link between the definitions and the MUs of the map. For each box and, therefore, for each MU the SUs are described. If one SU occurs in several MUs, it is useful to repeat its name and definition as often as needed. This is better than giving the legend a complex structure that makes it difficult to read.

Graphic and semantic information. It is convenient to distinguish on the soil map two kinds of information: *graphic information* (boundaries of soil delineations) and non-graphic information (numbers and qualifiers serving to characterize the soils and thus define the content of mapping zones. Non-graphic information is termed *semantic*, semantics being the study of the meaning of words. But some scientists prefer the term *tabular data* because of the way in which the non-graphic data are managed in the computer.

1.1.3 Constituent elements of the soil map (Brunet, 1987)

The map comprises five principal groups of elements, examined below.

Title and references

To say the title is necessary is to state the obvious. However, when a map is published in several sheets or even when several maps are assembled into an atlas, it often happens that the title and the authors' names are not reported in each section. Also, the map is almost always the product of collaborative work; it is only right to mention this and to display in a prominent place the list of collaborators indicating the sectors in which they have applied their mapping expertise (Fig. 1.3). It is also helpful and sometimes mandatory by law to indicate the source and the owner of the topographic base used to locate the soils in relation to roads, villages, rivers and other land features. On a reduced map it is important to properly locate the map in its geographic boundaries that correspond to the sections of a systematic inventory or represent the boundaries of an isolated study. Lastly, the date of publication should not be forgotten, as it also is an indispensable reference.

Legend

The legend gives the exact meanings of all the symbols appearing on the map, beginning with colours. It thus plays the role of intermediary between the map and its user. This legend should be easy to understand and appropriately presented, that is, structured, coherent and explanatory. We shall return to the question when the legend is to be constructed (Chap. 5, § 5.2.2).

Scale

By definition, scale is the ratio of a distance measured on the map to the same measured in the field, both in the same units. It is useful to memorize the following data that form the cartographer's multiplication table (Table 1.1).

Table 1.1 Relationship of scales, lengths and areas

SCALE (1/x)	1/1000	1/10,000	1/25,000	1/100,000	1/1 million
Length on the ground for 1 cm on the map	10 m	100 m	250 m	1 km	10 km
Area on the ground for 1 cm^2 on the map	1 are	1 ha	6.25 ha	100 ha	10,000 ha

Large scale is the scale with large 1/x, which would mean a small 'x', thus small coefficient of reduction. Unless it is for preparing maps of 'bedsheet' size, the large scale is useful only for small areas. On the contrary, *small scale* generally corresponds to large areas. These concepts are therefore quite contradictory to those that underlie popular expressions such as 'he makes a small-scale beginning' or 'he makes this on a large scale'. When

DEFINITIONS, OBJECTIVES AND CONCEPTS

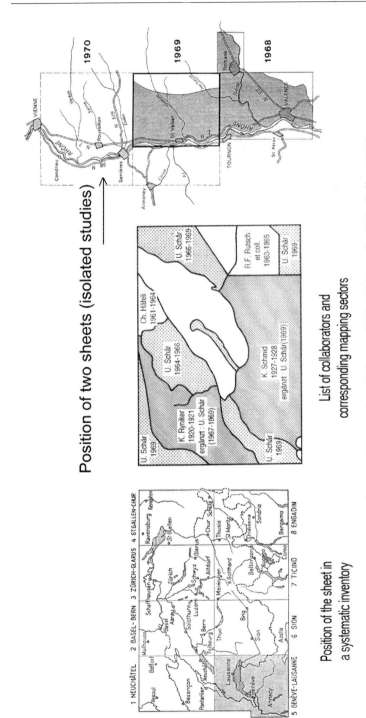

Fig. 1.3 Examples of proper mapping references (Bornand and Guyon, 1969; Schär et al., 1971; Collet, 1955).

facing a person for whom there is danger of ambiguity, it is useful to avoid the term 'scale' and speak of *detailed maps* or *synthesized maps*.

Actually, whatever be the objectives of the mapping and the survey efforts in the field, the map as a paper document is bound to present a certain degree of complexity. It will be stupid to produce a map with thousands of details too small to be legible or, on the contrary, a map uniformly covered with just one colour representing a single kind of soil. Therefore, a reasonable agreement has to be found between the scale of the map and the research effort allowed and degree of detail sought. Finally, each scale characterizes, a little exaggeratedly and by extension, a certain degree of precision. If a map is photographically or digitally enlarged or reduced, the scale (in the sense of precision) does not change. A scale of 1/100,000 remains 1/100,000 and this is fortunate after all is said and done. However, for this it is essential to draw the metric scale (graduated line) by the side of the scale ratio on the map, to enable the reader to visualize under all circumstances what a kilometre corresponds to on the map.

Coordinates and projection system
The data collected in the field are generally reported on a topographic base supplied by a specialized organization: for Switzerland the Federal Department of Topography, for France the National Geographic Institute, etc. In other words, the location of the data is primarily relative. For example, we observe what kind of soil extends up to a particular road. For the past many years, this has seemed adequate considering the objectives of mapping. Also, at the printing stage, the lines drawn on the edges of the map base to indicate the start of the geographic coordinates were often considered insignificant. Many soil maps have been published without these references to location on the globe. This is no longer acceptable today because of the possibilities of data management opened up by computerization of maps. With proper geographic references, we actually can:
- change the map projection system and, for example, adapt soil surveys to a new topographic base;
- match, with adequate precision, information of different kinds initially drawn on many maps of the same region but not necessarily on the same scale or in the same projection;
- easily and automatically join neighbouring maps for the same theme for the whole region.

All these possibilities can be combined. It will thus be necessary to be closely concerned with the coordinates and projection systems. Soil mappers still have to make progress to be at ease in this domain.

Graphic annexures
Mappers often have to provide by the side of the map, on the same sheet, supplementary information for visualizing the spatial position of the soils

or to show their morphology. This enables balancing of the graphic composition and to benefit without additional cost from colours whose use is indispensable for publishing the entire document. These supplementary elements may be many. But there are at least three principal types (Fig. 1.4).

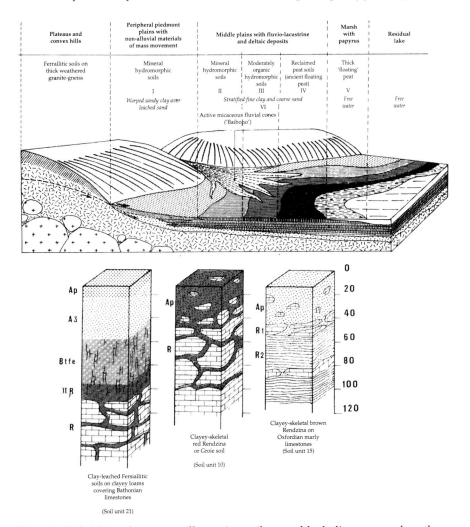

Fig. 1.4 Subsidiary documents illustrating soil maps: block diagrams such as those from Madagascar (Raunet, 1989) and profile drawings (Lagacherie, 1984).

Inset maps. For better understanding of the spatial organization of soils, many maps are accompanied by inset maps (small maps), prepared on a smaller scale, indicating how rocks, climate and other pedogenetic factors are distributed in the region. In particular, the maps on 1/100,000 scale of the Service d'Étude des Sols et de la Carte Pédologique de France (SESCPF) are always accompanied by four inset maps on 1/500,000 scale: pluviothermic, lithological, geomorphological and land-cover and land-use. These inset maps are sometimes prepared by the authors of the soil map, but the work is most often entrusted to specialists (climatologists, geologists, geomorphologists and botanists) who have worked in the region.

Block diagrams are executed in two dimensions but in perspective, and present the spatial organization of soils as a function of relief. Block diagrams at times represent an actually observed location in a given area. They can also be prepared automatically by calculation using a computer program that utilizes hypsometric data. But they often do not correspond to an actual landscape but rather to a theoretical landscape representing the typical regional organization. In such case, many sections of the map closely resemble it but none is perfectly identical to it!

Profile drawings. Illustrations of this type are less popular. However some soil scientists supplement their legend by drawings showing in concrete terms the appearance of the major soil profiles. At this level, many mappers have adopted the graphic conventions proposed by Duchaufour in the beginning of his book (Duchaufour, 1983). But the computer will no doubt give rise to other representations.

1.2 OBJECTIVES AND APPLICATIONS OF MAPPING

Soil mapping is a costly activity: it takes considerable time, it needs highly specialized personnel and it implies that many soil analyses will be done. It is justified only if its practical utility can be demonstrated. Here it is not a matter of listing all possible applications but of indicating which fields they cover. A deeper appreciation requires that we examine the functions of the soil considered as a *natural body*, a *component of the environment*, an *interface* regulating transfers, an exploitable *material* and a *support* for plant growth (Dumanski, 1994).

1.2.1 Knowledge of the natural milieu

Soil mapping has for primary objective the inventorying of soils and specifying their spatial distribution. Thus it represents a characterization of the physical milieu from a wide perspective and this can be used in all sorts of fields. For example, we meet a scientist who wants to establish the relationships between the kind of agricultural area and the socio-economic characteristics. Under these conditions, the author of a map will see in his department a parade of many users: geographers, phytosociologists, hydrologists and sometimes geologists, delighted to find at last a description and careful location of the surface formations of their study area. Over the past few years interest has been renewed in the characterization of soils. This is in particular the result of a better consideration of environmental problems.

1.2.2 Scientific research

The soil map is an excellent means of studying pedogenesis, that is, to link the soil mantle to the natural environment: morphology, climate, vegetation, time. This study of the relations between a soil and its environment is sometimes termed *chorological study* (Girard, 1983).

Some scientists, however, are somewhat disturbed. It appears to them that the soil map, prepared by using qualitative observations and some intuition, is not the expression of or the medium for a truly scientific procedure. They are mistaken, because the map is prepared by taking thousands of observations, For example, it is not unusual to take 2000 auger bores for a pedological study at medium scale. Therefore, when the authors of such an investigation subsequently maintain that such and such kind of soil is linked to particular environmental conditions, it is not just an impression but a clear expression of a precise statistical analysis with study of a sample located in large numbers and adequately structured. Also, all research pertaining to the soil must start from the field and return to it at the stage of validation of the hypotheses. This is the true basis of experimental research and means that we cannot always be limited to laboratory studies conducted *in vitro*. We will lose all contact with actual problems such as those existing in the natural milieu. Study of the land is therefore indispensable. But one should be sure of the representativeness of the sites where observations are made and sampling done. For this, mapping remains the preferred tool. Unfortunately, some scientists still select their soil samples with methods related more to excuse than to true scientific analysis. However, the time spent, the human and equipment cost of laboratory experiments, and their scientific and applied utility make this unacceptable.

1.2.3 Regionalization of agronomic problems

The links among specialists of soils, of plants and of agricultural production must be numerous. Actually, cooperation is possible in very many subjects. For example, agronomic experiments must be regionalized taking into account the nature of the soils. How are the production systems in a region to be analysed if it is not understood that there are three or four broad groups of soils each posing specific technical problems and each representing a defined production potential?

In fact, many cooperative efforts between Agronomy and Soil Science have already been accomplished. In consideration of their utility they will be covered in Chapter 9, § 9.1.

1.2.4 Planning the use of the soil

Settlement in developed countries is often dense. Industrial and transport infrastructures are many. In Germany for example, 120 ha of land were lost every day in 1997 *(sealing)* for locating highways, railways and various other constructions (EEA, 2000). Theoretically, the best lands should be reserved for agriculture and the poorest for other human activities. Unfortunately, things do not happen this way for two reasons, whatever efforts might be put in by champions of the rural life.

Firstly, good lands are the same for all people. Road builders, just as farmers, prefer accessible level areas, not too moist and without major obstacles. Thus there is competition for the same kind of area. Valleys and different levels of terraces are specially used. In France for example, the Rhône corridor, despite its small width between Lyons and Montélimar, has been striated with two railway lines, two trunk roads, one motorway, one lateral canal to the Rhône, not counting the High-speed Train.

Secondly, it is not fair competition between agriculture and the other branches of human activity if economic indicators are referred to. Establishment of a motorway or an airport is so costly that the price of the land is negligible in the project, even if the soils be excellent. Thus it is difficult the agronomic or ecological viewpoints prevail so far as occupation of the area is concerned. Only in some special cases (vineyards of high quality, protected nature parks) does this happen, sometimes indirectly after the intervention of the media.

In the general case, the mapper is not asked which soils should be by-passed! This makes very theoretical, not to say sterile, many studies conducted without sponsors and intended to define the farming, urban or industrial suitability of such or such sector. The only area where such studies can be done are in developing countries, in zones with very low population density.

On the other hand, the leaders of industrial and urban development are responsive to the impact on the natural milieu of the conversions they have decided to carry out. It is necessary for them to avoid various law suits later. Thus they do not hesitate to use the services of soil specialists even where the rules do not make this obligatory. Two specific examples can be mentioned.

In France, in the Rhône valley between Geneva and Lyons, the cutting of a canal lateral to the stream caused lowering of the groundwater level in the alluvium. Some soils, earlier hydromorphic, were found to have clearly improved. Others that had water at great depth were rendered too dry in summer. The INRA was asked to evaluate the damage caused, if any, in order to provide for adequate compensation to the owners of the land. The INRA also examined in which sectors provision of supplementary irrigation would be useful. All this was done based on accurate soil mapping along with measurements of groundwater levels and simulation calculations of the hydric regime of the soils (Bornand and Voltz, 1986).

In Switzerland, construction of motorways led to truncation of soils over a large width and great distances, thus on large areas. This resulted in removal of large quantities of topsoil that could be used to rebuild soils in old quarries or old industrial areas. Also, the cutting of tunnels provided enormous quantities of earth that had to be removed, with value addition if possible. The Agricultural Engineering Laboratory of the EPFL was consulted on several occasions for questions pertaining to these topics. The results provide good examples of the role that soils specialists can play in collaboration with project leaders; they will be presented in Chapter 9, § 9.2.

However, we will not be able to indefinitely let hundreds of thousands of hectares of very good agricultural land disappear from Europe every year. Villages were established several centuries ago in the most fertile zones: valleys and plains. Their expansion and transformation into large agglomerations progressively ate into an excellent and non-renewable soil resource. At present, agricultural overproduction hides this from us, if we forget that it is geographically limited and historically very recent. Will this last?

In the future then, we must wish for good integration of specialists of soil science when land-use plans (France) or land-assignment plans (Switzerland) or other equivalent systems are drawn up. In this framework, an in-depth practical study, village by village, could be done for preservation of the natural assets.

1.2.5 Agricultural suitability

The agricultural suitability of a soil is difficult to define because it is not merely a technical problem. Economic constraints also play a part. For example, studies on carrot (Nicollaud, 1988) showed that the best soils for this vegetable are not even-textured soils, potentially the best, but sandy soils. In fact, sandy soils allow quick uprooting even during the rainy season and thus enable supply to the wholesale markets of the cities in Europe without delay.

Furthermore, modern technical methods have enabled establishment of any type of crop. For example, in a country of the Maghreb, banana is grown in greenhouses. This apparently is profitable because of the locally adopted protectionist measures. However, the strong natural constraint in this case, lack of heat in winter, is suppressed at great cost. Above all, the economic conditions justifying costly development plans can change abruptly, leading to the ruin of those who have invested in it. What will happen to the banana producers of the Maghreb if the borders are opened up? Thus we should admit that *agricultural suitabilities* do exist, represented by a range of crops adapted to a given milieu. In this context, the role of mapping is two-fold. Firstly, the strong constraints that can diminish this range should be highlighted, for example insufficient soil depth for ploughing, presence of stones preventing production or harvest of tubers, presence of calcium carbonate leading to constraints in tree culture for rootstocks, etc. Secondly, the mapper must point out and spatially locate the hazards of change in the topsoil that may result from improper exploitation: danger of erosion, of salinization, etc.

1.2.6 Agricultural planning

Soil maps are often required by decision makers as preliminary to different types of agricultural planning. It is not possible to present here a complete list but a few typical examples may be mentioned.

Drainage
The utility of a preliminary pedological investigation prior to drainage is well appreciated by farmers. The cost of drainage is a function of the length of pipes laid parallel to each other. For the system to work, the porous pipes should be spaced not too closely (unproductive additional cost) or too far from each other (water is not evacuated and the investment becomes futile). Under these conditions the role of the soils specialist is useful and directly profitable. This is why, in France, pedological studies preliminary to drainage were very numerous between 1980 and 1985 under the impetus of the Ministry of Agriculture and the Interprofessional Department of Cereals, with the collaboration of the INRA and CEMAGREF. This case deserves to be retained as example (Chap. 10, § 10.6.6). However, this does not mean that all wetlands are to be drained, for some have essential ecological roles.

Irrigation
In France, the big regional development projects such as construction of the Provence canal or of the Bas-Rhône Languedoc canal were preceded by preparation of many soil maps covering practically all the zones liable to be irrigated by the canals in question. Actually, what use would it be to provide much water for irrigating soils in sectors where the farmers consider that the operation is technically too difficult and/or unprofitable? As the required soil surveys represent considerable work, the large regional devel-

opment corporations constituted their own teams of soil scientists and mappers in the nineteen-fifties. Today, the maps related to these improvements have been completed. However, soil mapping pertaining to irrigation has not completely ceased. It is continuing bit by bit as needs arise. For example, the possibility of irrigating the hillside lands when the EDF (l'Électricité de France) floods a valley by building a dam for producing electrical power is being studied. Pedological investigation becomes part of a compensation package with many components offered to villages affected by requisition.

Clearing and restoration of soils
Agricultural implements available to farmers are more and more efficient. They have enabled operations that were unthinkable just thirty years back. Thus, areas that had always seemed given over to moor, scrub or garrigue have been cleared and put under various crops. For reclaiming these lands, we have not hesitated to crush vegetation, pull out stumps, deep-plough the soil to increase its depth and even to crush stones to reduce them to acceptable size. But it is clear that these improvements cannot be done everywhere and under all conditions. There are stumps very difficult to pull out (box tree for example) and stones almost impossible to crush (crystalline rock or quartzose fragments). Sometimes mappers are called in as back up for defining the priorities in a given region, that is, the zones worth such clearing. However, clearing is often an individual action, executed at the level of the parcel. The farmer works alone or with the help of a specialist company that is entrusted with making the land suitable. In this case mapping is not useful, but investigations carried out on soil transformation can form worthwhile additional measurements. Thus, the evolution of organic-matter content subsequent to clearing of the Pyrenean Piedmont has been followed (Arrouays, 1995). Changes in active lime content of soils of Languedoc in which the large limestone fragments have been crushed have also been interesting. Presently, however, these development works are fewer because of the overproduction of agricultural commodities at the level of the European Union.

Redistribution
For redistribution of parcels between farms, it is necessary to exchange lands whose value and qualities are as similar as possible. Farmers often proceed among themselves, grouped into a committee. Mappers are sometimes called to provide a soil map enabling them to base the discussions on a solid scientific foundation. Such was the case in villages of Bardonneux and Plan-des-Ouates in Geneva canton. The soil maps prepared by EPFL served to draw soil-rating maps on which each parcel of land was given a quality rating between 0 and 100. These maps were then sent to the land surveyor in charge of redistribution. The surveyor then provided documents

as objective as possible for defining, before exchange was effected, the value of the lands. Things are quite complicated when gone into in detail. For example, a market gardener will not be seen offering two hectares of stony soil with 45 points in exchange for one hectare of loamy soil with 90 points! It will not work out well for him, and he will no longer be able to grow tuber crops. Therefore, quality ratings are useful groupings but are not sufficient for sorting out problems of exchange. They are generally used in instances where the soil is put to large-scale farming.

1.2.7 Protection of soils

For some years specialists have put out the concept of 'sustainable agriculture'. Just as applause or musical notes can be sustained long, sustainable agriculture is agriculture that can be continued interminably in a given milieu without causing modification in the characteristics of the milieu. It is thus a matter of conducting utilization of the soil with long-term views, to protect that natural resource. Unfortunately, this 'careful' management is not often practised in real life. Indeed, soil degradation is rarely catastrophic and rapid as described by John Steinbeck in 'The Grapes of Wrath' (Pillet and Longet, 1989). It often takes place slowly and insidiously. Among all kinds of examples, three are worth attention (Védy, 1990).

The battle against erosion
It is observed in the developed countries of the temperate zone that cultivated soils are becoming more and more susceptible to hardsetting (formation of a surface crust) and erosion. This is the result of the conjunction of three factors. First of all, mechanical working of the soils favours mineralization of organic matter and diminution of the structural stability of the soil aggregates. Then, disappearance of draught animals has reduced the quantity of manure available in areas with large-scale farming. Lastly, growing of crops that require weeding (maize, beet) has led to exposing the soil for a long time to action of raindrops and to the danger of surface run-off. The question has become cause for concern and has been appreciated by decision makers less because of erosion of soils than because of the consequences: silting up of hydroelectric dams downstream, pollution of water by phosphatic fertilizers washed down with solid particles, etc. Various studies have been conducted when demanded by the European Union. In particular, a map of erosion risks in Europe has been prepared on scale 1:5,000,000 (Chap. 10, § 10.1.3). This very general document has the principal value of locating the areas where the problem exists. But mapping research in greater detail has been necessary. Thus the manifestations of erosion have been spatially related to different characteristics of the milieu such as soils, slope, crops, parcelling, size of drainage basins, etc. (Chery,

1990; Chery et al., 1991; King et al., 1991). In some cases, the parameters responsible for the susceptibility of the soils to degradation are not always as expected. According to the authors mentioned above, slope is not specially the cause in France. On the other hand, the texture of the surface horizons and the type of crop seem to play a major role in the initiation of runoff. Such conclusions can typically be drawn in the case mentioned earlier, where mapping enables accumulation of very many observations and their statistical interpretation. Thereafter, once the causal factors have been identified, attempts can be made to minimize their effect.

Desalinization
Another example where mapping can help in understanding and restricting a process of soil degradation is that of salinization. In many dry regions, irrigation is done with water of poor quality. The water evaporates from the soil or is transpired by plants, leaving behind the salts contained in it. With time the concentration of salts rises and the soil becomes less productive before becoming virtually sterile. The method of desalinization is well known. It consists, in principle, of irrigating beyond the requirement of the crop so that the excess water prevents the salts from accumulating at too high a concentration in an aqueous phase of very small volume. There is also a leaching effect and it is imperative to provide for evacuation of the saline waters to the bottom of the soil by drainage. In some cases, calcium is added in the form of gypsum or plaster to prevent the dispersion of clay and impermeabilization of the soils. But familiar, known solutions are difficult to apply. Evacuation of salt-enriched drainage water to the river is to condemn the land-owners situated downstream to irrigate with water of still poorer quality. So it is necessary to view the problem spatially: to provide, as a function of the nature and geography of soils, channels to lead in fresh water and others, completely separated, for evacuating the saline water. Mapping is an excellent means for planning the structures. This is why many European research departments of soil mapping act in desalinization problems in the Maghreb or the Middle East. But the most spectacular example of this type of project is the exploitation of southern California. Water was pumped from the Colorado river, clarified of its mud, then transported nearly 280 km for being distributed to the plots. The drained water was evacuated to an endorheic depression. It has now formed a large artificial lake, the Salton Sea, the already considerably high salinity of which is continuing to rise slowly.

Fight against chemical pollution
Many forms of soil pollution, identified as worrying, pose problems in mapping. To begin with, there is excess fertilization with **nitrate** leading to pollution of groundwater. Some world-renowned sources of mineral water have been affected. Good agricultural practices have to be defined in the

corresponding watersheds in relation to the nature of the various soils present. Then there is accumulation of *heavy metals* in soils (Baize and Terce, 2002). These metals mostly enter through certain fertilizers or organic residues. Mapping can help in defining the zones in which sewage does not pose great problems. There is also enrichment of soils with *phosphorus*, which indirectly causes eutrophication of rivers and lakes. The spatial distribution of soils and crops as well as the structure of the hydrographic network often explains the features of the pollution (Bouchardy, 1992). Furthermore, *pesticide* residues are sometimes present in considerable amounts in surface run-off. To this already long list should be added radioactive elements, especially *caesium* and *strontium*..., if it is not believed that Tchernobyl will be an isolated accident.

1.2.8 Other applications

It is impossible to go through all the applications as they are varied (Purnell, 1994). Having prepared in France the soil map of Languedoc-Roussillon administrative region on the scale of 1/250,000 (Bornand et al., 1994), our team has been very practically faced with a variety of demands of users. Besides traditional questions on irrigation, salinization or land-use planning, we have from 1994 to 2002 dealt with:

- the distribution of calcareous and acid soils for taking lime dealers around,
- characterization of the soils of experimental parcels in the forest milieu,
- evaluation of soil quality for helping courts in settlement of disputes between people who want to demonstrate that a specific property could or could not be taken away from agriculture and used for buildings,
- the relationships among soils, hydric reserves and vine quality in the regional vineyards,
- study of soils suitable for asparagus,
- danger of pollution of groundwater by nitrates following establishment of industrial cowsheds and pigsties,
- suitability of selected areas to serve as zones for spreading composts or pig slurry,
- the role of soils in accumulation of phosphate in littoral lagoons,
- vulnerability of soils to accidental pollution from trucks carrying hazardous materials with a view to planning reclamation ditches in susceptible areas, where soils are porous and the groundwaters used for human consumption,
- geographic distribution of the red-legged partridge *(Alectoris rufa)* as a function of environmental characters.

On the whole, the experience showed that soil maps had many more small occasional users than large users posing just one major question before starting the intervention of specialists with additional studies and signing of a large contract into the bargain. But the small questions, legitimate and important to those asking them, are often costly to solve. What can you charge a person who takes a couple of hours of your time to present his problem and goes home only with a piece of good advice? In this situation, we responded by delegating to a technical organization—the *Hérault Chamber of Agriculture*—the responsibility of forming the interface between the soil map prepared and its users. At the same time, to limit the time devoted to advice of all kinds, a database was created on the Internet to enable the user to be as self-reliant as possible in getting information on soils. In this we follow the examples of the Americans, Canadians and others (Chaps. 6-10).

1.3 THE RUSSIAN SCHOOL OF SOIL SCIENCE

Historical aspects of soil mapping are discussed by country in Chapter 10. But we shall here make an exception of the Russian school of soil science. In fact, we owe to it some basic concepts of our discipline and it is appropriate to point them out in the first chapter of this book.

1.3.1 Dokoutchaiev (Margulis, 1954; Boulaine, 1983)

Heading the Russian school was Vassili Vassilievitch Dokoutchaiev (Dokoutschaeff, Dokuchaef or Dokuczajew...). The first of these spellings was used by one of the master's students, Agafonoff who knew Russian and French perfectly. The second comes to us from the person himself: it appears on the cover of a pamphlet published in French at St Petersburg in 1900. The third version is presently used by the scientists of the institute in Moscow that bears the name. The fourth appears in the presentation of the Pulawy Institute where he worked. Thus there is a choice....

A brief biography of Dokoutchaiev follows: he was born in 1846; in 1872 he was given the responsibility of teaching mineralogy and geology at the University of St Petersburg; between 1875 and 1879 he participated in the surveys for the soil map of Russia; from 1877 to 1879 he was asked to work on the *steppes with chernozem* at the demand of the Imperial Free Economic Society of St Petersburg, which was worried by seeing the disastrous consequences caused to agriculture by the droughts of 1873 and 1875 (it is said he traversed more than 10,000 km for this study); in 1883 he published *The Russian Chernozem*; from 1882 to 1886 he investigated the soils of Nijni-

Novgorod (Gorki) and published the study in 14 volumes with his students, thus accumulating unequalled experience in the world regarding soils and their mapping; in 1892 he was appointed Director of the first large Centre for Agronomic Research and Education established in occupied Poland at Novo-Alexandria (now Pulawy); he established a chair of soil science in this centre in 1894 and entrusted it to his student Sibirtzev, whose course demonstrated the thinking of the Russian school of science; in 1896 he published a new classification of soils in which he presented his theory of zonality; in 1899 he led a large soil survey in the Caucasus; he died in 1903, three years after Sibirtzev.

1.3.2   The contribution of Dokoutchaiev and his followers

Even considering things with hindsight, we are led to recognize that Dokoutchaiev's contribution is considerable. To begin with he invented, discovered or improved the basics of soil-survey techniques: study of soil profiles in place, identification of horizons, sampling and analysis of samples, demarcating soils, cartographic reporting.... However, the concept of designating horizons by the first letters of the alphabet is older and is partly owed to the German scientist Albert Orth who used it in a paper published in 1873 and partly to the Danish scientist Pieter E. Müller who published only after Dokoutchaiev (Tandarich et al., 2002).

And then, above all, the soil maps of various regions of Russia prepared in 1875-1899 under Dokoutchaiev's direction provide a harvest extraordinarily rich in observations.

Factors of pedogenesis

Thanks to his researches, Dokoutchaiev understood that soil is the result of the interaction of five soil-forming factors:
- climate;
- parent material;
- flora and fauna;
- relief;
- time.

We can then write:

soil = f(climate, parent material, living organisms, relief, time) and consider it a new paradigm, the *state factor model* [Jenny, 1941], the *soil factor equation* or even the *soil-landscape paradigm* (Hudson, 1992]. We can deduce that:
- in a given uniform landscape, the soils necessarily have common and stable characters; the soil mantle is thus spatially organized in a way that owes nothing to chance; it can be studied rationally;
- in landscapes that are comparable in all respects, the same kinds of soils will be found; this implies that the nature of the soil mantle can be predicted and thus mapped on the basis of what some scientists

term a *soil mapping model* (Hartung et al., 1991); in other words a deductive procedure is possible;
- lastly, if the soil observed is not what had been predicted, a scientific problem arises. Thus a most ordinary profile found in the most ordinary environment can provoke deep reflection, but it must be prepared by a suitable scientific culture.

The state factor equation thus promotes a deterministic procedure. Of course, all that is true only in broad terms and the Dokoutchaiev approach is only a model with all its imperfections. Soil characters show variations that are neither completely predicted nor fully understood. Some authors add a pinch of randomness to the system (Webster, 2000). Others even see in it the expression of a certain chaos (Phillips, 1998).

Dokoutchaiev made it clear that we do not know how to estimate the relative importance of each factor of pedogenesis. Thus he did not propose to give to the climate-vegetation pair a pre-eminent position in soil development.

The major pedogenetic processes were observed and analysed. Leaching of clay was mentioned; similarly, movement of organic matter and sesquioxides in the soils of the taiga. Excess of water linked to relief depressions was noted. The terracing of soils according to their age was observed and understood (the most developed soils were found on old, raised erosion surfaces). Of course the most spectacular advance of the era was the recognition of a latitudinal zonality of soils. Thus, the tundra, taiga, steppe, desert zone and tropical and equatorial forests are each associated with their characteristic soils.

Mapping and classification
This zonality is clearly seen when mapping is done. It is not the result of speculation. As far back as 1883, the map of the Chernozem strip of Russia in Europe, drawn by Dokoutchaiev and published posthumously (Margulis, 1954), demonstrated the presence of isohumic (same humus content) bands nearly corresponding to the latitudes (Fig. 1.5). It is noted in passing that the concept of isohumism was later distorted to now mean that isohumic soils are soils whose humus content is, if not identical from the top of the profile to the bottom, at least high to great depth.

Dokoutchaiev, in his enthusiasm, generalized a little too much. It seemed to him that the character of zonality should be applied to all things: 'It is remembered that the change and sequence of the five zones of soils of different colours in the northern hemisphere go along with increase in light and heat, with change in colour of humans and animals' (Dokoutchaiev, 1900). Thus the first attempts at classification would mention, along with characteristics of climate, rocks and vegetation, the most characteristic representatives of the fauna: soils with aurochs, bears, elk; soils with camels, dromedaries, etc. Later the author would come to consider only the soil

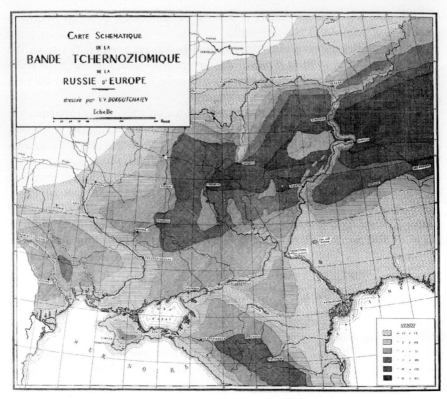

Fig. 1.5 One of the very first soil maps prepared in 1883 by the father of European soil science, Vassili Vassilievitch Dokoutchaiev (in Margulis, 1954).

fauna. The expedition to the Caucasus was undertaken because Dokoutchaiev wanted to observe the *vertical soil zones* which, according to the altitudinal variation in climate, should match the latitudinal zones demonstrated earlier. On land, one actually sees a strong similarity between the soils of summer pasture of high altitude (*eylagui* soils) and soils of the Arctic tundra. Also, 'typical chernozems' are found between 1500 and 2500 m, at least where the topography is level. In fact, this latter observation should be evaluated again in the light of current knowledge because writers of the 19th century had a regrettable tendency to mix up in the same category many different soils with black colour in common. But broadly, in spite of the changes caused by differences in relief and parent material, zonality is found with altitude and is mentioned in the map drawn by Mechtchersky.

In course of time, the concept of zonality has progressively been made clear. According to Dokoutchaiev and his students, three cases should be distinguished:
- the soil has '*responded fully to the normal union of physico-geographical and geobiological conditions of the soil region or soil zone in question*'; it is then qualified as *normal* or **zonal**;

- the soil has not responded at all to the law of zonality; in the extreme case there is no soil, but unconsolidated rocks outcrop: dunes, recent alluvia, loess, marine muds, etc; Dokoutchaiev talks of *abnormal* soils, later of *cosmopolitan* soils or even *skeletal soils* or *coarse raw soils*; for Sibirtzev they were *incomplete* or *azonal* soils;
- the soil is zonal, but corresponds to particular locations, where one of the principal factors of formation dominates, Such are, for example, patches of peatlands that speckle the podzol zone and correspond to wet depressions. In this case the soils are **intrazonal** or *transitional*.

The conceptual basis was sufficient for a disciple of the master, K.D. Glinka, to risk drawing the first map of soils of the world in 1908 (Fig. 1.6). This map was revised in 1929 by Agafonoff. Its accuracy is very limited and the map must not be considered in detail.

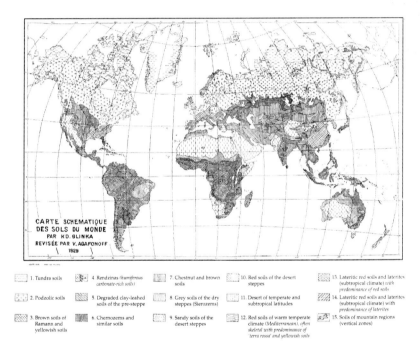

Fig. 1.6 The first soil map of the world drawn in 1908 by Glinka, student of Dokoutchaiev, revised and published in 1929 by another disciple of the master (Agafonoff, 1936).

1.3.3 Westward dissemination of the ideas

Towards the end of the 19th century there was total opposition in Europe to the concepts of Dokoutchaiev (Pedro, 1984a). At that time *Agricultural Geology* dominated thinking in Western Europe, viewing soil as almost exclusively related to the underlying parent material. Then the Russian concepts were progressively acknowledged because of three factors: the World Fair, translation of writings and soil science conferences.

Dokoutchaiev had already participated in the World Fair in Paris in 1889 and that in Chicago in 1892-1893, but it was the one held again in Paris in 1900 that revealed the magnitude of the progress in Russian work (Ototzky, 1900). In fact, the period from 1890-1900 represented ten years of intensive mapping in Russia that led to considerable experience being gained. Almost 45 reports of soil science investigations were prepared with soil maps and explanatory notes. The Fair of 1900 enabled presentation of 21 such studies conducted in all kinds of environments. An abstracted map, drawn by N. Sibirtzev, G. Tanfiliev and A. Ferkhmine took zonality into account. Another, drawn by Ototzky, summarized the studies done. The eight most typical soils were schematically presented in 'profile diagrams'. Many samples were exhibited in boxes or in the form of monoliths*. Nearly 30 photographs showed the typical soil environments of European Russia. The impact of this fair was great on the Western Europeans who discovered the methods of systematic inventorying of land and at the same time the scientific conclusions to which these inventories led.

Dissemination of the Russian concepts was also facilitated by the work of translation into foreign languages undertaken by the scientists of this school. Dokoutchaiev and his student Agafonoff were published in French. The works of another student of the master, Glinka, were translated into German and English. Many foreign scientists then became the disseminators of the new ideas: Murgoci (Romania), Stebout (Serbia), Ramann (Germany), etc. However, for political and personal reasons, young American scientists had some difficulty in obtaining Russian writings for translation and study (Tandarich et al., 2002)!

Lastly, the beginning of the twentieth century saw the appearance of a new form of communication exploited widely: the international conference. The first conferences concerning soil science were held in Budapest, Stockholm, Prague and Rome (Demolon, 1949).

Following the studies mentioned above, the first soil map of Europe was prepared in 1927 and presented at the First International Congress of Soil Science in Washington.

---

* Monolith: *a soil volume in the form of a right-angled prism with rectangular base, collected undisturbed and thin enough to be glued to a board for showing to the public in its natural orientation, that is, vertical. Preparation of moniloths requires great skill.*

1.4 THE ISSUES

1.4.1 Object, concept or image soils

In Soil Science, there are no strictly independent soil individuals separated from one another. This concept of individual is restricted to sciences such as botany (*a* plant) or zoology (*an* animal). The soil in nature is a sort of **continuum**, even though variations in its properties are sometimes abrupt enough to represent, in some way, boundaries. But these are essentially theoretical and conceptual, without real existence. In sum, we should distinguish (Baize, 1986):
- the *object soil*, the natural reality corresponding to the continuum;
- the *image soil*, the soil as we represent it to ourselves after having observed it (only the major features are seen and retained, somewhat like in a photograph where only the most visible external features are present);
- the *concept soil*, which has no real existence. It is a theoretical individual, a taxon of classification, most closely approaching our image soils.

Let us take an example. We are faced in the field with a soil (object). We describe it in a notebook (image). We reckon it to be a Vertisol (concept).

Viewing soils as distinct individuals is very efficient from several points of view. To begin with, it allows us to structure our knowledge in the form of classifications. Then, it makes possible the drawing of maps on which boundaries appear. Thirdly, the entities thus shown can be processed by means of computer systems suitable for the creation of categorial maps. But some authors judge this attitude rather harshly: *'The survey and classification of geology, soil or vegetation into exact, sharply defined classes or areas that meet abstract ideas of nicely defined, properly circumscribed units has been an enormous exercise in forcing continuously varying phenomena into exact moulds. Although much scientific evidence has been assembled to demonstrate that soils are not exact, homogeneous entities...'* (Burrough and Franck, 1995). We shall see in Chapter 2 mapping methods that, in their principles, favour different ways of understanding the soil.

1.4.2 Direct observation and indirect sensing of the soil mantle

Experience has shown that a scientist seeking to prepare a map on scale 1/100,000 is able to correctly conduct on average the survey of 700 ha per day. It is also observed that it is difficult for a mapper to dig and describe at this scale more than 20 **manual-auger bore holes** in one day (these bore holes are made with a manual tool the major types of which are described in Chapter 3, § 3.5.3). The auger has a maximum diameter of 7 cm and hence a cross-section of 38.5 cm^2. This amounts to saying that the map is prepared by directly studying an area in the ratio $(20 \times 38.5)/(700 \times 10^8)$ to the area to be characterized (approximately $1/10^8$). It will be objected that the mapper takes advantage of his bore holes to consider the soil around them and between them. Nevertheless the map is constructed without many direct observations. No opinion-poll company would want to understand a social phenomenon or predict the outcome of an election on such a slim basis. This is like, for example, polling two Americans for estimating the popularity of President Bush! However, mappers have to be satisfied with a very small amount of data if they wish to conduct their soil studies at economically acceptable cost. Under these conditions, there are many ways for achieving this. We shall briefly review them.

Indirectly viewing the soil
For such study it is essential to use aerial photographs. They give an overall view enabling generalization. They draw attention to image peculiarities that should be interpreted by reasoning or identified by field traverse with direct observation. They are also useful to 'see' what there is behind the hedge or hill. The contributions of photointerpretation are so many and so important that it is very difficult to produce a good soil map if aerial photographs are not available. We shall later discuss in detail the procedures for using photographs (Chap. 3, § 3.3).

All that has been said about photointerpretation is also true for remote sensing if mapping is done at very small scale (1/250,000 or smaller). Satellite images form a powerful means of generalization.

Consideration of pedogenesis
Great importance should be attached to analysis of the relationships that exist between intrinsic characteristics of soils and their three-dimensional distribution in the landscape (Girard, 1983). After Dokoutchaiev, all sorts of field and laboratory studies on pedogenesis have demonstrated the role of environmental factors in differentiation of soils. To characterize the environment is to understand what kinds of soils can be found. Also, the soil mantle has its own organizational logic. This eroded soil here on the topslope supplies the materials explaining the presence of that other soil lower down. Thus the mapper cannot be fully efficient without detailed knowledge of the laws and mechanisms of pedogenesis, therefore of the studies already done in the discipline.

Direct observation of soil
Nothing would be known without direct observation. A purely deductive procedure cannot be applied in mapping. We are not in the domain of the exact sciences. The mapper does not integrate in his analysis all the environ-

mental characteristics determining the modalities of pedogenesis; thus he cannot perfectly predict the nature of the soil. Moreover, the variability of the soil mantle is considerable. It has been shown in the French Jura that the probability of finding the same soil of a given convex relief on the immediately adjacent convex topography is at best 0.27 (Bruckert, 1989a). In such a context, analysis of the geological map and aerial photographs gives indications, not certainties. It is useful to conduct prudent verification at all places.

Lastly, the means being combined, a method of synthesis is necessary for tallying all the data to use them for drawing the map (Fig. 1.7); we shall return to this particular problem in Chapter 7 (§ 7.1-7.3).

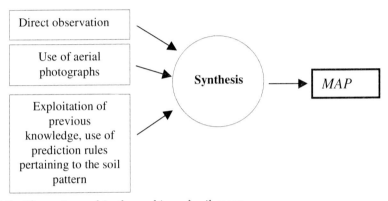

Fig. 1.7 Elements used in the making of soil maps.

1.4.3 Modelling the organization of the soil mantle

The soil is eminently variable in its three dimensions. The map will represent reality in a simplified manner. This simplification is the graphical translation of how the mapper 'sees the soil'. More specifically, the observer has available in the corpus of his knowledge some organizational models of the soil mantle and endeavours to find out which best suits the location studied. It could be said that this way of proceeding lacks objectivity. We will see moreover that there are other ways of dealing with the problem (Chap. 2). But the mapper does not have very much room to manoeuvre in. He must describe on a plane a three-dimensional object. Years of mapping directed by various specialists in all sorts of milieus lead us to think that there are four such organizational models of the soil mantle, presented below.

The profile model
Study of a soil section generally shows that the soil weathering mantle is organized in layers of varying thickness but parallel to the surface. These layers are termed **horizons**. They are not arranged in any which way but always occur in the same order, so much so that their succession constitutes

a characteristic *profile* that completely identifies the soil type. This organization in horizons, ordered and parallel to the surface is typical of level zones in all countries of the world. It is explained by the mode of development of the soil during pedogenesis. Its regular and also ordered character can be proved statistically. For this, investigators using a large number of observations perpendicular to the surface first inventoried all the horizon met with and classified them into a few principal types. Then, taking all their observations together, they drew up a proximity matrix (Table 1.2), indicating the number of times a particular type of horizon lay directly over another (Girard, 1983; Lahmar et al., 1989).

Table 1.2 Proximity matrix of horizons

		Types of horizon in 'underlying' position			
		H1	H2	H3	H4
Types of	H1	15	0	0	1
horizon in	H2	24	6	0	0
'overlying'	H3	15	40	7	0
position	H4	10	12	35	12

The table reads as follows: a type H3 horizon overlaid by a type H2 horizon was found 40 times.

It is then clear that certain vertically arranged associations of horizons are very frequent whereas others, though mathematically conceivable, are totally absent. Some scientists go even as far as to suggest a *'sociology of horizons'* (Brabant, 1989a). Thus the notion of profile, more precisely of typical profile, is justified both by genetic considerations and by repeated observations that show statistically that the concept is generally meaningful. Under these circumstances the first mapping method consisted of delimiting the portions of the area within which the soil profile is identical to that overlying a well-defined rock. This procedure, developed and codified around 1950 by the American Soil Survey is known by the term *series mapping*. The series were originally identified by a place name corresponding to the site where they were first found. They represent at the same time a mapping unit (object) and a taxonomic unit (concept) while acquiring a more or less vernacular name (Jamagne, 1993). Two remarks are necessary before going further.

In the strict sense of the term, the profile is a section with zero thickness. Therefore some scientists have felt that it is a poor model for representing the soil 'individual' inasmuch as the latter necessarily is a volume. This is why the American scientist Simonson introduced the concept of *pedon*, the elementary soil volume, characterized by a homogeneous profile and having lateral dimensions of the order of 1-7 m. Later on, and by misuse of language, the term 'profile' took on another meaning: it became the soil pit that serves to observe the profile itself. Thus it is customary to hear or read sentences such as 'three profiles were dug'.

Series mapping implies in theory that the soil is laterally uniform over large areas. In reality this is quite rare. Therefore, the concept of series leads to that of *impurity* or of *inclusions*. The American Soil Survey Manual originally considered that a good map had to separate series of purity greater than 85 per cent (85% of the pedons are identical). As this was often impossible, the Soil Conservation Service proposed a limit of 75 per cent in 1983. If the soils not included in this 75 per cent posed no specific problems of land use or development, the term *consociation* was used. The terms 'simple unit' (unit with a single soil) or 'pure soil unit', which obviously are incorrect, should be avoided. If the unit comprises very different soils, they are *dissimilar soils*. The above can be summarized as follows.

- Consociation = [(pure unit > 50% + similar soils) > 75%] + dissimilar soils < 25%,
- Compound map units = other cases.

Detractors of this series approach can claim that lateral variability is more or less smoothed out whereas vertical variability is preset!

'Horizon/soil volume' model

On slopes, horizons are often greatly variable and undergo great changes in nature and thickness over short distances. Indeed, it is theoretically possible to make sense of them by the profile/series method, that is, by defining the areas over which the profile does not undergo sensible change. But this leads to multiplying these reference profiles and to defining very small areas of applicability. It can be said that the compartments of the soil system become small and numerous. A typical example is presented in Fig. 1.8 (Brabant, 1989b).

The 'profile' model is obviously unsuitable. An attempt to apply it leads to imposing on the object of study a partitioning into subassemblies that do not directly account for its actual organization and might complicate study and interpretation. In this case, more particularly in tropical regions, we have to proceed in another way. The object whose lateral extent is to be found can no longer be the profile, it is the horizon in the broad sense of the term. That horizon is not necessarily horizontal: it is a layer and sometimes even a simple volume of soil material of any shape whatever. This approach using elementary volumes was developed by ORSTOM (now IRD) under the term *structural analysis* (Boulet et al., 1982). As the name indicates, this has for first objective the description of structures (in the sense of organizations) of the soil mantle. Its second objective, equally important, is to understand the evolution of these structures over time in order to determine how the system has been put in place and how it functions. The work related to structural analysis has been summarized by Ruellan et al. (1989).

28 MAPPING OF THE SOIL

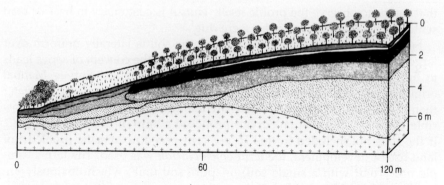

(a) this soil system is composed of an ordered combination of 8 horizons

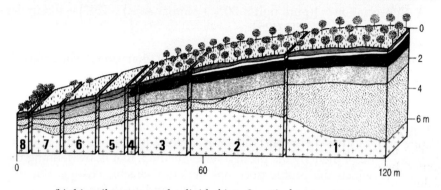

(b) this soil system can be divided into 8 vertical compartments

Fig. 1.8 Presentation of two methods for modelling the soil mantle: (a) horizon approach; (b) profile approach (Brabant, 1989b).

In cases that come under structural analysis, the complexity of the location makes representation of the soil mantle on a flat surface difficult. We shall return later to the partial solutions to this problem (Chap. 2, § 2.2.3).

However, the soil is not studied in trenches except in some particular cases (for example when a gas pipeline is to be laid). Lateral organization over moderate distances is not usually apparent. It is the result of a reconstruction. Yet, in science facts must be distinguished from interpretations. The concept of profile thus remains an observed element of soil along the vertical at a given location whatever the contributions of structural analysis might be.

'Associated soils' models
According to the American norms of 1983, there are three principal types of map units (MU) containing two or more soil units (SU) that are not separated on a small scale map:

- *Soil association.* Many soils present; they are organized in a manner that can be explained and is repeated regularly in space; these soils might be differentiated and separated on a large-scale map.
- *Soil complex.* Many soils so imbricated that it will be impossible to separate them even at a large scale, in spite of the fact their spatial organization obeys a certain logic, as in the preceding case.
- **Undifferentiated soil group**. The soils present do not have any regular arrangement. The proportion of the component soils can vary considerably from one end to the other of the unit in question. French scientists use the term *juxtaposition of soils*, which is less widely accepted.

Associations and complexes can be described and grouped using different criteria, the most traditional being:
- *morphology* of the organization (patches or strips of soils, etc.);
- *contrast* between the different constituents (see the work of Fridland, 1976);
- *size of constituents* (is it possible to distinguish them on a map and, if so, starting from what scale?);
- *origin of factors responsible for the imbrication* (topography, vegetation, etc.);
- *interaction between constituents* (do some of the soils considered give to or receive from the others constituents such as ions, clay, water, etc.?).

The terminologies relating to associations and complexes are not presented here. They, however, are of interest. Actually, on a map based on delimitation of map units (MU) containing many soil units (SU), the exact geography of each SU is not precisely defined. Then it is necessary to provide the user of the map with the means to find it. For this a precise apt vocabulary is desirable. For example, the map will permit the user to locate the association 'soil A + soil B + soil C', whereas the report will define that it is a catena (see definition below), with A on the topslope and soil C at the bottom. This is enough for proper field location. For further information the interested reader may consult various books (Fridland, 1976; Boulaine, 1978; Jamagne, 1993). Three very frequent types of association, however, deserve particular mention:

Catenas. The American scientist Milne, who worked in East Africa, introduced this term in 1936. Many soils are arranged on a slope in a regular sequence. The lateral differentiation is caused by one or more of the following factors:
- erosion (truncated soils on the topslope and colluviated soils lower down),
- lateral movement of water (dry soils at the top and moist soils at the bottom;
- movement of solutes and possibly of clay (leached soils at the top and enriched soils at the bottom).

Purists distinguish catenas from *toposequences*, the latter being observed on slopes without transfers having taken place. For example, in limestone terrain, outliers are often composed of a capping of calcareous rock over marls. Thus there is on the slope a succession of stony soils and clayey soils. But even in this case there are transfers (calcareous scree on marls). Thus mixing up of the two ideas is not forbidden. Catenas and toposequences are very frequent in hilly regions.

Climosequences. The soils are arranged in stepped bands according to altitude in relation to variation in the climate-vegetation pair (cf. climatic belts of phytosociologists). Climosequences are seen in all mountains of the world, particularly on acid rocks (granite, gneiss) that favour rapid pedogenetic development.

Chronosequences. Many neighbouring soils located in the same kind of environment (rock, topography) have very different ages. This is the most widespread case of soils of stepped alluvial terraces above streams and rivers. In Europe, this corresponds to a chronosequence of 330,000 years represented by four principal terraces (Bornand, 1978). Such locations are very useful for pedogenetic studies.

'Soil landscape' or 'soilscape' models
If the environment is heterogeneous and the map scale small, mappers cannot have for objective the systematic investigation of the boundaries separating two different soils in the field. Since greater degree of detail cannot be attained, they mainly seek to identify small natural units that appear homogeneous at a certain level, as in their geology, slope and any other element of the environment, and then to 'group' the soils appropriately. It is important to note that the term 'group' is a misuse of language because the landscape units do not result from aggregation of several soil units. Only later is their soil composition inventoried, if necessary. However, we generally proceed iteratively for precisely defining the container and the contents of what can be called a **soil-landscape unit** or **soilscape**. All of this corresponds to *physiographic mapping* or the *land system approach*. We shall return to it later. This procedure is attractive because it lessens the duration of field investigation and lends itself well to use of aerial photographs or satellite images. It leads to quick production of maps with sufficient precision for several applications. But we should be wary of the tendency to make soil maps without ever seeing the soils from up close!

It is possible to further push the consideration of physiographic mapping if it is found that the landscape units are naturally associated to form **small natural regions of soils** (SNR). We actually believe that ensembles of this type exist in nature. Unlike soil-landscape units, small natural regions are not characterized by homogeneity in their environmental constituents. Indeed, the soils, vegetation, rocks and even climates found in them can be many and varied. On the other hand, these natural components form a patchwork whose appearance is always the same from one end of to the other of the region in question. For example, in the Montpellier region an association of three constituents is seen: (1) hills of hard limestone with eutric soils and garrigues, (2) marly valleys with deep calcaric soils devoted to viticulture and (3) steeply sloping knickpoints forming the sideslopes of gorges through which streams flow. These last-mentioned correspond to superficial soils and are wooded. Thus the milieu is heterogeneous, but in an ordered manner conferring specificity.

Like all postulates, this is not demonstrable. The existence of natural regions has not been mathematically proved! But everyone knows that the Everglades in southern Florida, the oases of the Sahara, or Turkish Cappadocia have their particular 'look', so much so the whole world recognizes them. The concept of small natural region is thus validated by commonsense. As a consequence it is possible to do mapping at this level, at least for scales of 1/250,000 or smaller. The boundaries of natural ensembles of varying but definite soil content are drawn. But this ecological or environmental approach to soil geography needs to properly define the basic concept. Actually, the division of a territory into characteristic geographic units can be done at different levels. In the case of Switzerland for example, it is very easy to distinguish three or four such:

- the country is divided into three major natural regions: the *Jura*, the *Plateau* and the *Alps*;
- the Plateau, cut by valleys composed of moraines of varying thickness resting on different substrata, itself comprises different small natural regions of smaller extent, even within just the canton of Vaud: the *Côte*, the *Gros de Vaud*, the *Lavaux*, the *Orbe plain*, the *Rhône plain*...;
- the Gros de Vaud, to take just this example, is easily divided into two: in the west, a morainic sector consisting of loamy soils and, in the east, a sector consisting of sandier soils in which the molasse substratum is partly outcropped;
- lastly, in each of these regions, soil scientists are seeking to introduce some nuances although we are not yet equipped with detail sufficient to specifically delineate soil units. Thus in the moraine of Gros de Vaud, it is still necessary to distinguish the valleys with their succession of more or less sandy soils from the undulating interfluves in which the apparent monotony of the environment is broken by the presence of peaty depressions.

To sum up, a landscape can be characterized by ***different nested levels*** and corresponds to a land unit hierarchy. A single terminology for naming the levels, still not definitive at present, has been drawn up by various authors (Mabbutt, 1968; Bergeron et al., 1992; Brabant, 1992; Douay, 1994; Bornand et al., 1995; Bertrand, 1998a ...).

Let us take for example one of the attempts at standardization (Wielemaker et al., 2001):
- *Region* (example: volcanic mountains),
- *Major landform* (example: river floodplain),
- *Landform element* (example: terrace, backswamp),
- *Facet* (example: zone of silt deposition on a terrace),
- *Profile and its immediate environment* (7 m around).

It remains for the investigators to determine how to find the means for associating the concepts and the realities corresponding to these different levels without risk of error, whoever be the observer. More precisely, how are we not to confuse them from one region to the other? How are we to delineate each one objectively and precisely? Which definitive terminology is to adopted? Which hierarchical level is most suited for studying a given problem? Thus, soil-physiographic mapping, which is based on the delineation of natural soil regions and accepts as axiom that the soils are not to be individualized, has yet to make progress.

Summary

Mapping is therefore conceivable at different levels of approach, each aimed at the delineation of more or less bulky and complex objects. Table 1.3 summarizes the position.

Table 1.3 Mapped elements and types of mapping approach

Mapped element	Horizon or volume of soil material	Profile, pedon	Soil association	Small natural region, landscape unit
Type of approach	Structural analysis	Series	Association of series	Soil-physiographic mapping

In current mapping practice, the approaches are not always as distinct as the above schematic presentation may indicate. For example, there are soil associations in maps mostly containing consociations related to series. Moreover, these approaches are not spatially incompatible. Different types of mapping may be successively conducted in the same region by changing the scale. Some mappers have such a clear vision of all these nested levels that they evoke the concept of **Soil Landscape Hierarchical Model** (SLHM),in which their understanding of the natural milieu is ordered, from the level of the soil aggregate to that of the drainage basin! But this idea of SLHM should be more closely defined because, according to the writers, it represents the organization of soils in the natural environment, the cartographic image we wish to give to it or even the computer-generated structure enabling manipulation of the corresponding data. A well-documented article (King et al., 1994) considers all these concepts.

Lastly, it must be admitted that none of the approaches presented above provides a satisfactory solution to the problem posed by soils whose properties vary regularly from one end of the area covered to the other. An example would be the soil becoming more and more sandy towards the east of the Map Unit. Actually, all the arguments presented till now are based on the idea of a homogeneous body that may be the horizon, soil or landscape.

1.4.4 Similarity between soil types

It is essential for several reasons to be able to appreciate and measure the similarity between soil types:
- Firstly, every mapping approach rests on this notion of similarity between soils. Actually, delimitation consists of grouping pedons that are similar and separating those that are not. The human mind excels at this task: sorting is done intuitively, quickly and efficiently. But this approach is empirical and the process is not codified. Its precision is unknown if numerical means are not used.
- Secondly, defining similarity between pedons in the same unit is one way of measuring the homogeneity of that unit.
- Thirdly, most work bearing on verification of the quality of mapping leads to returning to the field (Chap. 5, § 5.3) in an area already mapped, in order to examine if the soil found agrees well with the indications in the map and legend. This means that it is necessary to compare the observed soil with the reference soil. This is rarely easy. Small differences are always seen, so much so that it is impossible to answer the question 'Is it the same thing?' with a 'yes' or a 'no'. It is therefore necessary to introduce a similarity measure to enable us give a qualified and quantified judgement.
- Fourthly, it may be useful to compare a soil observed in nature to all the reference taxa present in the databank to classify it automatically as a function of its maximum similarity. In addition, if it is different from all the know soils in the databank, it becomes a candidate for addition to the bank.
- Lastly, the mapper should, at a certain stage in his progress, consider the soil units corresponding to his provisional model and ask himself if some are not similar, in order to merge them. The human mind does this very efficiently and very quickly if there are up to 100 units corresponding to $[(100 \times 100) - 100]/2 = 4950$ possible combinations to be considered a priori. Matters become worse when the number reaches 300 or 400. This happens for example with mapping at a scale of 1/250,000 covering millions of hectares. The work of synthesis becomes very long and very difficult. Everything takes place as if the brain sees, at this stage, its performance diminishing, somewhat like a computer that has reached saturation (about 80,000 combinations to be

considered!). The use of automatic methods for detecting and quantifying similarity becomes very desirable.

We shall consider later (Chap. 7, § 7.4) the calculation of similarity between horizons and between profiles.

1.4.5 References and representativeness

As it is impossible to examine in detail every cm^2 of the soil mantle, it is necessary to study references and estimate their representativeness in the area. This is done at many levels.

Reference horizons and reference profiles
We will see later (Chap. 7, § 7.4) what efforts have been taken for recognizing, in a sample of real horizons or real profiles, the individuals that could serve as references, that is, as characteristic types. The others will be grouped around these as identical individuals or related individuals. The procedure can be generalized to references that are not objects but concepts (see above). The new French classification is based on this principle and is logically titled the *Référentiel pédologique*. English, Russian and Italian versions are available (RP, 1995).

Mapped reference area (RA)
The idea of reference individual may also be extended to an area (RA) in which may be found the whole range of soils of a given natural region. Various questions come to mind:
- Which is the optimum location of the RA so that it may really be representative and its size minimum in order to reduce as much as possible the mapping effort limited to it?
- What are the exact boundaries of the Small Natural Region of which the RA is representative? In other words, when would it be necessary to launch mapping of another Reference Area?
- To what extent would detailed mapping (at large scale) of the RA enable commencement of rapid mapping (at small scale) of the entire SNR in question, by profiting from the locally gained experience so that work would be better, faster and of greater accuracy?

We shall see later (Chap. 7, §§ 7.2 and 7.3; Chap. 10, § 10.6.6) the partial answers that have been given to these questions that are at the very core of mapping.

Conclusion and perspectives

The soil map thus seems to be a complex thing useful for planning many applications. Its preparation involves a dual modelling. For the one part, it is necessary to define which body is central to the delineations (profile, soil volume or horizon, soil association, landscape). For the other part, the correlations relating the soils to their environment must also be established. Without this effort at understanding, the mappers at the beginning of the twentieth century would have been unable to draw the first maps of soils of the world.

However, much progress has to be made in regard to the complete reviewing of concepts. Structural analysis and the soilscape approach are still in their youth. Potentialities are seen in them but there is a lack of works of synthesis presenting both. It is also necessary to overhaul the methods for description of soil associations and soil complexes. The variety of possible cases is infinite and one will not escape by a single descriptive approach associated with standardization of vocabularies. On the other hand, we are nearing the modelling of the organization of the soil units within the mapping units using their proportions and spatial dependence. Such attempts have been made (Goodchild et al., 1992). Thus, consociations and soil complexes could be precisely characterized by some mathematical parameters...

CHAPTER 2

VARIOUS KINDS OF APPROACHES

The various methods of mapping will be considered first in this chapter. Then all issues coming under publication of maps will be reviewed: modes of graphic representation, legends, scales, etc. In sum, the variety of possible options will be shown and the array of tools available will be presented. Thus mappers may deal with the diverse problems mentioned at the beginning of Chapter 1.

2.1 VARIOUS MAPPING PROCEDURES

2.1.1 Division of the real world

Mapping is the art of dividing the real world into subassemblies in order to make its organization easier to observe, analyse, understand, utilize and protect. There are many methods for realizing this division. Let us examine their principles before going into details. For this, we take the case of a soil property, clay content for example, and examine how it varies along a transect, which is an arbitrary line drawn in the field (Fig. 2.1).

It will be noticed that the variable studied is not observable in one portion of the transect for some reason or other (interrogation mark). Now we shall see which approaches can be applied to trace the distribution curve and to introduce the boundaries that will appear on the soil map.

The oldest and most traditional approach consists of introducing one (or several) limits in order to differentiate soil units that, in the chosen example, are richer or poorer in clay. This results in a categorial map, that is, a map comprising several discrete soil units.

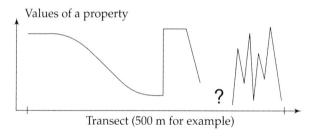

Fig. 2.1 Lateral variation of a soil property.

The problems with positioning limits are of two kinds. When the clay content changes gradually, the limit necessarily has to be somewhat arbitrary, relative to the threshold chosen (see point B in Fig. 2.2). When the clay content is not directly detectable (suppose it is the clay content at 1-m depth, which cannot be determined continuously), the boundary is vitiated by uncertainty as to its position. It is placed in the corresponding zone of uncertainty (point D). Otherwise, if the property varies rapidly in the portion DE of the transect, it is more practical to create a unit C corresponding to a soil complex than to try to position numerous boundaries more or less correctly. In other cases (point C), the boundary can be located with great precision. In all cases, details of variability of the property within the units are erased. The method is also applicable when the partitioning into categories is not based on numerical values but on qualitative observations.

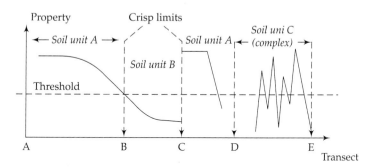

Fig. 2.2 Treatment of variability by introducing crisp limits and distinction of categories.

Another method, using geostatistics, consists of first characterizing the variation in the property studied across the entire study area. For this, samples are drawn from the field according to an often regular pattern (grid) and sometimes irregular pattern. Analyses follow, for numerical values are necessary! An appropriate mathematical function then enables interpolation of the value of the property at each point in the area and hence of the transect (Fig. 2.3). This leads to respect for the values determined at the

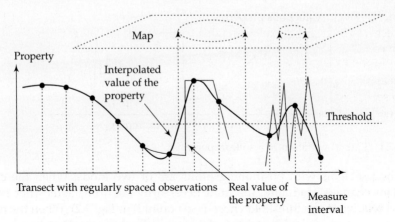

Fig. 2.3 Treatment of variability of a continuous function.

sampling points, to a proper representation of the zones in which the property changes gradually and to smoothing out abrupt changes in the property. Zones for which no values are available undergo similar, but less precise, interpolation because it is less supported by data.

The choice of thresholds corresponding to precise values of the property will then permit tracing of isovalue curves on a map.

Another method, which comes under *fuzzy set theory*, can also be used. This method consists of first separating core zones with unambiguously defined content and transition zones (dotted) whose assignment to one soil unit or other is debatable (Fig. 2.4). We shall take this further.

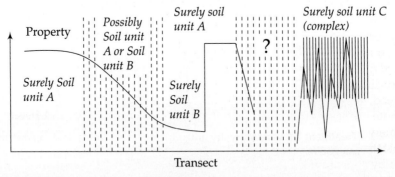

Fig. 2.4 Treatment of spatial variability in the framework of fuzzy set theory.

It is seen that there is uncertainty in the position of the boundaries (zones with dotted lines) as well as in the values of the properties of the units (zone with continuous lines). Thus we can, looking back, describe the approach in Fig. 2.2 as the *double crisp model* (Burrough et al., 1997).

One last approach consists of reducing the variability by dividing the transect into segments small enough to be considered homogeneous. This involves very detailed mapping of the area, and can sometimes be done using remote sensing. In other cases we are led to conduct what is called *grid mapping* of the land. For this, small squares are drawn on the base map (map serving as base for the soil map) and each one is studied before being assigned to a particular kind of soil (Fig. 2.5).

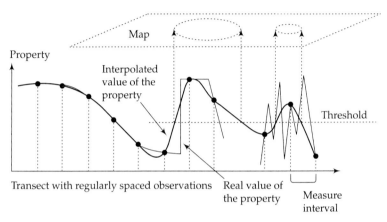

Fig. 2.5 Treatment of spatial variability with close sectioning of the study area.

2.1.2 Grid mapping

As mentioned earlier, the objective here is to present the method of arbitrarily dividing the zone to be mapped into small, similar elementary areas (squares or rectangles) and then identifying in the field the soil present in each of these areas. This leads to making observations (profiles or auger bores) at the nodes of a regular net. The method is therefore termed *grid mapping* by English-speaking writers. The elementary areas are 'sectons' according to earlier proposals of French-speaking scientists (Boulaine, 1980). Indeed, the term 'pixel' should be avoided, as it is too closely bound to remote sensing. Here, sampling is done systematically in the field: each elementary area is visited! Also, this method should not be confused with geostatistical mapping, which aims to identify the content of an elementary area by considering the contents of its surroundings.

Protocol to be followed
The procedure for grid mapping consists of three or, occasionally, four stages:
  1. Observations (auger bores or profile pits) are taken at the nodes of a regular net, often square-mesh (Fig. 2.6). This is *aligned systematic sampling*.

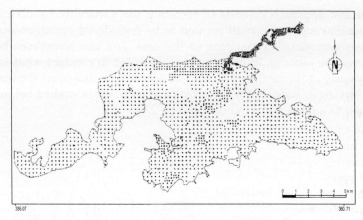

Fig. 2.6 Location of sampling sites in grid mapping; study of the Rochefort marsh (King, 1986). The white patches within the marsh correspond to water bodies, which are zones without soil observations.

2. The data from the nodes of the net are treated by a clustering process to try to arrange them into coherent soil subassemblages, each identified by a numeral or letter. At this level, the concept of similarity between soils is used. This concept, mentioned earlier (Chap. 1, § 1.4.4), will be discussed in detail later (Chap. 7, § 7.4).
3. An identifying *label* derived from the classification process is placed at each point of the net. Then the boundaries are introduced as follows: if two or more adjacent points have the same identifier, they are included in the same unit. If two adjacent points have different identifiers, they are separated by a boundary passing midway between them. Thus, in the final cartographic representation the point information is traced as an elementary area of geometric shape. This means that the observation point is presumed a priori to be representative of its nearby environment.
4. A smoothing step can possibly be applied to suppress the sectons that might be found isolated within a large homogeneous area. Contrary to what happens in remote sensing, this smoothing is normally useless in soil mapping. Adjacent points are often closely related and are naturally located in the same unit, which eliminates any marbled appearance and no artificial intervention is needed. Actually, the soil mantle is a thin layer within which abrupt variations are rare but at the same time significant (continuum concept). Thus there is no reason to reduce their number.

There are methods for determining the optimum spacing of grid points considering the variability of the milieu (Bishop et al., 2001). But they go from consequences to causes. We determine by hindsight, using the information received earlier, to what degree it was excessive! The opposite, lack of information, is the more difficult to prove the more an excessively open net erases variability so well that it cannot be noticed it had been ignored.

Disadvantages of grid mapping
The first problem lies in the difficulty of defining the similarity between two soils. In stage 2 for example, are two profiles having small differences in many properties as similar as two profiles very obviously differing in a single property? In the field, the observer is generally capable of deciding, and choosing the best solution to this kind of problem, while being aware of the limiting cases for which his own diagnosis is debatable. If a computer is used, all cases that might occur must be foreseen from the start!

The second difficulty in grid mapping is that the method is never applied in all its rigour. Actually, the regular location of profiles or auger bores in the field is often impossible. Regular location will result in making observations on the tarmac of a highway or in the water of a river. When required, the observation sites are moved from their theoretical positions or are suppressed (Fig. 2.6). When such displacement is required, it is done in a manner that does not require the subjective interpretation of the operator. For example, ten paces are taken to the north.

In the third place, the sampling points can constitute particular cases and may not be representative of their immediate environment. This is devastating in the field, but if observations are many, statistical compensations occur.

The fourth disadvantage stems from what we have just seen: some profiles having been displaced, the grid loses its regularity and the process of delimitation (stage 3) becomes more complex.

Lastly such a method gives useful results only if the observation points are many. Therefore it is prolonged, costly and tedious, and cannot be routinely applied.

Advantages of grid mapping
This method has not been abandoned because it has advantages sometimes similar to those offered by geostatistical mapping.

Firstly, as said before, it suggests relying on systematic sampling. Thus it is amenable to sustained statistical analysis as, for example, calculation of the level of impurity in a mapping unit or estimation of the spatial variability of a soil property (Chap. 5, § 5.3.6).

Secondly, this method is suitable for large-scale mapping, when the soil boundaries are not clearly marked on the landscape and there is no evident connecting thread to find them based on vegetation, topography, etc. We can cite the instance of inextricable tropical forests where mappers open up parallel tracks with the machete and take observations every twenty metres. So too cultivated alluvial lands that often are uniformly flat and where the environment hardly provides any information to help in demarcating the soils.

Thirdly, grid mapping corresponds to a particularly logical procedure. It should be noted that all observations are made before attempting the process of delimitation. This will not always hold true, as we shall see later.

Fourthly, this method allows us to take into account, at least in theory, all available information on the soil, be it quantitative or qualitative.

Fifthly, whether it is really an advantage or not, grid mapping lends itself to employing poorly qualified labour, used for digging the pits that the specialized mapper will then quickly describe.

Lastly, since delimitation is done after taking observations, and by automated techniques, simulation is possible. For example, one can visualize the map that will be obtained with one-fourth the number of observations. This enables us determine the point beyond which additional effort will be unproductive in terms of the quality:cost ratio.

Examples of grid mapping
European countries that have conducted grid surveys or at least have organized a systematic characterization of their land through profiles dug in a grid are:
- England (5×5 km),
- Austria (4×4 km),
- Denmark (7×7 km).

In Greece, mapping of 5000 ha has been done by defining 110 sectons. Principal component analysis was used to find, out of the 48 soil properties observed at each point, the 12 that carried the most information and sufficed to support the classification process (Theocharopoulos et al., 1997).

In France, grid mapping was done of the Rochefort marsh (King, 1986). It has served as basis for simulations aimed at determining the effect, on the maps drawn, of change in the criteria used to group the sectons. We will return to this question later (Chap. 5, § 5.3.2).

In total, grid mapping is a useful tool for research and of analysis of mapping precision. Rarely employed because of its cost, it is the best in some particular cases.

We can think of improving this kind of mapping by taking into account the relations between neighbouring sectons. This involves introduction of 'spatial constraints' in the procedures of multivariate classification. Suggestions have been made by Olivier and Webster (1989). In other words, it is a matter of finding the interface with geostatistics (see below).

2.1.3 Geostatistical mapping

We can barely touch upon geostatistics, a discipline that is the subject of whole volumes. Basic books can be referred to (Matheron, 1970; Journel and Huijbregts, 1989; Webster and Olivier, 1990). The last-mentioned has been written by authors knowledgeable about soils. It gives many examples from our discipline. For further detail, the reader may also refer to the voluminous special issue of Geoderma devoted to the subject (Vol. 62, 1994). It presents the papers of the Working Group on Pedometrics of the International Association of Soil Science. Lastly, there are review articles as well (Goovaerts, 1999).

Brief review of basic concepts
Let us consider a certain number of points in geographical space (for example auger holes). The values of a property (clay content, say) are determined at these points. Geostatistics has the principal objective of examining in what manner and in which proportion these values of clay content can be correlated among themselves. More precisely, we try to test whether they are more similar the closer they are. If this were exact, the application is immediate: it is possible to predict the value of clay content for each point by considering the values found in its neighbourhood. Thus a method of interpolation can be defined. The steps in calculation are described below.

(1) Definition of variograms. Let us take the pairs formed by all the points taken two by two and group them into say 15 groups according to the distance separating their two members. The mean variability, the *semi-variance* as it should be termed, among all the points located at a distance h from one another is calculated from the following equation

$$\gamma(h) = \frac{1}{2m} \sum_{i}^{m} (x_i - x_{i+h})^2$$

where x_i is the value of clay content at one point, x_{i+h} the content at another point located at distance h, and m is the number of pairs at the distance h in question.

Then the curve, termed *variogram*, linking the values of $\gamma(h)$ and the values of h is drawn (Fig. 2.7).

The distance beyond which there is no link between the values taken by the points is the *range*. Beyond this is found the maximum dispersion corresponding to the *sill variance*, which represents the variance of the population in conventional statistics. Two merged points are obviously characterized by the same clay content. In other words, the curve theoretically passes through the point (0, 0). However, if the heterogeneity at very short distance is very high, there is a *nugget variance* effect, marked by a nearly vertical increase in $\gamma(h)$ when h increases, giving the impression of a non-zero ordinate at the origin.

Of course when there is no spatial link a flat variogram (horizontal line) would be obtained.

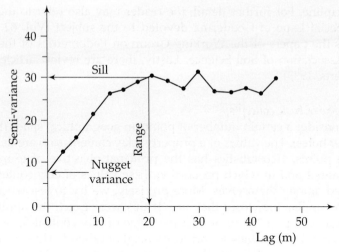

Fig. 2.7 Example of experimental variogram (Walter, 1990).

(2) Modelling of variability. In the second step, geostatistics suggests characterization of the experimental curve obtained. For example, it can be reproduced in the portion affecting the range by a second-degree function chosen from the small array of those having adequate mathematical properties. In this step, the variability seen in the field is modelled more or less completely.

(3) Interpolation. The value to be estimated at a given point will be a function of that measured at each of the known points but the latter will have a significant and specific influence only insofar as they are located at a distance less than the range. The precise contribution of the neighbours or, more correctly, the weightage given to each will be the direct result of the type of function used in the preceding step of the procedure. Interpolation methods are many; the best known is *kriging*. It satisfies several statistical constraints aimed at restricting error.

Advantages of geostatistics
In soil mapping, geostatistics has two essential purposes: first, it helps to describe the variability in soil properties in the soil units and second, it enables drawing of maps by a process of interpolation between isolated points. It is particularly useful when the property studied does not correspond at all to what is seen in the soil or the environment (for example, maps of Pb or Cd content). The results obtained are then presented in the form of isovalue curves (isopleth maps).

Conceptually, the contribution of geostatistics is very important because it represents an alternative to other kinds of mapping. Actually, it starts with the postulate that the milieu is a continuum and the changes in it are gradual. On the contrary, conventional mapping, as will be described later, seeks to recognize and spatially separate objects having their own individuality, be they horizons, pedons or landscape units. Thus it is greatly desirable and very purposeful to contrast these approaches (Voltz and Webster, 1990; d'Or and Bogaert, 2001).

Some believe that geostatistics is the only valid statistical method in mapping. This may be summarized as a syllogism: (1) geostatistics studies spatial relationships and finds them; (2) conventional statistics presumes independence among the variables; (3) therefore conventional statistics is worthless and only geostatistics has a sound basis. An important paper that appeared in *Mathematical Geology* showed that the reasoning is totally wrong and is based on confusion (de Gruijter and ter Braak, 1990). This does not prevent this error from being propagated anew. Indeed, conventional statistics requires that the points or control samples be selected at random or at least without bias; it surely also wants the different variables to be treated as independent. But it does not postulate that the different values of a single random variable have nothing to do with one another! It is completely the reverse if the Gauss law or Poisson law is referred to. Thus the reader need not be perturbed. When he reads a paper that manipulates data by geostatistics and also by descriptive statistics, he does not presume the simultaneous dependence and independence of spatial data! But perhaps, through this misleading problem and subconsciously, one comes across the fundamental question of continuity or fragmentation of the soil mantle into separate pieces with their own identity... .

Limitations of geostatistics
Geostatistics is not without theoretical shortcomings. To begin with, like all statistical methods, it presumes that various hypotheses relating to the variables and the methods by which they were acquired have been fulfilled. In particular, it postulates that variability is similar in different portions of the area, which is almost always untrue. But it is fair to say that there is a particular kind of kriging to resolve this problem and avoid excessive simplification. Secondly, geostatistics is barely applicable where abrupt discontinuities are observed in soil because this challenges one of its fundamental hypotheses. Thirdly, it is tedious to use because it must rely on at least 150 or 200 observation points. Moreover, to use it as a means of determining the variability of the soil units obviously presumes that this variability is fully taken into account. For this, it is necessary that the breadth and length of the study area each represent at least 2-3 times the range (Gascuel-Odoux et al., 1993). Furthermore, it is applied above all to quantitative variables, though methods suited to qualitative variables have appeared (Walter, 1990). Lastly, it will be naïve to think that it can, under all circumstances provide results as good as those of the mapper who uses, besides

his auger bores, the characters of the environment, his knowledge of the laws of pedology, his experience of the milieu and his aerial photographs.

This said, geostatistics is a useful tool. Also, it raises interesting questions at the conceptual level. In our discipline, it has to be coupled more closely in the future with classification methods, as pointed out earlier (§ 2.1.2). It will also be necessary to use it jointly with analysis of the organization of the soil mantle (Chap. 1, § 1.4.3) to render objective the interpolations between observations made along the vertical and to reconstruct the exact location of soil volumes in space. The way has been cleared (Gascuel-Odoux et al., 1991; Bierkens and Weerts, 1994).

2.1.4 The fuzzy approach

This method is particularly suited to investigations based on remote sensing. The mathematician Zadeh originated the method and published it in 1965. In our field of activity, land use is the primarily concern even though specialists have cleared the way for the fuzzy approach in soil science (Odeh et al., 1992; McBratney et al., 1992). A special volume of *Geoderma* is devoted to the problem (Vol. 77, 1997).

Fuzzy content of areas

Let us take the example of the remote-sensing specialist who draws a land-use map with three classes: residential (R), forest (F) and agriculture (A). Each point in space has a specific nature and may correspond to the roof of a house or to the foliage of a tree. But at the scale of the information unit, the pixel, the specialist often sees intermediate cases, as for example a forest dotted with houses with a ploughed field in one corner. This is termed a *mixed pixel*. If, say, 60 per cent is forest, 30 per cent habitation and 10 per cent agriculture in the pixel, we can write

$$P = [0.6, 0.3, 0.1]$$

P is the *fuzzy vector*. It is more commonly denoted by $P_x(y_i)$. It also gives the *degrees of membership* or *fuzzy membership values* of the pixel x at each of the categories y_i of the map [i here varies from 1 to 3, in the general case from 1 to c].

In the example chosen, it is logical to allocate the pixel to forest and similarly assign the others to their *dominant membership*. This is what some scientists term the *defuzzification* procedure. But then the *inexactness of classification* will be 40 per cent. Some prefer to call it the *attribute ambiguity*. This opens the door to various possibilities of graphic representation.

- Prepare a map of degree of belonging to forest of each pixel by splitting the interval [0-1] into classes; on this basis we can obtain three maps, one each for F, R and A;
- Choose a limit, say 0.7, and say that the membership in a category i is validated for a pixel only if $P_x(y_i)$ exceeds the limit chosen. The pixel studied above will not be considered to be sufficiently in the forest category and will remain unclassified;
- Launch a smoothing procedure; for instance, our pixel normally grouped under F could be assigned to R or A if it is isolated within a core of pixels of either of the latter types on condition that its resemblance to R or A reach a certain limit, say 0.3 or 0.4.

More generally, instead of retaining a value for characterizing the content of a pixel, we keep c values. The approach is then less simplistic.

Actually the fuzzy vectors may correspond to other things than mixed pixels and degrees of membership. Sometimes, they conceal probabilities (Lark and Bolam, 1997). Some authors mention *probability vectors*. But the concepts are quite different. Let us return to our example 'R, F, A' and let us assume that the numbers given above correspond to these probabilities. In the first analysis nothing seems to change: our pixel taken as example remains assigned to forest. Actually this changes everything: the hypothesis according to which it will correspond to a 100 per cent residential zone is then of low probability (0.3), but it is not excluded. Worse still, in the framework of the theory of possibilities, it is appropriate to answer the question 'is it possible that the pixel belong to class A?' with 'Yes, the possibility exists; that it happen is barely probable but even this, as a possibility, is a certainty and has a value of 1 over 1'. We see thus that we must be very wary and examine carefully what the data identified as 'fuzzy' conceals. The more the vocabulary used varies widely from author to author the more this is necessary.

Various authors use fuzzy vectors to represent the degree of affinity of map units or pixels of the area to the taxa of a soil classification. These taxa are generally represented by median or mean values of the variables characterizing them (Zhu, 1997).

Fuzzy nature of boundaries
On a map prepared using fuzzy vectors, let us consider two categories, say Forest (F) and Residential (R) and then examine (Fig. 2.8) the transition from the one to the other along a transect (straight line) linking two internal cores, one purely forest and the other purely habitation. We know that all the pixels of the map are grouped into the categories F and R with *degrees of membership P* that go from 0 to 1. We can thus construct curves showing how P changes for the two categories R and F along the transect.

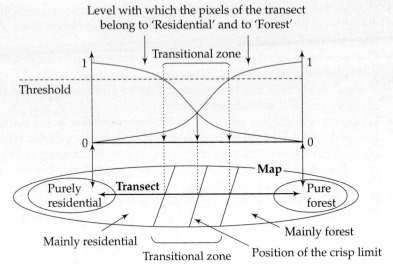

Fig. 2.8 Characterization of the fuzzy limit between two categories.

It is then easy to trace a threshold below which the pixels are neither in F nor in R. This is the transitional zone or corridor of transition. The two curves intersect at the *crisp limit*, which a field mapper will undoubtedly be able to find intuitively.

Indeed, the curves may be of different kinds. If they come from remote-sensing procedures, we can consider them as measured data. They can also correspond to a mathematical model, for example if the fuzzy vectors have been established in the framework of geostatistical investigations. They can also form an empirical model the simplest of which corresponds to the straight segments (Fig. 2.9).

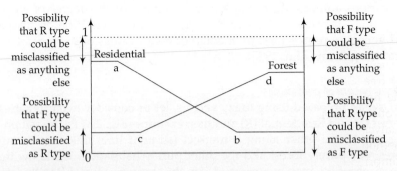

Fig. 2.9 Modelling the transition between units (Davis and Keller, 1997).

In the example in Figure 2.9, unlike that in Figure 2.8, the value 1 is not reached for the units R and F (no core zone). A risk of error exists as indicated. This empirical model can be supported by using an already prepared map. We go to the field to refine it. The curves linking the frequencies of occurrence of R and F units outside the boundaries originally assigned to them are drawn as functions of the distances to the pure internal cores of R and F. This enables us determine the slope of the straight segments ab and cd in Figure 2.9 (Cazemier, 1999).

Instead of directly using the values of $P_x(y_i)$ to characterize the transition zones and to establish the boundaries, we can calculate slightly more sophisticated indices (Zhang and Kirby, 1999):

- *The confusion index* (introduced by Burrough); it is given by $1 - P_x(y_i) - P_x(y_j)$, where i and j are the two classes corresponding to the greatest values of P for x. The confusion is zero if, for example, $P_x(y_i) = 1$ and obviously $P_x(y_j) = 0$. The confusion in the grouping is large if the group j has almost the same value as i and if the two correspond to small degrees of membership. The confusion zone is then assimilated to the transition zone. As the confusion is zero for easily identified regions, it serves to construct the boundaries starting from a map of pixels.
- *The entropy*, not very well named because its homonym in physics has a quite different meaning. Here it means a measure of the disorder traditionally used by botanists; it is given by

$$H(P(x)) = \sum_{i=1}^{c} P_x(y_i) \cdot \log_2 P_x(y_j)$$

When the disorder is large, the boundary is near.

To state that the boundary corresponds to the maximum disorder or to the zone of greatest confusion indicates that we are in the framework of models that presume a priori fuzzy and gradual boundaries. Yet there are abrupt and perfectly clear soil boundaries: cliffs, banks of a marsh, etc. Fuzzy set theory, though very useful, thus cannot cover all situations.

2.1.5 Free survey or categorial mapping

Principles

By free survey is meant the method in which the mapper works mainly in the field and is absolutely free in choice of location of auger bores and profile pits. Of all the methods this is the one requiring the most effort in analysis and understanding of the milieu. It is also the most commonly used method. It culminates in preparation of categorial maps. In the field, it comprises the following principal phases:

Choice of mapping criteria. During a preparatory field tour, which may last several days, even more than a week, the mapper makes contact with the milieu. He determines which characteristics of the soil are sufficiently variable and appropriate to serve as basis for the mapping considering its objectives. The ideal is to find mapping criteria that may easily be observed, for example surface stoniness. But it is at times necessary to refer to soil properties more difficult to observe, such as presence of rust-coloured mottles below a depth of 70 cm. The mapper must take all the time during this stage of the investigation. Beginners, worried because they do not immediately understand the modalities of spatial organization of the soils, actually tend to wish to proceed very rapidly to avoid what seems to them shameful hesitation. In France, during the mapping of the Haut Vivarais, more than six days were required before the actual investigation was started in suitable conditions. This was the period needed to make hundreds of preliminary observations and to properly understand that topography played a greater role in the spatial organization of soils than differences in the parent material (granites, gneisses, metaschists). Choice of criteria is indeed a fundamental step in mapping on which the quality of the final study depends. The aim is to draw up a rough plan for a method to partition the area and to refine and enrich it as the survey progresses to culminate in the final elaboration of the map legend.

Delimitation of the soils. The soils are delimited on the basis of selected criteria, with the help of surface observations, auger bores and inspection of cuts found during the survey (quarries, fresh road cuts, building foundations, etc.). Features clearly separating various soils are obtained, features that can often be traced directly on the aerial photograph. These are the well-known *crisp limits*. Every evening, after field work, these features are checked and refined under the stereoscope. The features are then transferred every day to a base map to make a sketch, that is, a provisional map or a 'minute'. At the same time, an embryonic legend is drawn up, which will be completed, improved and restructured as the work progresses. The procedure for drawing the delineations in the field will be described as precisely as possible in Chapter 5, § 5.1.

Fine characterization of the soils. The soils having been delineated, they are characterized by judiciously digging a limited number of profile pits. These serve to describe the properties that are not visible in the auger bores or in natural cuts and, also, for drawing soil samples for analysis.

Correlation tours. Explanation of the procedure would not be complete if these tours with two objectives are forgotten. If the mapper works alone or with an assistant, it will be useful from time to time to rapidly retraverse the sectors already surveyed (mapped) to verify that he does not deviate and that he continues to name the soils of the same type in the same way and groups them in the same unit. If many teams work in parallel, it is absolutely indispensable for them to do these correlation tours together. Thereby complete agreement will be reached on mapping criteria and vernacular, mnemonic or scientific terms to name the various units provisionally or finally.

Advantages and disadvantages
Mapping of the Free Survey type is a widely used method. Based on direct observation and executed in the field, it leads to reliable but rather costly results. It could be criticized as a bit empirical because it is not far from expertise. Also, the mapper should be especially competent. For example, he may encounter problems of **bias** because the area has not been systematically surveyed. In the forest zone of the Limousin in France, the following adventure befell us: observations had been made using a car. They were taken along rural roads, of course avoiding parcel borders and ditches; but this precaution was not enough. Indeed, the roads in the region tended to skirt the bogs located in the depressions. In short, we were in danger of not finding soils with redoximorphic features whereas they were predominant in the area. Aerial photographs were of little help in this case because the zone was completely wooded. Constant attention is therefore necessary to be certain of the representativeness of the information gathered.

Morphopedology
Morphopedology is a particular kind of Free Survey. It has the objective of putting observation of soils in the framework of regional geomorphology. A superficial analysis could lead us to think that the geomorphological approach to mapping, like the climatic approach (zonation in the Russian school) is a sector-based attempt to observe soils by arbitrarily giving preference to one of the factors responsible for the spatial organization of the soil mantle. Thus the same object soil will be considered from different points of view. But this is not quite correct because each of these methods has its own merits and its favoured domains of application. Soil zonation is well suited to schematic mapping of soils of the world. The geomorphological approach is very useful at the regional scale. It started with studies conducted since 1970 by ORSTOM, the organization that became IRD, and by scientists of l'Institut de Recherches Agronomiques Tropicales (IRAT) in Montpellier. A special issue of the journal l'Agronomie Tropicale was devoted to the question (see Kilian, 1974). The first totally morphopedological map (and termed so) was published in 1970 by R. Bertrand.

Morphopedology proposes an original view concerning time and, simultaneously, space.
- In time, pedogenesis (soil formation) is considered in its relationships with morphogenesis (formation of relief). When pedogenesis prevails over morphogenesis, the soil becomes deep. In the opposite case, it becomes thin and may even disappear completely through erosion. This means that we are concerned with the rate of ablation in relation to hydric regime and wind action.

- In space, in a given area, certain landscapes appear to be linked, to the extent they seem to represent the different stages of transformation by weathering and erosion of the same original geological structure.

To observe all this it is necessary to view the natural milieu with sufficient detachment, that is, to consider areas covering several tens, hundreds or thousands of km² and at times to envisage the continental scale (Bertrand, 1998b). One must also take into account development periods counted in tens of millions of years. This enables us imagine the original situation and to reconstruct the morphological evolution that has taken place for finally determining the placement of a give landscape in the chain of transformations. Figure 2.10 was constructed on the basis of studies conducted in Africa from the dry Sudanian zone to the humid tropical zones (Tardy, 1993; Bertrand, 1998a).

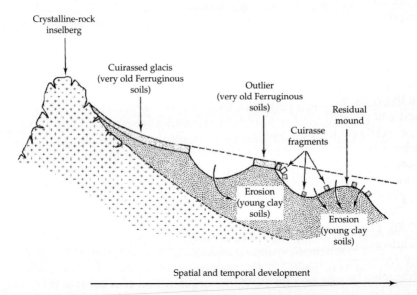

Fig. 2.10 Example of the relationship between soils, morphology and age: evolution of cuirassed glacis in Africa.

The structure of the landforms leads us to believe that an iron cuirasse existed originally and totally covered the entire region. It was formed in dry climate. This cuirasse was dissected in a more humid climate by a hydrographic network that sank deeper leaving behind small plateaus aligned to the same altitude. Then some of the plateaus were attacked from the flank by backward erosion. Their cuirasse was then dismantled from its edges until it completely disappeared. At this stage, the plateaus gave way to hills with rounded summits. Study of the aerial photographs thus permits us, starting from the shape of the interfluves, to deduce the presence or absence of the cuirasse and normally associated soils. The soils are thus replaced in space, time and in the framework of climatic changes. From this we can better make out their origin, age, and mode and speed of transformation. Application in cartography is obvious since it is practically impossible to visit all the landforms. In very concrete terms, the breaks-in-slope that line the sides of the landform separate landscape units associated with specific soils and serve to locate the boundaries.

The same approach was used for the soil mantles of Brazil (Pellerin and Queiroz-Neto, 1992). More generally, morphopedology shows that it is necessary to know how to be detached and handle the 'macroscope'. Some of the results obtained by this method are fascinating. Various tropical landscapes are revealed to be related; they carry thick regoliths; they have evolved over considerable time by enduring climatic changes that should be placed in the category of great upheavals undergone by the Earth (continental drift, climatic changes from the Tertiary to the Quaternary). Such a synthesized view enriches mapping and soil science.

Phytoecology

As the name indicates, phytoecology has the aim of studying the natural milieu by going about it starting from the vegetation. It introduces the idea of indicator role of vegetation and seeks to put plants back in their climatic, soil and geomorphic environment. It gives great importance to soil-vegetation relations in particular and is thus a method of soil mapping even if scientists may at times grumble that the best way of studying the soil is to observe it directly. Phytoecology is distinguished from the pedoclimatic zonation of Dokoutchaiev and his followers by its ambitions: it claims to be applicable to large-scale mapping and not just continental-level mapping.

Developing fast when inventorying was fashionable in Europe (1950-1970), phytoecology has since become subject to some disaffection. But its future seems rather bright for at least two reasons. Firstly, modern computer techniques allow it to develop to a more quantitative approach. For example, the validity of prediction of soil pH from vegetation in the Vosges could be quantified (Gegout, 1995). Secondly, it has necessarily to be done when one wishes to tackle soils through remote sensing.

2.1.6 Synthesis

Mapping methods are very varied. Also, some can be combined. To exaggerate a little, each mapper has his own method. The presentation we have made is necessarily schematic.

It is striking to see that each method starts with a model, kind of relating a priori to the organization of the area to be subdivided. Each approach has its features, advantages and disadvantages. According to the characteristics of the milieu studied, the means available and the aims pursued, one or other will reveal itself the most suitable. The most recent (geostatistics, fuzzy set theory) are evidently the most attractive at the level of scientific thinking. Their conceptual contributions are great. We shall see this again when we touch upon the problems of precision. But that does not mean that these methods could replace the good old one that consists of primarily working in the field. We shall see why.

Conventional mapping versus remote sensing
Examples of good soil maps prepared entirely by remote sensing have been rare and always have pertained to very specific milieus since the method came into being. Of course, one would like to do without a costly soil survey and to keep to some site checks in the field. But experience has shown that this is well-nigh impossible! At the worst, the advent of remote sensing has been, in a way, damaging for soil mapping. It has allowed sponsors to believe that they can have better results for less cost. College exercises have indeed multiplied. But the number of hectares actually mapped in Europe has stagnated because, although remote sensing is useful, it is not an alternative to conventional mapping. On the other hand, we shall see in Chapter 7, § 7.3.3 that remote sensing can help in proper choice of field observations. The sampling operation can thus be reduced and optimized. But there is no question of its being reduced to almost nothing!

But in some respects remote sensing has proved its great usefulness in investigations of 'regions'. The 'region' concept is not purely pedological. It integrates other characteristics such as landform, plant cover, agricultural land use by man, all things under the sun. In the French vineyards of the Côtes-du-Rhône for example, the regions that give the Grenache and Syrah grape varieties their special quality could be identified through remote sensing (Vaudour, 2001). When all is said and done, for us remote sensing is a useful tool in soil mapping, and not a soil-mapping method.

Conventional mapping versus fuzzy set theory
On many points scientists set fuzzy maps against categorial maps prepared conventionally by matching crisp limits.

On the one hand, fuzzy maps are more precise. This is an evident truth that needs no calculation to be established, in spite of various, sometimes tedious, attempts at quantification. Indeed, a categorial map gives a schematic view, often useful, but necessarily a bit removed from reality. On the contrary, an uncorrected aerial photograph or a raw remote-sensing image is a minimal distortion of reality but is useful only to the expert if no effort is made at synthesis of what can be seen. From all evidence, the fuzzy map is positioned between the two (Fig. 2.11). This is when the adage 'All that is simple is untrue and all that is complicated is of no use' should be remembered.

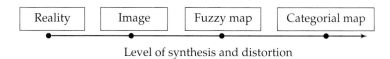

Fig. 2.11 From reality to categorial map.

In the literature on the subject there is a spectacular and interesting example (Gruijter et al., 1997). Even if it be slightly in error, the (categorial) soil map of the Wesepe II area (Netherlands) in colour on scale 1/50,000 teaches us something about the distribution of soils. On the other hand, the continuous soil map of the same area in contrast resembles a painting by Georges Seurat (1859-1891). From a distance of 30 cm, no organization is seen!

Fuzzy maps are particularly useful to demonstrate the limitations of the conventional approach. Let us consider an area in which we find, say, 26 kinds of soils, designated A to Z. The mapper put in the field seeks to identify the soil at a given point. He is going to find it is C, which amounts to excluding all other possibilities considered as 'equally and completely wrong' (Woodcock and Gopal, 2000). Indeed the human mind struggles to do better: it can recognize a category or say 'I do not know', but it cannot mentally calculate a degree of membership of such zone to each category. A fuzzy map, on the contrary, suggests a more precise and qualified diagnosis. But to return to the expert working in the field, an attempt was made to pass from the Yes/No test to a more refined diagnosis (with five levels). The context is that of verification of a cartographic document prepared earlier (Woodcock and Gopal, 2000):

- Absolutely right, the match is perfect (score 5/5),
- Good answer (score 4/5),
- Reasonable or acceptable answer; may not be the best possible answer (score 3/5),
- Understandable for such or such reason, but globally wrong (score 2/5),
- Absolutely wrong (score 1/5).

Conventional mapping versus geostatistics
Geostatistics has its eulogists. As all soil boundaries are gradual, at least when viewed from up close, some users came to think that geostatistics was going to supplant all other mapping methods and to support a new paradigm, founder of a revamped approach. This would finish maps with crisp boundaries. Some would go even so far as to humorously dispatch conventional mapping to the 'Dark Ages' while comparing geostatistics to the 'Age of Enlightenment'. We shall not follow them completely in recommending their luminous synthesis (Heuvelink and Webster, 2001). In our turn, let us handle the exaggeration: it is perhaps because soil boundaries are scarcely

visible in nature that the human mind claims its perceptiveness by being able to locate some of them conceptually!

Above all, boundaries are not the most important thing in cartography, even if they are a means to separate categories. The contents are more important. For example, it is well known that the boundary between Ferralsols and soils of desert zones on a soil map of the world does not correspond to anything tangible in the field and, besides, this is not of importance.

Also, geostatistics uses a lot of numerical data; so it is costly in survey time and requires many analyses. It obviously cannot be routinely used over large areas.

Lastly, conventional mapping is the only one of all the methods presented that had been conceived for easily handling qualitative variables: soil colour, texture, etc.

An attempt at correlation between the methods

The methods can be regrouped on the basis of two principal criteria (Table 2.1).

Table 2.1 Different kinds of soil or related mapping

Mapping Technique	Mapping Purpose		
	Soil alone	Soil seen through its environment	Other natural characteristic
Extrapolation (inductive method)	Free survey	Morphopedology, delineations based on climate or physiography	Agricultural geology, phytoecology (+statistical display)
Systematic approach	Grid survey	Fuzzy approach based on vegetation	No known example concerning these approaches to soil mapping
Interpolation (deductive approach)	Geostatistics	No known example concerning these approaches to soil mapping	

To begin with, the methods are distinguished according to the element that is mapped. Some clearly have for objective the delimitation of soils, and that too directly. Others propose a totally indirect approach. The soil is presumed to be closely linked to another constituent whose spatial distribution is then studied. *Agricultural geology* (Chap. 10, § 10.6.1) falls in this category. In the 19th century, the existing geological maps were interpreted in terms of soils and agronomy. From this point of view, phytoecology, associated with remote sensing, is comparable. Its aim is not to delineate soils but it claims, justly no doubt, to give indications of the nature of the soil mantle. There are, between these two groups of methods, intermediate methods in which the objective is to separate the different soils, but indirectly. The pedoclimatic zoning of the pioneers of the Russian School of Soil Science is in this category (researches of Glinka and Agafonoff). The same is true of soilscape zoning whose significance and philosophy were discussed at the end of the preceding chapter. Lastly, fitting in this category is morphopedology about which has been written 'it reaches soil science through morphology' (Decade, 1984).

Furthermore, it is necessary to distinguish the methods according to their operating procedure. We have seen in Chapter 1 that the main problem was going from point observations to characterization of the entire area to be mapped. For this, there are three principal ways of proceeding. The first suggests selecting observation sites, one by one, benefiting in each case from the experience acquired till then (the 'step by step' approach). Thus, selection of one site is often the result of what had been learnt from the preceding one. Free survey relies on such an approach. It enables us determine the distribution of soils based on a logical reasoning using a very small number of observations. Grid survey corresponds to exhaustive survey of the entire area to be mapped. Observation sites, therefore, are very numerous and selected according to a systematic sampling procedure (grid sampling). The third approach, using geostatistics, consists of studying the soil at points and applying mathematical methods for interpolation and generalization over the entire area. In this case, the location of observation sites is sometimes similar to that in a grid. But it is often obtained by a system adapted from random sampling (Chap. 4, § 4.6.2).

2.2 GRAPHICAL TREATMENT OF PEDOLOGICAL OBJECTS

In Chapter 1 we saw how specialists have schematized and modelled the different kinds of organization in soils they could find in nature: horizons, soil material volumes, soil series, more complex organizations... . We shall now see what graphical representations have most often been suggested for presenting these organizations on the map, to wit, in two dimensions.

2.2.1 Spatial representation of point data

In many cases, mappers use point information (soil analyses, vegetation census, etc.) referred to x and y axes. It is possible to represent this information directly with no attempt at interpolation or indication of the relationships among sites (Fig. 2.12). What some term a *cartogram* (King, 1986) is obtained. If the observation sites are many and their mode of representation

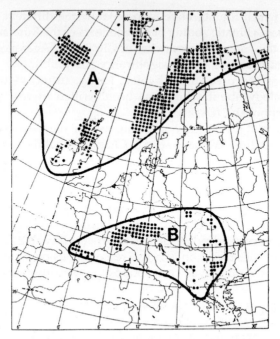

Fig. 2.12 Example of a cartogram: Subarctic region (A) and Alpine region (B) where the plant *Oxygria digyna* is found (Ozenda, 1986).

is judiciously chosen, visual synthesis can be immediate. Suitable computer programs are available for getting good results at this level.

Some English-speaking authors use the term cartogram for what French speakers prefer to call *cartographic anamorphosis*. This refers to maps in which the pattern is modified so that, say, the area of each country in Europe appears proportional to its population or gross national product. Such a pattern would be useful only if the reader knows well the appearance of the same map without distortion.

2.2.2 Graphical treatment of map areas

The different kinds of boundaries and contents
On a categorial map, for example one showing series (Chap. 1), the pedons that belong to impurities and do not match the general definition are not represented. Consequently, the elementary soil areals composing this series are 'homogeneous', at least at the graphical level. In addition, their boundaries are represented by a line even if they are gradual or not definite. These conventions have long been accepted unanimously by specialists. Then suggestions to proceed differently were made, because the new techniques enabled us go further and to show uncertainties in the contents and

boundaries. This leads to the idea that there are four cases that can occur according to whether the contents are well defined or not, and whether the boundaries are well defined or not (Fig. 2.13).

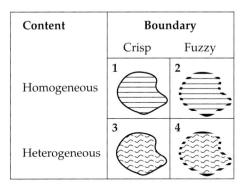

Fig. 2.13 The four types of elementary soil areal that may be found in nature.

For some years when cases 2, 3 or 4 were encountered in the field, they were treated and represented as if they were cases of type 1. No means was available to do otherwise. Also, the procedure did not have disadvantages alone: after all, the main thing was put on the map with synthesis in mind. Today, fuzzy set theory has enabled conceptual progress which, however, has scarcely been translated into modes of representation if we except the possibility of replacing the line representing the boundary by a *buffer*, a sort of corridor whose width could be the image of the uncertainty in the boundary. However, such a representation, while appropriate in a given case, would make most soil maps needlessly cumbersome.

Totally arbitrary boundaries
In some cases the boundaries drawn on maps pertaining to the soil may be entirely arbitrary. In Brittany (Aurousseau, 1987), analytical data were available for 20,000 samples of unknown location but referenced by their belonging to different villages (administrative districts). The data could be used to calculate and graphically represent, by administrative district, the mean levels of P_2O_5, carbon, etc. This sort of exercise, the utility of which will be shown in Chapter 9, § .1, is termed *statistical mapping*. The terminology is not quite appropriate, because there is no phase of cartographic delimitation in it. This approach has both advantages and disadvantages (Mathieu et al., 1993). On the one hand, to express mean values by administrative district is to have a firm grip on the concerns of the agricultural world, because agricultural advisers can set about their activity of promoting better controlled fertilizer application at exactly this level. But on the other hand, these groupings on administrative basis are done without evaluating the geographical and pedological reality; they could thus be very artificial

and barely precise at the same time. We shall touch upon calculation of the corresponding error in Chapter 7, § 7.5.

Various computer programs make the work easy. The main thing is to use digitized map bases representing for example, French administrative districts, Swiss cantons or various countries of the world. So, it will suffice to fill in a table indicating the value assigned to each prerecorded administrative unit, then to use the program to organize the graphical representation. The method is very popular in politics when it is urgently necessary for a magazine to draw the map of results of an election.

2.2.3 Graphical treatment of three-dimensional soil volumes

Although the horizon is the smallest mapping object, its cartographic representation for all that is not the simplest, quite to the contrary. Actually it is scarcely used. But it has its place in our synthesis.

Representation of the depth of occurrence of horizons
An isopleth map can be drawn of the depth of occurrence of such or such a horizon. The map corresponds to the curves showing the level (negative altitude) of the top of the horizon in question (Fig. 2.14, left half, plan view). These maps can be generated following exploration with radar (Collins and Doolittle, 1987). The vertical sequence of horizons can also be represented in perspective by making use of the computer (Fig. 2.14, right half). This representation obviously pertains to small areas examined in detail.

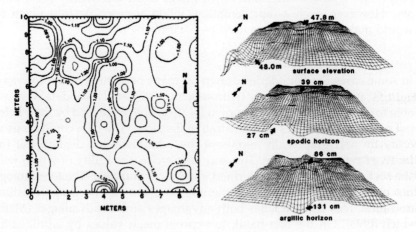

Fig. 2.14 Representations suggested by Collins and Doolittle (1987). The drawing on the left corresponds to the argillic horizon shown in the right hand diagram.

Tomograms

A *tomogram* is a view of the soil mantle from above obtained as if the soil has been scrapped off to a determined depth by closely following the relief irregularities on the surface. Figure 2.15 presents an example of the vertical section (profile) and one of the horizontal section (tomogram).

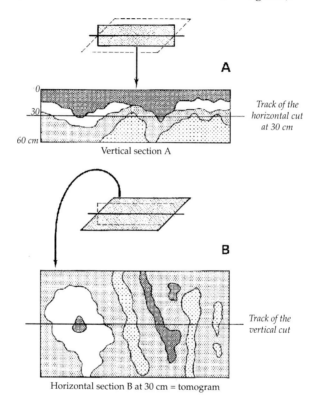

Fig. 2.15 Example of tomogram.

Two perpendicular planes intersect along a straight line; the reader can verify this by making a photocopy of Fig. 2.15 and folding it suitably that the two drawings are similar at the single level of this line represented in two sections (the author's pride prevents him from suggesting the page be torn out for making the same examination). All this might appear to be an amusement for the specialist, but it is nothing like that. The human mind, which finds it hard to play with complex volumes, is forced to resort alternately to one and then to the other of the two-dimensional views proposed in Fig. 2.15. Thus the map is an 'area tomograph' while the description of the soil horizons in the map legend furnishes the third dimension (the vertical). At the same time, the report generally accompanying the maps mainly presents profiles and then recreates the lateral dimensions by text and explanation.

So-called 'analytical' display

Analytical display can be contrasted to the *synthetic display* in categorial maps. This analytical display enables us draw the vertical variability of horizons within a soil map. It was used, for example, during the mapping of the department of Aisne on scale 1/25,000. The bases for this analytical display were:

- the texture is shown by colour: sandy soils in red, loams in green and clay soils in blue;
- the depth at which a particular texture appears is linked to the orientation of lines: flat colours on the surface, lines at 45° for a texture appearing between 60 and 80 cm, vertical lines for a texture appearing below 80 cm, etc.;
- a series of specific symbols serves to indicate redoximorphic features, presence of calcium carbonate, proportion of coarse fragments, etc.

Figure 2.16 shows schematically an analytical display comparable to that of the Aisne maps, which cannot be properly reproduced in black and white.

Analytical display is rather rare and might disappear. Indeed, use of the computer enables us quickly obtain graphical solutions on demand, on which only useful information is given. There is then no room to want to put

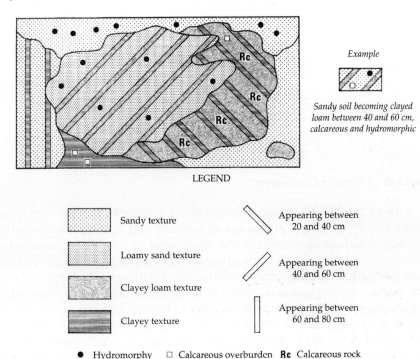

Fig. 2.16 Display of the analytical type.

everything on the map at the risk of making it illegible. Actually many published maps correspond to a predominantly synthetic display but lightly mixed to the extent that black hatching serves to characterize the underlying rocks. It is elsewhere noted that use of soil formulae (§ 2.3.3) amounts to selecting a synthetic display of map boundaries, associated with an analytical naming of the content of units.

2.3 TYPES OF LEGEND

2.3.1 The concept of taxonomic references

At least in theory, there are two ways of delineating soils *on large scale*: with or without direct reference to an existing classification (see 'image soil' and 'concept soil' in Chap. 1, § 1.4.1).

The first procedure consisting of making the 'mapping-classification' amalgam has long been preferred by some scientists. With passage of time, this way of working has been severely criticized (Pedro, 1989a). It leads to presentation of maps in which the units have esoteric names, cutting off soil science from its users. Later, it leads the mapper to arbitrarily chop up the soil continuum according to a classification system not necessarily suited to local distinctive features. The characteristics satisfying the requirements of arrangement and division are no longer drawn from real life. Lastly, the procedure above all introduces much confusion in the map legends when the classifications are modified.

So we realize that large-scale mapping should be done on the basis of dividing the area into typical units—homogeneous if possible—without first worrying about finding out if this would correspond or not to what such or such classification proposes. For example, we will without further ado group together 'red gravelly soils of high wooded terraces'. Americans proposed their series with this idea, each one drawing its name and definition from the local features of real soils. Attribution to an official classification is proposed at a later stage, to facilitate dialogue with the rest of the scientific community. So this attribution has various features (RP, 1995):

- *It may be perfect or imperfect*. It is perfect if the soil exactly matches in each of its characteristics a type in the classification serving as reference. The attribution is imperfect in other cases; they are *taxadjuncts*.
- *The attribution may be unique, dual or multiple*. The attribution is unique if there is only one matching taxon in the reference classification. It is dual or multiple if there are two or more matching taxa. Obviously a dual or multiple attribution is, by definition, imperfect with respect to

each of the taxa involved. If not, this will otherwise signify that these taxa have the same definition or even incomplete definitions not based on the same properties. The classification system will then be gravely in doubt. Far from being theoretical, the concept of dual or multiple attribution is, on the contrary, operational and easy to use, in particular if genetic grouping systems are referred to. Indeed it is often quite easy to realize that a particular soil has a genesis that places it midway between several other typical and well-known soils. If the attribution is not unique and perfect, the soil studied is said to be *intergrade*.

Soil classifications such as *Soil Taxonomy* of USA (USDA, 1998), WRB of the international community (Deckers et al., 1998) and RP of French-speaking countries (Baize, 1998) allow an adapted designation for these intergrade soils.

On the other hand, **on small scale** (1/500,000 and smaller), description of the soil mantle implies that the division proposed relies on general concepts. Direct reference to a classification is required.

All this has been summarized by Gaucher (1977) in the expression *Is it a question of knowing the soils or of identifying them*?

2.3.2 Structure of legends

Legends are not disorganized lists of soil units. Each legend groups the units observed on the basis of some criteria judged essential. The principal structuring elements useful in legends are the following:

- *Important features of the soils.* For example, some legends of studies done earlier on drainage highlight texture and hydromorphy (Favrot and Bouzigues, 1975).
- *Features of the environment.* Many legends emphasize the organization of soils within large geomorphic units.
- *Soil classification.* The legend may present the units in the order suggested by a particular classification, even if the soils had been delimited in the field without direct reference to that classification. It is usually enough that the units observed are labelled in accordance with it.
- *Combination of the above elements.* In order to better inform the reader, an attempt is often made to present in the legend the soil environment, domain by domain, in tabular form. Given below is a very brief simplified extract from the legend of the *Soil Map of England and Wales* on scale 1/1,000,000 (Table 2.2). But the table in certain other maps is excessively detailed and reading it becomes difficult.

Table 2.2 Extract from the legend to the *Soil Map of England and Wales* (SSEW, 1965)

1: Name	ALLUVIAL GLEY SOILS	GLEY-PODZOLS
2: Map symbol	Map unit 5	Map unit 65
3: Associated soil group	Brown alluvial soils	Stagnogley soils
4: Parent material	Marine alluvium	Glaciofluvial
5: Characteristics	Deep loam, locally clayed, ground water levels controlled by ditches and pumps, unstable topsoil structure under cropping	Deep sandy soils with high groundwater and with subsurface pan (hardened horizons) associated with clayed soils with impeded drainage
6: Relief	Level; <10 m	Level to gently sloping; 10-150 m
7: Land use	Horticulture	Arable

2.3.3 Standard mapping symbols

Some authors feel that legends are not sufficiently synthesized and that they do not permit instantaneous understanding of the main soil properties. They suggest *standard mapping symbols* for describing a given soil in a compact and simplified manner by highlighting what is important in land-use planning or in pedogenesis. This has at least two advantages: first, mappers are thus led to push their synthesis to the limit, which compels them to clearly distinguish the essentials from the non-essentials; second, the symbols proposed are compact and so can be entered on each elementary soil areal to facilitate reading and interpretation of the map. Many systems are followed. We may mention in particular that of the Department of Land Management of the canton of Vaud (Switzerland), the one proposed by the Canada Agricultural Research Branch (Canada, 1981) and that used in France by the Soil Science Laboratory of the Faculty of Science in Besançon. The last will be presented by way of example. It has been gradually perfected (Bruckert, 1987, 1989; Lucot and Gaiffe, 1994, 1995). Thus it has undergone a few changes and adaptations to specific conditions. In principle, the expression has three components in the form $(X/Y) \cdot Z$.

- the term X serves to record the characteristics of the zone explored by roots (texture, presence or absence of calcium carbonate, whether the soil is transported or not);
- the denominator Y is used to indicate the position and nature of obstructions to root penetration (depth, rock type, level at which excess water appears);
- The third component, Z, is used to indicate the presence of various constraints: stoniness, susceptibility to erosion, soil workability… .

The four panels in Table 2.3 serve as a guide to using the system. An example is also given, the representation of a noncalcareous silty clay soil on a hard limestone layer between 30 and 60 cm and strongly affected by excess water.

Table 2.3 Symbols for the soils of the University of Besançon

Code	Texture
a	clay
l	silt-loam
s	sand
h = rich in humus	
* = calcium	
' = transported material	

Code	Constraints
i	Susceptible to drought
e	Susceptible to erosion
g	Lack of oxygen (from 1 to 4)
pi	Stoniness (from 1 to 9)
t	Workability limitations
b	Sealing soil
k	Boulders

Code	Type of obstruction
K	Hard calcareous stone
J	Powdery $CaCO_3$
M	Marl
O	Clay
G	Sandstone
S	Coarse alluvial deposit
T	Compacted horizon
E	Siliceous silty horizon
Y	Redoximorphic mottles
W	moraine

Code	Depth to obstruction
1	From 0 to 15 cm
2	From 16 to 30 cm
3	From 31 to 60 cm
4	From 61 to 90 cm
5	More than 90 cm

Example: $\dfrac{\text{la} \cdot \text{g4}}{\text{K3}}$

In this system, suited to local conditions, the summary given for the soil is oriented towards highlighting constraints that limit the possibilities of agricultural use. Other systems, on the contrary, are turned more to soil classification and genesis.

Such individual characterization of each map zone can, if necessary, substitute for all other legends. More precisely, we go from a *closed legend* (list of soils numbered from 1 to *n* and linked to as many boxes) to an *open legend* (symbols allowing the description not only of the soils observed but also of all others likely to be present). However, decoding of the symbols by readers of the map may be difficult and may lead to errors when working without the help of a computer. A test was done with geography students at the University of Calgary (Canada). It proved to be largely negative. Thus it is better to use the symbols with the backing of a conventional legend.

2.4 SCALES

2.4.1 Evolution of scale in the course of mapping

We must distinguish scale of survey, scale of provisional maps (still called 'draft maps' by some) and scale of the final map version. A good solution consists of dividing the linear scale by about two at each stage of the work. For example, when survey is done on a map base at 1/5,000, the provisional map is at 1/10,000 and the published map is at a fixed scale of 1/20,000. This is exactly what was done in the soil map of Belgium in 457 sheets. In this way, precision is not lost when the boundaries are reported. The examples given below (Table 2.4) represent the working methods very conventionally used.

Table 2.4 Evolution of scale during mapping

Survey	Provisional or draft map	Publication
1/5000	1/10,000	1/20,000
1/30,000	1/50,000	1/100,000
1/50,000	1/100,000	1/250,000

We should always avoid making surveys at a scale too much detailed compared to that of the final document to be printed. Indeed, this will be an encouragement for execution of very precise mapping, without reference to what is demanded. In such a case, the mapper tends to put too much data on the final map which becomes difficult to read. This fault, corresponding to a lack of synthesis, is very widespread. The optimum is indeed very difficult to define. Actually, the same map may appear simple or complex according to whether the choice of colours is judicious or not.

D.G. Rossiter, in his course freely available on the Internet, proposes retrospective calculation of the true scale of a map by enlarging or reducing it until the boundaries in it represent a suitable density of features (search for site *ROSSITER SOIL RESOURCE INVENTORY ITC*).

2.4.2 Principal mapping scales

At least five principal scales are distinguished:
- *Survey at 1/10,000* and more detailed. These are the very large scales for which applications at the level of parcel or group of parcels can be envisaged. Instances are survey before drainage, land consolidation, etc.
- *Survey at 1/25,000*. This scale is typically used in plains or valleys before organization of irrigation. It is also a large scale but can be used for regional planning.

- *Survey at 1/100,000*. The scale of 1/100,000 was chosen for the systematic inventory of France under the Service d'Étude des Sols et de la Carte Pédologique de France (SESCPF). It is an intermediate scale quite suitable for characterization of the natural milieu.
- *Survey at 1/250,000*. From this scale on, considered small, we abandon soil mapping in the strict sense for orienting ourselves to physiographic mapping with the aim of delineating and characterizing Small Natural Regions (SNR) of soil. This scale is thus useful for providing a rapid inventory of soils at the scale of a small country.
- *Survey at scale 1/1,000,000 or 1/5,000,000*. This corresponds to surveys with essentially didactic and pedogenetic significance. For example, the soil map of France (on 1/1,000,000) or the soil map of Europe (1/5,000,000).

In English-speaking countries or those influenced by English, many maps have been published on scale 1/63,360, corresponding to one *inch* (2.54 cm) to the *mile* (1609.34 m).

2.4.3 The concept of virtual scale (Boulaine, 1980, modified)

Consider a given map of area S' (measured on the paper). For this a definite number Ob of observations have been made with description, be they auger holes, profiles or pickaxe cuts. The ratio Ob/S' then represents the observation density estimated on the map, that is, in a way allowing us to be free of the scale or, more exactly, allowing comparison of many maps done at different scales. If:
- E is the scale (say 1/10,000),
- L' is a length measured on the map,
- S' is an area measured on the map,
- L is the corresponding distance on the ground,
- S is the corresponding area on the ground,

we have
$$L' = LE$$
$$S' = SE^2$$
or even
$$S' = SE^2 \cdot 10^8,$$
if the area on the map is expressed in cm² and that in the field in ha.

In principle, it is accepted that it would be necessary to obtain one observation per 0.25 cm² on the map, that is,

$$\frac{Ob}{S \cdot E^2 \cdot 10^8} = 4 \quad \text{or} \quad \frac{Ob}{S} = 4E^2 \, 10^8$$

Actually this ideal density is very high and **the required density** is only

$$\frac{Ob}{S} = \frac{4}{K} E^2 \cdot 10^8 \tag{1}$$

where *K* is the **mapping efficiency**, ranging (Boulaine, 1980) from 1 to 20 according to the following factors:
- accessory documents available: aerial photographs, topographic and geological maps…,
- personal efficiency of the mapper,
- readability of the landscape.

Therefore, the mapping efficiency can be minimal for a virgin-forest landscape, in a region devoid of topographic and geological coverage, with a beginner as surveyor. Indeed, we shall see in Chap. 3, § 3.2.2, that even excellent mappers hesitate to claim a mapping efficiency greater than 10. Be that as it may, this approach is worthwhile from various view points:
- It shows the considerable part played by factors difficult to take into account and the level of approximation (from one to twenty times).
- It enables calculation of the number of observations to be actually taken:

$$Ob = \frac{4}{K} E^2 10^8 S \qquad (2)$$

- It serves to compare the actual observation density to the required density and thus to define the **screening degree** of the milieu, D^0A. The screening degree is generally a number smaller than 1 and, at the limit, could be equal to zero:

$$D^0A = \frac{Ob/S}{4E^2 10^8/K} = K \frac{Ob}{4E^2 10^8 S} \qquad (3)$$

- Lastly, one can deduce from all this the **virtual scale**, that is, the scale for which the degree of screening will be equal to 1. The equations are:

$$D^0A = K \frac{Ob}{4E_r^2 10^8 S} \qquad (4)$$

$$1 = K \frac{Ob}{4E_v^2 10^8 S} \qquad (5)$$

where E_r = real scale
  E_v = virtual scale.

Therefore,
$$E_v = E_r (D^0A)^{1/2} \qquad (6)$$

This equation (6) is useful. For example, if one has only half the information theoretically required taking into account the mapping efficiency (D^0A = 0.5), the virtual scale will be 70 per cent of the real scale, which gives, say, 1/35,000 instead of 1/25,000. From another point of view, it is misleading to the extent the virtual scale is independent of the real scale as shown by its definition (cf. equation 5 above). Thus, in a given case, whatever the announced scale, we will always have the same virtual scale, which is logical!

Conclusion and perspectives

Chapter 2 has shown the great variety of mapping methods: variety of aims of zonation, variety of survey techniques, variety of modes of graphic representation and of legends. We still have not drawn up an exhaustive catalogue of the possibilities, which indeed are unlimited to the extent we can combine in all sorts of ways the options proposed for surveying or drawing of the map! This would mean that many means are available for treating all sorts of scientific or applied problems.

Progress will be seen in the future. We can expect mapping to be aided more and more by the computer (Chap. 7). We are permitted to hope that possibilities will increase without prohibitive rise in complexity and costs.

CHAPTER 3

PREPARATION OF A FIELD WORK PLAN

Having seen in the preceding chapters the fundamental concepts and wide range of options possible in soil mapping, we can now touch upon practical mapping starting with preparation of the project. We envisage here all the phases of the work that precede going to the field, principally establishing conditions of contract, preparing aerial photographs and collecting various documents and equipment.

But before going further, it is suggested that the reader examine carefully the drawings prepared by the SOL CONSEIL Society, Strasbourg (Fig. 3.1). We owe them to the talents of D. Schulthess and H. Duchaufour (Duchaufour and Party, 1986).

3.1 CONDITIONS OF CONTRACT

3.1.1 Generalities

Preparation of a soil map brings together the following major players:
- the Project Leader is, as the title indicates, the person who executes the mapping work;
- the Contracting Authority sponsors the investigation, that is, orders and pays for it (the administrative departments, for example);
- the Scientific Management Committee is often a third important partner. It acts when the work is not done independently but is part of a broad mapping programme comprising many studies of the same type. This committee then has the responsibility of setting up the programme and monitoring it. Thus it ensures uniformity in methods of study. Its

Fig. 3.1 Summary of the mapping method of SOL CONSEIL, Strasbourg.

experts, nominated for their varied competence, often belong to, or are selected by, the administration.

The project leader and the contracting authority are made to sign an agreement the terms of which are aimed at ensuring the quality of the study and to protect the interests of both parties. But before coming to the signature stage, a dialogue should be established so that the objectives of the study are clear. It must be known what is desired and at what scale the work can be done suitably. In addition, the project leader should not agree to do anything without verifying the existence of documents likely to greatly facilitate the mapping work: detailed geological maps, photographic coverage, topographic bases. In some cases, one is forced to make provision for completing the topographic base before any soil mapping is taken up.

Let us examine, for instance, the technical specifications drawn up in the very different cases of mapping at scale 1/10,000 and mapping at 1/250,000. It will only be necessary to present summaries emphasizing the principal points. In reality, the technical conditions of contract are documents covering at least ten pages.

3.1.2 Conditions of contract for studies at scale 1/10,000

What follows is based on the specifications in the general technical clauses developed by the INRA-CEMAGREF Scientific Management Committee for the project 'Reference Drainage Sectors' of the French Ministry of Agriculture (Favrot, 1990). This contract deals with study objectives, technical specifications, financial aspects, delays and ethical problems.

- *Objectives.* The first objective in each case is generally preparation of a large-scale map on which the units drawn should have adequate purity (Chap. 5, § 5.3.6) and sufficiently accurate boundaries (30-40 m). The second objective is preparation of an explanatory report of defined contents, structure and size. Soil profiles described and analysed are to be brought together in the adequate appendix. Each of these profiles should represent a minimum amount of descriptive and analytical data.
- The *technical specifications* are precisely detailed. First of all the type and number of observations must be well defined. In general, for obtaining adequate precision at scale 1/10,000 one auger bore every two ha and one soil profile every 20 ha are accepted with, in any event, at least one profile per identified soil unit. The profiles should be described on standardized forms, for example those of STIPA (Chap. 4, § 4.4.2). The analyses must be done in an approved laboratory that gives sufficient guarantee regarding the quality of the data. Clauses for editing the map are fixed. If the project leader supplies an unedited document, it must be on a topographic base of 'stable support' type that will not distort.
- *Particular technical specifications*, if needed, may round off the technical specifications. For Reference Drainage Sectors, it was necessary to detail the specifications for obtaining supplementary data pertaining to the hydric functioning of the soils (hydrodynamic determinations, inquiry among farmers, setting up field experiments).
- *Financial clauses* should be unambiguous in the agreement. In many cases the cost is fixed all-inclusive per hectare, it being understood that the project leader bears all the expenses before ending with the final map and report. In other cases, the cost of survey, analysis and publication are detailed.
- The *work timetable* is fixed and late delivery penalties are imposed. Naturally a certain time overrun should be tolerated if it is justified (for example, serious vagaries of climate).
- *Conditions of utilization of the documents* are also specified by detailing all that comes under the concept of ownership. The contracting authority and sponsor is obviously the owner of the publications ordered by it. But will this ownership go as far as confidentiality and prohibition of the project leader from presenting papers or scientific

communications on the subject? Will the name of the project leader appear on the map and report? If so, where and with how much legibility?

It would be better to deal with all these questions in the agreement rather than have problems, sometimes difficult to resolve, later between the parties.

3.1.3 Conditions of contract for studies at scale 1/250,000

Most of the items presented in the preceding section could be repeated here: problems of delay, quality, confidentiality, etc. We will simply emphasize now what pertains to studies on small scale as part of a systematic coverage of the territory, which therefore would involve numerous public and private partners. For this, we shall rely on the conditions of contract proposed for the coverage at 1/250,000 of France under the programme 'Inventory, Management and Conservation of Soils' (IGCS, 1993). The following items are detailed:

- The *role and responsibility of the different participants*. The 'National IGCS Committee' guides the operations. The 'Regional Management Committee' organizes the financial arrangements and the work calendar. The 'Regional Scientific and Technical Committee' looks after the execution of the work according to the rulebook and guarantees its scientific quality. The 'Contracting Authority Responsible for the Region' has above all the task of constituting and administering the soil database that brings together the data from the mapping. These data should be entered in the DONESOL format (Chap. 6, § 6.3) and a copy should be provided to the Service d'Étude des Sols et la Carte Pédologique de France (SESCPF). The same responsible Contracting Authority organizes the availability of these data to the benefit of various users. The 'Project Leader'... has only to carry out the work demanded.
- The *nature of the agreements to be entered into by the different partners* (Ministry of Agriculture, INRA, Regional Contracting Authorities, Project Leaders) is also detailed; typical agreements are suggested.
- The *conditions to be fulfilled for obtaining different quality-labels* are defined. As indicated in Table 3.1, this label depends on the observation density obtained as well as the area that had been mapped earlier on large scale and has to be synthesized at 1/250,000. In other words, the fact that one part of the map has been established on the basis of documents having a precision exceeding what is normally required is appreciated.
- The *ownership, distributorship and price of the data* are also envisaged in the conditions of contract. These are delicate questions considering the number of partners involved. A price has been fixed for each kind of data: soil map, various derived thematic maps, semantic information....

Table 3.1 Observation density for scale 1/250,000 (IGCS, 1993)

Area already mapped at a scale larger than 1/100,000 prior to the new mapping programme	Maximum area per auger hole	Maximum area per profile	**Resulting quality-label**
10%	800 ha	8000 ha	'Basic'
30-50%	400 ha	4000 ha	'Standard'
More than 50%	200 ha	2000 ha	'Optimum'

3.2 ESTIMATION OF TIMES AND COSTS

3.2.1 The viewpoint of experts

The following items (Table 3.2) pertaining to quality norms are mean figures given by experts. They are the result of experience gained in France and Switzerland over three or four decades. The costs (in Euros, base year 2002) take into account the vagaries of a temperate, not too humid climate. They also integrate various causes that provoke small delays. Thus it is possible to work faster for less cost but it will be dangerous to make a commitment to do it! The paying organizations want a lot of results for little expense. In consequence, the quality norms tend to decline a little. When the figures had to be corrected downward from the first edition of this book, the old norms (1996) have been kept in parentheses.

Table 3.2 Observation density for drawing up the conditions of contract[1]

Scale	A[2]: Area mapped per day in ha	B: Number of ha per auger hole	C: Number of ha per profile	Cost/ha, base 2002, Euros[3]
1/10,000	40-80	1-5 (0.5-3)	20-50 (10-50)	40-80
1/25,000	100-250	5-20	50-200	20-30
1/50,000	250-500	20-50	200-500	5-7.5
1/100,000	500-1000	50-100	500-2000 (300-1000)	2.5-5
1/250,000	3000-9000	100-400	2000-4000	0.2-0.3

[1] This table was drawn up with Mr J.P. Party (SOL CONSEIL) taking into account the opinions of Messrs R. Bertrand (IRAT), M. Bornand (INRA), J.C. Favrot (INRA), M. Gratier (Service d'Aménagement du Territoire du Canton du Vaud), Isabelle Boutefoy and J.M. Vinatier (Soil Information Rhône-Alpes).
[2] This corresponds to just reconnaissance and delimitation of mapping units by a team of two persons.
[3] This is the all-inclusive price including salaries and is therefore higher than that sometimes practised by the administrations that do not charge for the time spent by their officials. This price also takes into account the time in the office needed for the work of synthesis and editing.

3.2.2 An attempt at generalization

It was seen in Chapter 2, § 2.3.3 (equation 1) that:

$$\frac{Ob}{S} = \frac{4}{K} 10^8 E^2$$

or, by inverting both sides,

$$\frac{S}{Ob} = \frac{K}{4 \cdot 10^8 \cdot E^2} = \frac{K}{4 \cdot 10^8} (1/E)^2$$

From this,

$$\left(\frac{S}{Ob}\right)^{1/2} = \left(\frac{K}{4 \cdot 10^8}\right)^{1/2} (1/E) \tag{7}$$

In this expression, $1/E$ is the reciprocal of the scale or, for example, 25,000 for 1/25,000; besides,

$$\left(\frac{S}{Ob}\right)^{1/2}$$

is a quantity expressed in hectometres (since S is in ha). Thus we have a linear relationship between the reciprocal of the observation density (one observation every x hm) and the reciprocal of the scale. This can then be graphically displayed to link between them the figures given by experts in Table 3.2. This method in principle applies to the three columns A, B and C of the table because all of them express areas in relation to scales. But only the reciprocals of the observation densities corresponding to the auger holes (column B) are considered here.

Examination of Figure 3.2 leads to the following comments. To begin with, the opinions expressed by experts are consistent: the limiting densities proposed for different scales are virtually aligned, which means that they all correspond to the same values of the mapping-efficiency coefficients (K). It is seen that experts propose norms that correspond indirectly to low values of this coefficient (between 1.2 and 4). The line corresponding to $K = 2.56$ has been drawn on the figure. It is approximately in the middle of the estimates. Lastly, it is noted that the experts assign a slightly higher mapping efficiency to the detailed scales (up to 12.8 at 1/25,000). It may be asked if this must be followed. Actually at small scales, taking the environment into consideration improves the mapping efficiency and reduces the number of observations required.

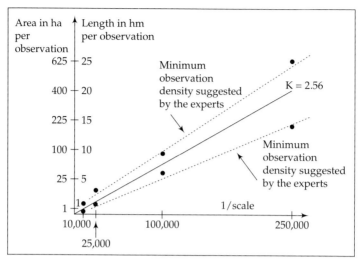

Fig. 3.2 Comparison of scales and observation densities.

Let us return to Table 3.2 and compare the first two columns. The experts do not expect that a *two*-mapper team will do 30 auger holes every day, which it is theoretically capable of sinking and describing, based on the time available and taking into account the presumably normal physical strength. For example, at 1/100,000 one auger hole is enough for 100 ha, while it is expected that only 1000 ha need be surveyed per day. This obviously corresponds to 10 auger holes. Thus it is not valid to estimate the time to be spent in the field from the observation density required (column B) and from the hypothesis that 30 observations could be taken per day. Indeed, especially at small scale, we are not satisfied with 'boring'. It is very necessary to look, move around and think.

It can also be stated that there is an approximately linear relationship between the square root of the area covered per day and the reciprocal of the scale. After calculation we have found that:

$$(S_{day})^{1/2} = \frac{1}{4000}(1/E) + 5.6$$

It is, therefore, easy to deduce from this the time required for survey on the basis of a single team of two persons working. On the whole, we will retain the following norms generalizing those in Table 3.2:

$$\text{Auger bores} = \frac{4}{K} 10^8 E^2 S_{ha}$$

$$\text{Profiles} = \text{Auger bores}/10$$

$$\text{Time in days} = S_{ha} \Big/ \left[\frac{1}{4000}(1/E) + 5.6\right]^2$$

For example, to survey 1000 ha at 1/10,000 the following will be necessary for a two-member team: about 400 auger holes (at $K = 10$), 40 profiles and 15 effective working days involving 27 auger bores per day. It goes without saying that these formulae do not give very accurate numbers. They may be modified in accordance with particular conditions. The method of calculation proposed has the sole aim of fixing the order of magnitude of efforts needed for characterization of the soils.

3.2.3 Exercise in planning a field programme

In the Appendix will be found an exercise of calculation of time and cost relating to the execution of a soil survey (Appendix, § A1). A suitable calculation sheet is proposed.

3.3 ORDERING, PREPARING AND UTILIZING AERIAL PHOTOGRAPHS

3.3.1 Generalities

Recapitulation of wavelengths

Figure 3.3, taken from Dent and Young (1993), summarizes the situation pertaining to the various wavelengths that can be utilized in mapping.

When we go beyond the visible wavelengths (0.4-0.7 nm), the information recorded can only be seen when digitally converted or displayed in false colours. The following display is conventionally used in false-colour infrared photography (Table 3.3):

Table 3.3 IR display in false colour (Dent and Young, 1993)

Wavelength	Displayed as
Blue	Filtered/suppressed
Green (0.5-0.6 mm)	Blue
Red (0.6-0.7 mm)	Green
Near infrared (0.7-0.9 mm)	Red

But the entire difficulty arises from the fact that the objects do not emit radiation of a precise wavelength; their reflectance may be high in many visible or invisible wavelengths. Thus, on a good false-colour infrared photograph, broad-leaved trees are bright red or dark red, conifers are more blue, roads are white or light blue and water and wet areas are black.

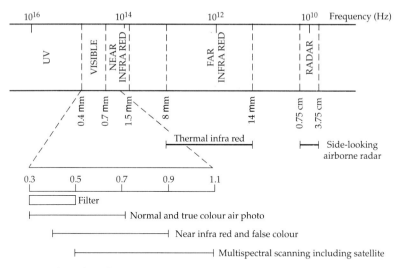

Fig. 3.3 Wavelengths of remote-sensing systems (from Dent and Young, 1993).

With the present state of the technique, black-and-white or false-colour photographs are often the best tool the mapper could use for large-scale studies because of their excellent resolution. But the future belongs to natural-colour photographs. Doing a portion of the map using them will convince the user of their great superiority. Colour enables us see the colour of the bare soil and eliminates confusion with the vegetation be it dominantly green or yellow.

Satellites whose recordings are useful in systematic soil mapping have a much lower resolution (Girard and Girard, 1999). Thus they are preferred for reconnaissance studies at 1/250,000 scale or smaller. Then purchase and handling can be avoided of a very large number of photographs with too broad a precision, that is, inadequate precision.

Role of aerial photographs
Aerial photographs play many important roles in soil mapping.
- In some cases, when a topographic base is not available, they first of all help photogrammetry specialists to prepare a base map useful for soil survey.
- Certain mapping departments, such as those in USA and Canada, use enlarged aerial photographs for display, overprinting the soil boundaries with, within them, the numbers identifying the soils.
- Very often, moving around the area is made easy if aerial photographs are used as guide. This, however, is not true in mountains because the photograph does not always enable correct location of the trails and changes in altitude unless the stereoscope is used, which cannot be done on the road. It is better in the mountains to be guided by the

topographic map, not forgetting that the version called 'tourist map' has the advantage of showing the layout of identified trails.
- Lastly, aerial photographs directly help in soil mapping: locating sites that must be visited and extrapolation from them, which are usual photo-interpretation procedures.

3.3.2 Procuring aerial photographs

Procedure
Generally, it is necessary to proceed in two stages. First, the specialized Institute is asked to provide an *index map* for the area in question, that is, a small-scale topographic base showing the location of the areas covered by available photographs. This enables us in the second stage to order for the photographs needed: those exactly corresponding to the study area. The precise location of photographs on the index map is found because the centres of photographs are indicated by corresponding numbers, or the flight paths of the aeroplane are drawn, or the exact boundaries of the photographs are displayed for one photograph in two or one in three. The overlap in photographs is large (Fig. 3.4).

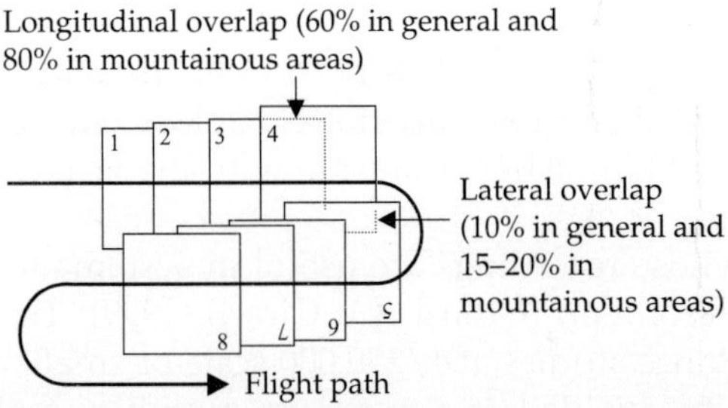

Fig. 3.4 Line of flight and overlap of photographs (Poncet and Lalau-Keraly, 1984).

The total overlap in each strip exceeds 100 per cent so that a stereoscopic view is ensured. Overlap between strips is large. Theoretically, it is overall three frames in four! But the pilot of the aeroplane may face difficulties, such as clouds or abrupt changes in relief obliging him to make several passes over the same area. Some index maps thus suggest up to ten times the theoretically required number of photographs for ensuring coverage without overlap! This represents the price one has to pay for obtaining aerial characterization of high quality.

When ordering photographs, one must think of observing in relief the edges of the area to be mapped. It is useful to broadly calculate and procure some photographs partly outside the area for enabling proper stereoscopic viewing of the margins.

The index map also provides the principal data pertaining to the flight. The mapper, to understand correctly what is seen on the photographs, should remember the date they were taken.

Options
All sorts of options are possible when ordering photographs.

The first option pertains to the type of mission: France, for example, is fully covered by black-and-white missions. But for certain sectors, colour infra-red photographs are available. In other regions, totally special missions have been flown. This must be found out.

The second option relates to the date the photographs were taken. The same regions are photographed at short or long intervals. In some countries the missions are repeated every five or ten years. A succession of five or six since the Second World War can be counted. The latest is obviously the most accurate in terms of condition of sites: use of land for roads, crops, etc. But this is not necessarily the best for the objective pursued. Thus for an important mapping study, it may be useful to compare the most recent missions on some photographs before selecting which to order. In certain cases it is essential to procure all for a small area to do diachronic studies. For example, a scientist of CEMAGREF in Grenoble studied with this method the 'abandoning', that is, the retreat of human influence in the valley of the Maurienne with appearance of fallow lands and secondary forest formation (Delcros, 1993).

The third option concerns the type of photograph we wish to have. It is possible to order for paper prints with or without enlargement relative to the original negative. The paper may be glossy or matte. Glossy paper always gives a print of better quality but it is impossible to write on it with a blacklead pencil. For this reason many mappers have for the past several years ordered photographs on mate paper. Nowadays, some felt-tip pens enable writing on the plastic base of glossy prints. Such prints are, therefore, preferable. But if it is desired to preserve the original photographs they can obviously be overlaid with a transparent tracing film.

It is often desirable to procure not only paper prints but also copies of the original negative. This can be useful if one has high-quality draughting equipment available or if it is desired to get several photographs of the same area, say at various enlargements.

3.3.3 Preparation of aerial photographs

The consequence of large overlap is that the mapper in most cases is forced to work with many photographs. He risks losing considerable time (and

frequently so) if he does not array his photographs in such a way that the appropriate ones can be found rightaway. The following preparatory operations are recommended:
- Draw on the edge of each photograph an arrow to indicate the north, so that orientation is easy.
- Add beside the arrow the letter R (for right) or L (for left) to indicate whether the photograph bearing the highest number is found on the right or the left for stereoscopic examination, with reference to the flight of the aeroplane in the W-E or E-W direction. Some photographs carry notations on the margin to indicate this and to avoid the above operation. But the steps are useful for the ease of use they provide.
- When needed it is possible to create, for the area surveyed, a layout map drawn at the scale of the final map by using the appropriate topographic base. *'Time spent in sorting, labelling and locating the photographs pays off in time saved and irritation avoided at later stages of the survey'* (Dent and Young, 1993).

3.3.4 Principles of photo-interpretation

Information furnished by the photographs
The soils are not identifiable directly on the aerial photographs unless three conditions are satisfied: vegetation is absent, natural-colour photographs are used and the soil has a characteristic colour. It is therefore necessary to proceed indirectly by finding on the photographs visible elements to which the soils are linked. Indeed on a photograph three types of information are mainly seen:
- first of all, everything pertaining to *land use of the area* (natural vegetation, crops); for example, it is relatively easy to identify on the photographs: orchards, grasslands, broad-leaved trees, conifers, etc.;
- then, using a stereoscope, the *morphological organization of the landscape* is marked out starting from the different landforms present: valleys, terraces, alluvial cones, etc.;
- lastly, the photographs enable observation of *'infrastructures of human origin'*: parcels, farm boundaries, etc.

The aims of photo-interpretation
On the one hand, the objective is to identify and delineate approximately the landscape units that exhibit some specificity and sometimes internal uniformity. Thus will be distinguished a completely wooded high plateau or a flat-bottom valley covered with grassland, etc. These landscape units will become soil-landscape units if the field survey enables associating characteristic soils with each of them. In any case, this prezoning defines compartments that will be studied one after another. It thus helps to regionalize

the problems and to organize the survey. It corresponds, if one prefers, to study of *landscape organization*.

On the other hand, for each unit thus identified, photographic analysis should be done in order to link the constituents to understand their interactions. But the analysis should not be restricted to research on correlations between one character or other of the environment and some property of the soils. This will be, for example, detecting wet soils (Gleysols) going by poplars that can be easily identified by their tall shape. This procedure is not very correct: look out, for example, for the cypress of the Mediterranean zone for they also have tall shape but actually indicate drier soils! This kind of correlation can thus be useful in specific cases but are anecdotal in nature. Indeed, one must proceed with a more holistic and deeper analysis of the photographs to understand the *functioning of the landscape* that is right in front of the observer (Forman and Godron, 1986).

Functioning of the landscape
As stated earlier, it is not a matter of investigating correlations between the elements taken two by two, but between groups of elements taken as a whole (*cf.* Principal Component Analysis of statisticians). To return to the example considered above: presence of poplars is an index (questionable) of presence of soils with redoximorphic features. But the diagnosis becomes more reliable if it is noted that these trees are located in depressions favourable for water to accumulate in, and that they correspond to parcels where signs of drainage are seen. The presumption is further reinforced if the soils are black on colour infrared photographs and if there are no houses in those areas. It is thus seen that identification of Gleysols uses at the same time observation of land use (poplars), landform (depression) and human activity (ditches). In other words, all the elements the photograph shows are examined for determining if they confirm the hypothesis 'the soils are wet'. Figure 3.5 summarizes the procedure.

At least three factors are likely to help in this systematic analysis. In many areas of the world, human occupation is very old and in the end closely related to the environment. If one would take the trouble to think, very many correlations will be found between landform, land use and soil. The second asset is the experience of the mapper who, in a given case, will think of finding on the photographs some indicators that are probably or at least possibly present. The third point is fundamental: the main thing in this work of photo-interpretation is not done in the laboratory but in the field. Remaining within the framework of the preceding example, it can be said that the mapper, though he has not thought of poplars, will see them at a bend in the road, then will seek for them on the photograph and will examine whether they form a useful indicator or not. In total, the mapper simultaneously resorts to two views of the natural environment to draw the soil boundaries: direct observation, and indirect observation through aerial photographs.

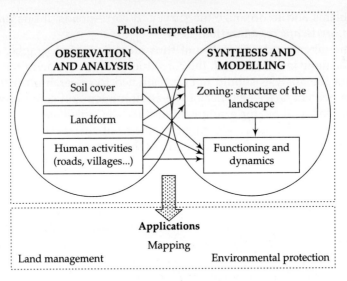

Fig. 3.5 Use of photo-interpretation in soil mapping.

Even if he is advised, before going to the field, to closely examine the aerial photographs, there is no separation between the photo-interpretation phase (in the laboratory) and the validation phase (in the field). Rather there is an alternation of the two phases and only for the sake of clarity are they presented separately!

Discussion
This systematic analysis is productive. The presumptions being many, the quality of delimitation of soils increases. Above all, it is still possible when one of the indices normally considered is lacking (Chap. 5, § 5.1.3). Thus, presence of Gleysols can be suspected even when poplars are absent (depressions, black soils, ditches, etc.).

Certainly there are difficulties. Firstly, there is the poorly experienced or unlucky (difficult terrain) mapper who does not find on the photographs clues that enable him speed up and refine the delimitation of soils. Then what is left is the auger and the possibility of multiplying the number of direct observations! The rate of progress of the mapper slows down. Soil-environment correlations are then tested a larger number of times. The mapper is forced into a more intense effort of thought for being allowed to bore less often. Slightly better, subtle deductions become possible. At last, everything clicks, a model of spatial organization is built up and the rhythm takes off again at the same time as morale returns! There is also the case, quite rare in real life, when a soil-environment correlation is found but without the ability to explain it. For example, in the Massif Central of France we mapped a canton (administrative subdivision) in which Gleysols were concentrated in three sectors composed of *smaller* and *more regular* parcels. It was thought that drainage would have forced an organization of the parcels in accordance with the structure of the system of outlets. But no ditch or drain, functional or abandoned, could be found. The explanation was something else altogether: these poor lands had been left as common grazing lands till the French Revolution, at which time they were divided. This explains the regular shape and small size of the parcels because the cake to be divided was small and many portions had to be cut out of it! The use of criteria with unclear significance is obviously awkward for the conscientious mapper...

Many portions of the region often have the same features. This is very important because Nature then gives the mapper replications to better learn how to read the landscape and to understand the spatial organization of soils. Identical milieus are sometimes very numerous. Then the mapper can define the soil-landform relationships on the basis of such a large number of instances whereas no laboratory scientist could ever be provided a sampling so voluminous for basing his experiments on!

3.3.5 Examples of interpretation

An exercise at photo-interpretation for mapping purposes is given in the Appendix, § A2. It illustrates, in a specific case, the concepts presented above.

3.4 PHOTOGRAMMETRIC DOCUMENTS

Importance and limitations of prezoning
Digital documents are accessible in developed countries. They can be purchased or even drawn up by the organizations producing soil maps, especially:
- *Digital Elevation Model* (DEM) giving the altitude of the land at the nodes of a uniform-mesh net.
- Satellite (remote-sensing) data treated to give, in particular, vegetative cover (for example, the *Corine Land-cover Map* of Europe).
- Manually drawn maps that are relatively easy to digitize, geological maps in particular.

Topography, vegetation and parent material are the main factors in pedogenesis. These factors are, therefore, related to the geography of soils. It is thus tempting to combine them to do a prezoning so that the later work of the mapper in the field will mostly consist of identification of the contents of units delineated beforehand by automated combination of data related to the environment. References are provided later (Chap. 7, § 7.3.3).

The procedure has advantages. It can speed up fieldwork and thereby result in substantial savings. It helps to optimize the selection of observation and sampling sites. It brings rigour to an automated treatment done the same way throughout the area in question. It encourages much refinement in the layout. The last point poses a problem of principle: what does the user desire? Does he want an accurate map crammed with details or rather does he count on the mapper to make a synthesis and bring out the main things? If he wants to use the map for a series of small farms he will prefer detail with the danger of using the document at a level of accuracy it does not always have (for example, which soil is there in such or such exact farm?). On the other hand, if he wants to plan the use of the area or simply to understand the organization, the user will prefer a synthesis that shows him the simplified spatial distribution of the major soils. Certainly one can transform a detailed map to a more synthesized map but not the other way around. But this is the work of the specialist and the user cannot do it.

Besides this, the combination of environmental information hides various pitfalls that ought to be avoided. The boundaries linked to the prezoning are not all valid: we must know how to review them. At the same time, some important soil boundaries may not be predicted; they must be accounted for and added. Thus much time should be spent in the field. Without this the procedure may result in a soil map with uneven accuracy. We have seen this particularly in mountains. In such regions the geological maps used give all sorts of detail on slopes and cliffs where the geological strata are clearly seen with their folds, thrust faults and overthrusts. On the contrary, these geological maps describe valleys laconically as bearing 'recent alluvia'. Under these conditions, the soil map arbitrarily made from the geological map without sufficient field survey will contain all kinds of useless detail on slopes for which it would be enough to mention 'thin, eroded stony soils'. But it will be practically silent about alluvia, while the soils in them, important for agricultural planning, are very varied in texture, organic-matter content and hydric functioning. Moreover, the procedure could help to conceal the incompetence of certain private teams that are quite capable of putting out very lovely coloured maps obtained by combining environmental information but are incapable of validating these maps in the field from a soil-mapping viewpoint. The worst is that that the technical experts of these automated and sometimes superficial approaches can claim lower cost and products enjoying the aura that touches all that emerges from the computer. They practise unfair competition against true mappers.

Strengths and weaknesses of remote sensing
After several decades, there is enough objectivity to make a first assessment of the remote-sensing approach to geography of soils:

- Remote sensing, which takes into account the relief and organization of the landscape, is very efficient for delimiting territories (called 'terroirs' in France) that yield products (wines, cheeses) of specific and recognized characters (Vaudour, 2001). More generally, remote sensing is extremely useful at small scale when the objective is not to establish soil boundaries but landform boundaries; it enables consideration of large areas at one go (use of the macroscope). By smoothing procedures it is possible to work at a chosen level of detail.
- Remote sensing can make a very useful contribution to soil mapping in the context of a landscape approach at small scales (large areas). Besides, it has been shown that it is possible to find out almost three-fourths of the segmentation of area drawn by a mapper for land in the cultivated zone (Girard, 1995).
- Locally, some features of the environment can be observed by satellite and at the same time related to the nature of soils. For example, patches of salt or of wetness can be seen. Unfortunately, the relationships brought out between phenomena and wavelengths are empirical and limited in time and space. It is the same with aerial photographs. But the comparison stops there. On the one hand, it is normal to use these tenuous and empirical connections locally, moving around in the field, photograph in hand, because one calls into question the system of interpretation every moment. On the other hand, it is abnormal to use correlations so volatile for establishing a general system of interpretation applied in the laboratory for using many satellite images!
- At medium and large scale, the contribution of remote sensing in soil mapping is reduced further because it is not possible to see below the surface and through the vegetation from a satellite.

In any case, remote sensing is a discipline in its own right that cannot be treated or attacked here. Reference to the book *Application de la Télédection à l'Étude de la Biosphère* (Girard and Girard, 1975) is recommended. Written by experts in vegetation and in soil, it presents various examples in the domain of interest to us.

For some years, the use of satellite images has been very different from that of aerial photographs for at least two reasons. Firstly, the resolution of satellite images was much poorer; they were thus suitable only for studies on very small scale. But improvement in resolution increases possibilities. Secondly, satellite recording is digital, so that it only allows use on the computer. But with the advent of efficient, low-cost scanners this difference is also blurring, even though a digitized photograph provides only one kind of information while images taken by satellite correspond to recording several channels (wavelengths).

3.5 DOCUMENTS AND EQUIPMENT

3.5.1 Basic documents and bibliographic research

In the first place, cartographic documents pertaining to the natural area should be procured: topographic maps, aerial photographs, satellite images, digital terrain models (Chap. 8, § 8.6.1), geological maps, etc. Then, regional publications pertaining to soil, geology, etc., should be searched for in bookshops as well as university libraries near and far (such as for former colonies).

3.5.2 Office equipment for use in survey

- *Field sheets* for entering description of auger bores, profiles and soil units (Chaps. 4, 5 and 6);
- *stereoscopes:* there are many kinds of stereoscopes with varying ease of operation, price and quality;
- *desk lamp* with flexible or fully articulated shaft with extension cord because hotels and other places on tour are not equipped for stereoscopic study of aerial photographs and light is insufficient almost everywhere;
- *assorted notebooks, paper sheets, pencils, erasers,* pencil sharpeners, colour pencils, ball-point pens, felt-tip pens for writing on aerial photographs, scissors, adhesive tape, etc.

3.5.3 Field equipment and tools

Munsell chart
A *Munsell colour chart* is used to characterize the colour of the soils examined (Chap. 4, § 4.2.4).

Augers
There are five types of *augers*. All of them in principle enable digging holes to 1.20 m.
- The *conventional auger* (Jarrett auger) is made of metal, is heavy and rigid, and comprises three parts. First is a vertical shaft on which marks or notches indicate depths. Second, a hollow portion for digging; its edges are sharpened along the cut. This hollow portion usually has a diameter of 7 cm, so that considerable soil material can be brought to the surface quite quickly. Freeing this soil is quite difficult. A 'digger' or spatula is often used, taking care not to hit the sharp edge of the auger, which would reduce its efficiency. The third part of the tool is a bar welded at a right angle to the shaft, to help turn the

assembly for digging. Rubber sleeves are often fitted where the hands hold the bar (indispensable in severe cold).
- The *Dutch auger* is made of aluminium alloy. The hollow portion is a little narrower than in the conventional auger (6-cm diameter). It is much lighter and is tending to come into general use. It is more efficient than the preceding in stony soils.
- The *screw auger* is not hollow and resembles a wood screw. We have not tested it.
- The *bucket auger* is actually a hollow tube that is pushed into the soil by turning it and pressing it to push down the teeth or sharp edge it has in front. This auger is very suitable for sampling peats. Some models can be used for drawing undisturbed samples (Stolt et al., 1991), especially the chisel auger consisting of a hollow tube with a longitudinal slit.
- *Motorized augers* are of many types. The augers with arms are of little use. Of course less effort is needed for digging, but the machine has to be pulled along in the field, which requires a sort of yoke lifted by two persons. The use of this machine is, therefore, limited to cases where many bores have to be dug in a very small area. Augers mounted in front of all-terrain vehicles may be useful. They allow digging of large-diameter holes. The problem of their transportation to the desired location remains.

Pickaxe
A pickaxe is useful to scratch the soil on the surface or in profiles, and to loosen the soil brought up by the auger.. The best tool is that with a flat, gently curved part on one side and a solid metal hammer on the other. With this one can scratch the ground or crush coarse elements to examine the degree of weathering. Unfortunately, soil mappers are few in number and manufacturers have not really put any specialized tool on the market. One can find in sporting goods shops a *speleologist's hammer* with the requisite qualities although a bit costly, a little heavy and made of a metal that wears out too fast. The *slate worker's pickaxe* is partly satisfactory in that it consists of a suitable peen joined to an almost useless kind of tomahawk (as in the case of American Indians).

Hydrochloric acid
A bottle of dilute hydrochloric acid is required for determining what is called effervescence (Chap. 4, § 4.2.7). The acid is used diluted to one-half, one-third or one-tenth. The FAO and ISO recommend a three-fold dilution. One should not forget that the substance is corrosive. Some mappers use glass droppers; others prefer a very small plastic phial. These containers are attacked by the acid and end up leaking; they must be replaced frequently. They should be transported in a small airtight container labelled 'Danger:

acid' because, although this must be avoided totally, the container could be lost in the field.

Plastic bags, labels

The samples drawn are transported in sturdy plastic bags. *Thermostable* bags are available and are very useful when determination of water content is necessary. In such case, the samples brought to the laboratory are weighed and then placed directly in the oven with the bags open. The weight of the bag is negligible compared to that of the soil or of the water. The bags are weighed again after drying. The method is quite fast without loss of accuracy.

Labels and string should be taken to the field for closing the bags and identifying them.

Instrument for measuring pH

In the early stages of a survey, it is useful to determine the pH of the soils studied. There are many ways of doing this. The most popular is a mixture of coloured indicators, a drop of which is mixed with a soil aggregate. The liquid, separated from the solid fraction, takes on a colour that gives the pH to within half a unit. The popularity of the method comes from its speed, simplicity and the fact that it avoids transporting expensive and bulky equipment to the field. Of course pH meters with sturdy compact electrodes are available. But their use in the field demands precautions. Buffer solutions, distilled water and washing equipment have to be carried. Above all, it is necessary to wait for the solution and soil to come to equilibrium, which takes considerable time. In other words, the pH meter is potentially more accurate, but if it is not used in proper conditions, it is better substituted by equipment more easy to handle.

Knife, spatula

Study of a soil and drawing samples from it involves use of a knife or spatula. The knife, shaped like a dagger, has as broad a blade as possible, flat and not curved. The handle should be smooth so that the knife can be pushed into hard soils through the day without injury to the hands. To find such a tool is not easy. When one is found, it is desirable to immediately remove the sharp edge, which is dangerous and of no use in this work. Some workers prefer a trowel or putty knife, which could be good options for soft soil.

Metre rule

A metre rule is necessary for measuring the depth of horizons. One would prefer a folding wooden rule or, better still, a rule made of flexible fabric with coloured strips, with a nail for fixing it to the profile. The profile can be photographed with an appropriate scale.

Clinometer
A clinometer is used to measure slope, a useful value for several reasons (Chap. 8, § 8.6.1). Excellent compact and sturdy instruments are available. A plumb bob hooked to an axis marks the vertical while the observer sights the line of the slope the inclination of which is to be measured. The best instruments, oil-immersed, show two graduations (in degrees and in per cent). Thereby there is no scope for error and the appropriate value can be entered on the field sheets.

In general, measurement in per cent is preferred because it can also be directly obtained by calculation, using the topographic base and the contours. To go from a measurement in per cent to a value in degrees, the angle with known tangent must be found out. This is one of those instances where the naturalist mapper is astonished to find that inverse trigonometric functions have some use after all.

3.5.4 Special appliances

Each of the appliances presented in this section would require long explanations. There are many articles and theses pertaining to them and we shall be satisfied with just mentioning their existence and use.

Microcomputers
Microcomputers are indispensable if direct recording of description of profiles and auger bores is planned (Chap. 4, § 4.4.1).

Positioning by satellite
Positioning of a point by means of satellites is done with the help of a GPS *(Global Positioning System)*. This apparatus has been used in soil mapping in USA since 1986 or 1987.

The principle is the following (Puterski et al., 1990; Louchard, 1994). Twenty-four NAVSTAR satellites have been placed in circular orbits at an altitude of 12,000 miles (about 20,000 km). Their movement is such that four to six of them are always visible from any point on the Earth. The satellites emit every second (and the same time through synchronization with the atomic clock) a complex signal on the frequency of 1575.42 MHz. This enables them to transmit their position. The GPS receiver also generates this signal that it also repeatedly receives with time-lags related to the distance of each source. This information is enough for calculating the distance between each satellite and the receiver, and for calculating the position of the receiver on the Earth. However the signals of four satellites not in line have to be received.

The accuracy of the positioning is limited by various factors: reflection of the signal at the forest canopy, atmospheric disturbances, inaccurate measurement of the time on the ground and also degradation of the signal by the *U.S. Department of Defense* for military reasons. A discrepancy of one microsecond represents an error of 300 m.

The accuracy may be improved if the coordinates of one of the measuring points are calculated at the same time and known accurately. For this, apart from the mobile station used in the field (receiving antenna + electronic device), a fixed station is also used in parallel. Both can be linked together by telephone or post-synchronized. This is the technique of *differential GPS*.

Another solution to improve the accuracy of positioning is to repeat the measurement continuously a number of times, for example 300 times, that is over several minutes. The errors then tend to cancel out. Some years ago, the error in positioning obtained with a popular instrument was 73 m without correction and 6 m with correction, however with overrun in 5 per cent of the cases (August et al., 1994). Improvements have been seen since then. Without correction, an error of 14.4 m was attained, and with correction, 1 m by repeating the measurement 90 times (Gao, 2001). The user may have the impression of a still better result to the extent that the determinations done one after the other may be reproducible, all measurements being a little wrong in absolute value...

Under these conditions, the GPS is a good tool for locating profiles and auger bores. Above all, it can be coupled with instruments that will be described below so that the corresponding determinations may be correctly positioned.

Portable radar

This is Ground Penetrating Radar (GPR). It has been in use in USA since 1979. The equipment comprises a radar source with a 12-volt battery or 50-watt generator, a series of antennae capable of sending and receiving signals from 80 to 500 MHz, a microprocessor and a graphic unit. The whole thing is arranged in a trailer pulled by a vehicle. The measurements are done discontinuously, but the vehicle advances at 4 km per hour.

The electromagnetic wave emitted is partially reflected when it meets a discontinuity in the soil. This generates a signal that is processed by the receiving antenna and sent to the graphic system. This system converts the strong reflections to black and weaker reflections to shades of grey. The depth of the discontinuity may be determined by the following formula (Olson and Doolittle, 1985):

$$P = \frac{ct}{2E^{1/2}}$$

where P is the depth reached in m, c is the speed of light in m s^{-1}, t the time in s and E the dielectric constant of the soil (about 13). The method should be calibrated because this dielectric constant is a function of water content, salt content and type and abundance of clay minerals.

Several tests have shown that GPR has a penetrating power barely exceeding 3 m. It works particularly well in sandy soils. It is an excellent tool for detecting occurrence of a compact B horizon and for estimating the thickness of soft material over hard rock (Collins et al., 1989). The error may be less than 6 cm (Schellentrager and Doolittle, 1991). In USA, GPR is used to determine transects (aligned, linked observations) and to evaluate the exact composition of complex mapping units when updating old soil maps.

Electric sounding
The resistivity meter has been much used in soil science, as for example in France (Bottraud, 1983; Dabas et al., 1995). The instrument is a sort of comb with four teeth (four electrodes in line). The two outer electrodes carry a current I. The two inner electrodes, the closest, measure the potential difference resulting from the current. Ohm's law, $V = RI$, is applied. In other words, the resistance to the current can be easily determined. From this, assuming a simple geometry for the system, the electrical resistivity of the soil expressed in ohm-metres can be deduced, and also the conductivity, that is, '1/resistivity'. The electrodes are pulled along by a vehicle and scratch the soil as a rake would or are even replaced by metal casters that turn while remaining in contact with the soil ('rolling electrodes'). Thus measurements can be made continuously.

Resistivity depends above all on water content, salinity, texture, organic matter content and temperature (Banton et al., 1997). It is used for locating shallow water tables (Gras et al., 1997) and for drawing maps of salinity generally presented in the form of isovalue curves. A review of the question is available (Rhoades, 1993).

Earlier, these systems were also used for studies on organization of the soil, for example to detect the depth of occurrence of a petrocalcic horizon. Accuracy is very good (error less than 5%). The advantage is that by inserting the electrodes the depth explored can reach 5 to 10 m. But with the advent of GPR, this use has tended to vanish.

Electromagnetic induction (EMI)
Electromagnetic induction is another method for determining soil resistivity. The principle is as follows: the instrument consists of a coil through which an alternating current is passed. A magnetic field is created. It induces Foucault currents in the soil, thence a magnetic field that in turn generates a current in a second coil in the emitting appliance. This current is then measured. Thus there must be no contact between the soil and the instrument that may be pulled along about 25 cm above the surface. When the coil is vertical, the soil is explored to 1.5-m depth. When the coil is horizontal, the soil explored is half less deep. The contribution to horizons to the reading recorded depends on the position of the coil and its height above the soil. The instrument should be mounted at the end of a 3-m long

wooden beam because it is sensitive to masses of metal within a distance of one metre. Metal detectors also work on the same principle except that one tries to nullify through calibration the effect of soil so that the electrical and magnetic effects of only the objects looked for are found. The vehicle carrying the instrument moves along and transects are studied.

Electromagnetic induction seems to be a possible alternative to GPR, which is costly and time-consuming. This is why many papers try to compare the two methods (Doolittle and Collins, 1998; Inman et al., 2002).

Radiomagnetotelluric determinations
These again measure soil resistivity. But this time the signal is composed of long radio waves, for France: those of Europe 1 (183 kHz) or of Monte Carlo (216 kHz). One may be astonished by the ease with which earthworms pick up their favourite musical pieces! Although it may be so, the transects done at right angles to the direction of the signal enable us to find out the depth of the soil or water content of its surface horizons (Chaplot et al., 2001). This may help in drawing up a drainage plan for a parcel (Dupis et al., 1991).

Seismic refraction
Seismic techniques are old and much used in geophysics. They have been applied in soil science since 1966, at least in USA. Their use in France is more recent (Meyer, 1984). For study of very superficial layers the basic equipment consists of a metal plate and a 4-kg hammer. The operator uses this to send a shock wave into the soil and to send to the measuring system an electrical impulse determining zero time (when hammer and plate come in contact). One or several geophones are also required (to detect when the signal reaches different points of a transect) and a series of connecting cables. A specialized unit records the times and processes the signals.

In a soil composed of two superposed layers, the compression waves are propagated in all directions. But three particular cases deserve attention. A direct wave advances on the surface. Another passes through the first layer of the soil and returns upwards after having undergone reflection from the second layer. A third is refracted downwards to a limiting value of angle of incidence. This value depends on the relative speed of transmission through the two layers (Snell-Descartes law). At this limiting angle, the wave advances to the level of the contact without crossing it before emerging on the surface at an angle equal to its angle of descent. In total, the itineraries followed are known and the speeds of transmission in the two layers are obtained through calibration. Thus it is enough to measure the transmission time for calculating the depth at which the second layer of the system is located.

The method functions perfectly up to 10 metres. Its accuracy, generally high, depends on correctly measuring the propagation time. But it is not easy to use. For tracing a 500-m profile with a plate-geophone distance of 25 m, it is necessary to take into account one working day for two persons and 1500 blows with the hammer! The fatigue is considerable. The interpretations require an additional half a day. Consequently, seismic refraction is not often used in soil mapping.

3.6 PROBLEMS OF PRACTICAL ORGANIZATION

3.6.1 Relations with residents and authorities

Mappers are obliged to move on fields, meadows, woods…, all of which have owners. In principle it is necessary to ask their permission for digging auger holes and more so for opening up soil-profile pits.

For auger bores, permission is not generally requested, to save time. For profiles, mappers go from farm to farm, by the side of the sites where they wish to dig and take time to explain what they are doing and what they want. In general, the farmers accept on condition that the pit is filled up again quickly. It is then useful to note down the address of this farmer to take to him when possible the results of the analyses that would have been done. However, permission is not always easy to obtain. For example, the head of the farm may be absent and wife or children do not dare take the responsibility of permitting unknown people to dig a pit in a field. Sometimes it happens that permission is given but the mappers err by a few metres and dig in an unexpected location. Lastly, in some isolated places, woods particularly, seeking the owners is difficult. Mappers at times open profiles without permission. In short, it may happen for various reasons (error, anxiety to work fast) that pits are dug on the farms of people who have not been warned. If these people find it out, they may react with concern or even legitimate anger. Thus it is essential to take as many precautions as possible.

All the authorities must be informed in writing in advance: mayors (France), mayors or syndics (Switzerland), police, etc. If there is a minor problem with an owner, the mapper could explain to him that it is useless to run to look for the police, which is already aware of the question! This is psychologically important. If there is a major problem, it is obvious that one will be very happy with having informed the authorities in advance. In the same way, it is useful to get in touch with the local press and publish short articles in newspapers explaining what will be done in mapping: where, how and why.

3.6.2 Health hazards

Very fortunately, mapping is generally not a dangerous activity, but accidents may happen. So precautions are necessary. The chief ones in our opinion are the following:
- Considering the kind of activity, it is essential to be vaccinated against tetanus. It must be remembered that the vaccination lasts for ten years.
- The risk of snakebite is not totally negligible. In Europe, the bite of the viper (*Vipera aspis*), is not lethal for a healthy adult who knows to keep calm. The medical fraternity is not sure of the usefulness of the serum. For most specialists, this product has absolutely no efficacy but it is wise to inject the patient who asks for it for the sake of reassurance and to demonstrate that the doctors have done everything possible! There is thus no need to travel in the field with the serum (tropical countries apart!), considering its doubtful efficacy and the difficulty in preserving the product. On the other hand, it is said that calciparine…. But we cannot go further because of the risk of being accused of illegally practising medicine. In short, the doctor must be consulted before and after the accident.
- The greatest risk in mapping is generally that of travelling in motor vehicles. It is dangerous to work alone and, for example, drive a vehicle even over a short distance and at the same time look at the soils that pass by and monitor one's itinerary on the topographic map!
- When one goes on foot in the mountains it is essential to form teams of two persons each, particularly so in remote areas.

3.6.3 Formation of teams

For practical reasons it is useful to work in pairs for describing profiles or auger bores. Also, mapping relies on memory and it is often useful to share memories and thoughts in the field. Lastly, mapping by oneself is demoralizing: the work progresses slowly and evenings in the field hotel can be dreary. This must be avoided as much as possible. The ideal mapping team is thus composed of two persons, in all circumstances. It is formed of a scientist and an assistant with a driving license. Unfortunately, some mappers work alone to bring down the cost. This is particularly true for the youngest of them, when they set up their own business.

3.6.4 Location of teams in the field

The number of kilometres travelled by a team of two surveyors having one vehicle is estimated by:

$$KM_{tot} = J(km_1 + km_2)$$

where J is the number of days worked, km_1 the number of kilometres necessary on average to travel between the living quarters and the place where mapping is done, and km_2 the number of kilometres covered during the actual mapping. The value of km_1 can be calculated in the following manner:

$$km_1 = Nt(\alpha D)$$

where:
- Nt = number of trips (2 or 4) according to whether the team returns to the living quarters for lunch;
- D = distance in km, as the bird flies, between the base camp and the place of work; with slight simplification, $D = (d_{min} + d_{max})/2$ with d_{min} the minimum distance to go to the field (= 0 if the base camp is in the study zone) and d_{max} the distance to the farthest point in the area;
- α = the coefficient for statistically converting the straight-line distance to distance by road. Our tests have shown that α varies from 1.2 (excellent road network, level land, base camp chosen at the centre of the network) to 2.4 (sparse network in mountains).

But km_2 must also be estimated. On very large scale, one can work on foot, but 'must follow the vehicle'. At medium or small scale, one can 'work by vehicle'. More correctly, the vehicle is used to proceed as close as possible to the sites one wishes to study. We suggest, as a first approximation, assigning to km_2 the following number of km: 5 at scale 1/10,000, 20 at 1/25,000 and 50 at 1/100,000 or smaller scales.

The day having only 24 hours, time lost in travel is also lost in fuel, fatigue and hours of work. This is why the practice of the mapper returning to his distant place of residence is very bad. The ideal thing is thus to find lodgings in the immediate proximity of the area to be surveyed or, better still, in its centre. In Europe, the hotel is made the base camp. Many attractive solutions (camping, lodging in a tourist hostel, etc.) are not advisable because they cause loss of precious time in buying food, preparing meals and cleaning the room. To realize the magnitude of this, it is enough if we estimate what an hour lost costs a team of two scientists by way of salaries, travel allowance, arranging for a vehicle and being obliged to work only part of the week (Monday, they must reach the study area and Friday evening or Saturday, they have to return home).

3.6.5 Recapitulation of documents, equipment and precautions

The mapper should not forget anything if the survey is to be organized properly. A check-list is generally very useful, above all if one is sent to remote regions where it will be impossible to purchase the material lacking (Table 3.4).

Table 3.4 Check-list for preparation of a field programme in Europe

Administration	Basic documents	Personal needs
— letters to administrative authorities — letters to police authorities — form letter to farmers — hotel reservation — getting vehicles ready — (assignment orders)	— aerial photographs — geological maps — base maps for sketches and topographic maps for moving in the field — other maps and scientific literature pertaining to the area — road maps (way to the study area)	— boots, heavy shoes — raincoats, hat — medicines: — mosquito repellents, sun screens, alcohol, cotton — rucksack — means of payment

Field equipment	Office equipment	Special equipment
— augers, picks, knives, magnifying lens — field pH meter, wash bottles for water, HCl — plastic bags, labels, flexible metre rule — mattocks, spades, rakes — storage boxes for small bore and various documents	— cameras, film, batteries (flash (profile photographs) — lamps, stereoscopes, extension cords — field diaries and notebooks, auger bore sheets, profile cards, Munsell colour charts — notebooks, pencils, pencil sharpeners, ballpoint pens, felt-tip pens, erasers, scissors, adhesive tapes and rulers (for measuring coordinates)	— base-camp microcomputer, field microcomputer — resistivity meter — magnetotelluric apparatus — seismic refraction euipment — GPS, GPR, EMI — clinometer, compass, altimeter

When the project is in lands where living conditions are difficult, the precautions to be taken are more numerous and the check-list should be expanded, because the least oversight can be very annoying (Dent and Young, 1993). It is necessary to provide in addition for:
- obtaining all the necessary administrative permissions including (at times) the right to take photographs;
- vaccinations and other health precautions (antimalarial);
- purchasing or renting all-terrain vehicles;
- supply of water, food, currency and petrol, possibly obtaining priority for buying fuel;
- local recruitment of assistants, guides, interpreters;
- arrangements for transport of soil samples to the analytical laboratory;
- means of communication with headquarters;
- names of persons (friends, relations, scientific colleagues, important people), who can be contacted in the country in case of need.

Conclusion and perspectives

Many implements are well designed but expensive; they are not used regularly. So we can dream of a survey in which the mappers will benefit from a modern 4 × 4 vehicle, fast on the highways and efficient in mountains. The vehicle will be endowed with a motorized auger mounted in the rear. An appliance of GPS type will be carried and will enable positioning by satellite. A simple knob will serve to lower the mast for generating an EMI profile continuously with automatic recording. The on-board computer will provide on demand all desired data on vegetation, rocks and landforms in the area. Photographs available to the team will be in natural colours. Altimeter, compass and clinometer will form part of the basic equipment stored in the map compartments of the dashboard. In short, the era in which show-offs (excessively chrome-plated 4 × 4s), amateur sailors (GPS) and Sunday gardeners (motorized augers) are better equipped than soil scientists would have ended.

CHAPTER 4

DESCRIPTION OF SOILS IN THE FIELD

In this chapter we shall first indicate how to describe soil profiles. The problem is not exhaustively covered; only some essential features of the horizons are reviewed, because there are books specifically devoted to this issue. We will only show that descriptions of soil features are precise and based on science as well as on its application. We examine later in detail, except from a purely computer point of view, how to build a system for recording in a data bank the information pertaining to these profiles. Indeed, all specialists of the natural environment are faced with problems of this type at some time or other in their work. After that, we shall present the methods for checking the quality of the work. Lastly, we end the chapter by explaining practical rules that may be usefully observed for organizing the profile description in very correct terms.

4.1 GENERALITIES

Field characterization of soils is chronologically not the first task to be carried out in soil mapping. But this aspect of things is introduced quite early so that useful definitions can be presented in the chapter and used later.

4.1.1 Bibliographic orientation

The quality of mapping depends on the quality of the observations taken in the natural environment and particularly on the quality of profile descriptions. These descriptions are very complex, at least if one refers to the mass of observations that was agreed to be done and recorded with a view to

using them later. In these circumstances, many investigators and institutions have judged it necessary to publish papers explaining how soils are described and how the corresponding information can be stored in a more or less standardized manner. The best-known international publications are:
- First of all, the *Guidelines for Soil Profile Descriptions* (FAO-ISRIC-UNESCO, 1990). This is the most recent of an entire series of publications in English and in French.
- Then, an international attempt at standardization of methods for characterization of soils is presently led by the ISO (International Standards Organization) under *Technical Committee 190* with the help of national standards organizations, mostly European. The work planned is aimed at harmonization and standardization of description of soils, sampling methods and methods for physical, chemical and biological characterization of soil samples. Numerous provisional or final documents are already available. Unfortunately, there has hardly been progress regarding the question of devising a norm for the detailed description of soil profiles. In this, each group defends its own knowledge and its software that it would have to revise in order to use new norms. International agreement has been reached only on some points (Table 4.1) till 1 May 2002.

Table 4.1 Available ISO standards pertaining to soil characterization (http://www.iso.ch)

ISO Ref.	Year	Title
11074-1	1996	Soil quality – Vocabulary – Part 1: Terms and definitions relating to the protection and pollution of the soil
11074-2	1998	Soil quality – Vocabulary – Part 2: Terms and definitions related to sampling
11259	1998	Simplified soil description
11074-4	1991	Soil quality – Vocabulary – Part 4: Terms and definitions related to rehabilitation of soils and sites

- Thirdly, various international publications have introduced concepts that have become de facto standards. Such is the case of the *Legend of the Soil Map of the World* created under the sponsorship of the FAO and the UNESCO. This legend has been improved through many successive approximations. The latest edition of the text (1989) is in French and English. It offers a unified system for defining and naming soil horizons and soils. More recently, the WRB (*World Reference Base*) has been proposed to improve and thus replace the earlier publication. It has been achieved through various documents (Bridges et al., 1998; Deckers et al., 1998; ISSS et al., 1998).

Along with the above, sometimes earlier, each country has formulated manuals for describing soils in the field. Some of them are very close to the

publications mentioned above, others present very original aspects. It is impossible to cite all the documents. We shall only mention:
- *American publications.* The first edition of the famous *Soil Survey Manual* was published by the U.S. Department of Agriculture (USDA) in autumn 1937. Several editions followed. The latest, of 1993, is quite comprehensive. It is found on the Internet (search for site STATLAB IASTATE EDU). It was followed in 1996 by the *National Soil Survey Handbook* (NSSH) accompanied by a *Pedon Description Program* (PDP), both since withdrawn. The official document recommended by the USDA is the *Field Book For Describing And Sampling Soils* (version 2). It is available on the Internet and can be downloaded (search for site NSSC NRCS USDA NEBRASKA).
- *British publications.* They have been realized through publication of a glossary (Hodgson et al., 1976) and also by publishing a more clearly explanatory book (Hodgson, 1978). A more recent book has reviewed and updated the principal concepts (McRae, 1988).
- *Canadian publications.* The system has been perfected as part of the CanSIS *(Canada Soil Information System)*. It comprises manuals in English and French for description of soils, with many successive editions, the latest in 1987 (ECSS, 1987). A powerful computer system has been developing for handling soils data, consisting of descriptions, analyses and mapping surveys. Canadians were pioneers in this area. Their contribution is appreciated and recognized at the international level.
- *French and Francophone publications.* Started more than 25 years ago (Jamagne, 1967), they have been continued in the 'Information Technology and Biosphere' Society, and later under the sponsorship of 'l'Agence de Coopération Culturelle et Technique' (ACCT). As this agency works in the context of French as a spoken language, the publications have been directed with the participation of French-speaking countries of Africa, Belgium, Canada and France. Switzerland, which is not a member of the ACCT, was not represented. Various publications have been put out by the group and also in different French-speaking countries. Belgian scientists in particular have published a glossary and conducted research on soil data banks (Kindermans, 1976; Delecour and Kindermans, 1977). In France, a vocabulary of soil description, known as STIPA, has gone through three editions (Bertrand et al., 1984). The INRA has also published a guide for description of soils (Baize and Jabiol, 1995). Besides, the book *Regards sur le Sol,* published by AUPELF, is a good introductory manual for characterization of soils (Ruellan and Dosso, 1993). Also, the Référentiel Pédologique (RP, 1995) is an efficient system for naming soils. It has gone through an edition in English (1998), Italian (2000) and Russian (2000). Lastly, AFNOR, the French Department of Standards, established the standard NF X32-003 in 1998 for describing soils in detail (57 pp), but this standard does not bear the ISO hallmark.

4.1.2 Content and structure of a profile description

The purpose of describing a soil profile is to preserve the image of the soil, but the latter is not bound to be the most complete and most detailed possible! An effort at synthesis is necessary. The description of a soil profile can be organized in six principal parts.

References and geographic location
The pit number, location with reference to administrative units, coordinates, date of observation and observer's name and affiliation are given. For the coordinates, a GPS is used when possible and useful. But it is never desirable to rely completely on electronic recording. It is advisable to mark the location of the profile by a cross drawn on the base map or on the aerial photograph and to keep the list of sampling sites up to date (Chap. 5, Table 5.1). All this is essential to easily find out to which site the data pertain and to avoid errors.

Profile environment
The natural environment around the profile pit—climate, geology, geomorphology, hydrology, vegetation and land use—is described succinctly. Pedology should not be forgotten in the list! Indeed, a typical profile in a certain milieu is not necessarily representative of its immediate surroundings; such is the case with a soil described in a silt-filled crack between two bare slabs in high mountains. Description of the environment is therefore important; it details the setting in which the observation is taken.

Site and area description
The environment having been defined, it is useful to detail exactly where the profile was found (on a mound, beside a tree, etc.). The area of the terrain around the pit must also be described because the sectional view (first horizon) does not substitute for a view from above for an entire series of characters (spacing of shrinkage cracks, presence of signs of erosion, etc.).

Description of the horizons as such
Other than the upper and lower boundaries, the major features to be observed in the horizons are related to colour, mottles and redoximorphic features, organic compounds, reaction of the soil to applied HCl, coarse fragments, structure, surface appearance of peds and voids, texture, mechanical properties of the material (consistence), organization of voids, and root growth and other forms of biological activity. To this already long list it is appropriate to add secondary formations, that is, accumulations within the horizons from various weathering processes followed by migration or relative accumulation (salt crystals, iron accumulation, etc.).

It may well be asked if there is an order to be followed for describing the various characters reviewed. On this subject, habit often prevails. One scientist starts with colour and ends with texture. Another does the opposite! This is not of much importance. Above all, it should be decided whether all the characters (texture, structure, colour, etc.) of one horizon will be described first before proceeding to the next horizon or the texture, then the structure, then the colour, etc. of all the horizons will be described in that order. Nowadays the second option is preferred. This is not so much the manifestation of a desire to immediately compare the different horizons for each of their characters as the consequence of the way in which the field sheets are presented for the sake of the computer treatment that will follow.

But it is imperative to leave till the end of the description the diagnoses and syntheses, especially the horizon designations (A, B, etc.).

Particular characters of the surface and subsoil horizons
Characteristics occurring only once in a profile (for example, depth of water table) may be separated from characteristics that have to be detailed for each horizon (for example, colour or structure).

Synthesis
To avoid burying the essentials in minor details, the observer is invited, at the end of the description, to leave the pit, move away ten paces and summarize what is important. This is the moment for handling the macroscope. Four principal points are to be considered:
- *Soil moisture regime*. This is deduced from morphological observations associated if need be with hydraulic measurements and data from farmers' survey.
- *Summary of soil profile morphology*. The summarized presentation of the profile is structured in the following manner. First, the properties that hardly change from the top to the bottom of the soil profile are described. For example, it may be stated that the soil is yellowish throughout. Then the properties that change widely from one horizon to another are listed because it would be impossible to carry out, as far as they are concerned, an overall diagnosis. For example, silty then clayed. In such a case, it may be said that the horizons are differentiated by texture or other property that varies widely from the top to the bottom of the profile. In other words, the main features of vertical development are listed.
- *Classification* (in the sense of reference to a national or international system of soil nomenclature) and *allocation* (in the sense of positioning compared to the map legend). Chapter 5, § 5.2.1 will show how to indicate which Map Unit (MU) and which Soil Unit (SU) the profile described belongs to.

- *Management and protection:* mention of the constraints to management that could result from unfavourable soil properties or the immediate environment, observation of degradation hazards.

4.1.3 Remarks concerning auger holes

It is obvious that descriptions of auger holes are much simpler than those of soil profiles. There are two reasons for this. First, it is not possible to devote more than a few minutes to an auger hole. Second, use of the auger causes disorganization of the material making observation of certain properties (structure, location of roots, etc.) impossible. The properties discussed below can be described in auger holes.

4.2 DESCRIPTION OF CERTAIN PRINCIPAL PROPERTIES

It is not our objective to explain here how to describe each of the properties that may be observed in soils. They are too many. The STIPA sheets (§ 4.4.2) allow us to describe up to 188 of them for a horizon! Very fortunately, one does not find in nature a soil displaying all the possible characteristics at the same time. For example, if the soil is not calcareous, there is no need to describe the shapes of lime accumulations. In this section, we shall simply draw the attention of the reader to the essential points and present a vocabulary to be used later. To know more, one has to refer to the glossaries mentioned above.

The observer who has gone down into a profile pit finds himself facing the soil profile. What one should do first is to restore to freshness the vertical face of the section with a knife or pick in order to distinguish the horizons. The profile is alternately smoothed (to observe mottles) or scratched (to expose the natural cracks of the soil).

4.2.1 Horizon boundaries

The first objective of the observer is to define the upper limit and lower limit of each horizon. Distinction of horizons requires a quick preliminary inspection of the section with study of colours, textures, structures and reaction to HCl (§ 4.2.7). Transitions, that is, gradual passing from one horizon to the next are often seen. In some places of the vertical face, one is sure that the material seen belongs to the horizon itself. In other places, one would find it hard to decide whether the material belongs to the horizon or the one above. It is generally estimated that each horizon extends from the middle of one transition to that of the next. But the extended transition may be described in the same way as a homogeneous horizon, if very accurate work is desired.

In general, creation of pointless transition horizons must be avoided. They complicate the soil description without always clarifying or detailing it. Figure 4.1 shows how one generally proceeds (example A) and how one is sometimes obliged to proceed (example B). In the latter case, a soil with three horizons is described as if it had five, and five soil samples are taken! Of course all of this must be explained in the sheets for later use.

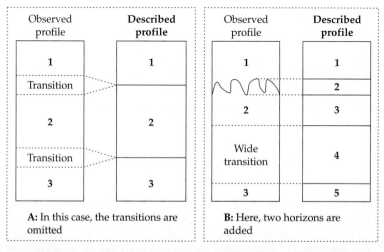

Fig. 4.1 Schematic diagram of two profiles for the purpose of description.

Actually, everything depends on the aims of the description. The suggested simplifications, valid when the mapping has an applied objective, does not give a full account of the way in which one horizon passes into another. Thus they are not suitable for pedogenetic studies. Besides, the process of dividing the profile into horizons (this is the first task to be completed) must be distinguished from the process of identification of horizons (as A, B, etc.). This identification is an interpretation, hence a synthesis resulting from the totality of observations. it should be attempted only at the very end of the description.

Transitions, even those eliminated arbitrarily when horizons are separated, are not always completely forgotten because their thickness is measured and written down properly. Besides, these boundaries, when thick, diminish the contrast between horizons as we pass from one to another. For checking the relevance of the division proposed, it is often useful to take a few grams from the middle of the various horizons and place them in the hollow of the hand, arraying them in order, one facing the other. The contrast between horizons is then reproduced in the best manner.

Some observers consider that the depth value '0' is not of the surface but of the beginning of the horizons that are not strictly organic. For example, they record a 4 cm thick litter on top of the soil with the references '–4 to 0 cm'. This system leads to complications when the data are processed. It would be better to place zero at the surface of the soil. The error introduced by the fact that the position of 'zero' could vary during the year is slight and probably smaller than the uncertainty in visually locating the boundaries between horizons. Of course one may err in particular cases. This is as true for description of depths as for the rest of the characterization of the profile. In addition to the conventional indications, it is strongly recommended that the exact situation be properly described in the spaces reserved for *comments* in the field sheets.

The base of the pit could correspond to a horizon boundary (for example, occurrence of hard rock), but this is not the general case. Many observers have the habit of characterizing the depth of the last horizon in the form '60-120+ cm'. The + sign indicates that the lower boundary of the horizon in question has not been reached. Some computer systems provide for management of this special feature.

4.2.2 Particle-size distribution and texture

Particle size limits

Many properties of a soil depend on the proportion of particles of different sizes it contains. Particularly affected are the mechanical properties, cation exchange capacity, hydraulic conductivity, etc. The names given to particles of different sizes found in the soil are not quite the same in all countries. Table 4.2 presents the widely used systems.

Table 4.2 Examples of particle size scales

FAO-ISRIC, 1990		SAND					
CLAY	SILT	very fine	fine	medium	coarse	very coarse	GRAVEL
2 µm	63 µm	0.125 mm	0.2 mm	0.630 mm	1.25 mm		2.0 mm

USDA, 1993		SAND					
CLAY	SILT	very fine	fine	medium	coarse	very coarse	GRAVEL
2 µm	50 µm	0.10 mm	0.25 mm	0.50 mm	1.00 mm		2.0 mm

F S	LIMONS		SABLES		GRA-
Argile	Fins	Grossiers	fins	grossiers	-VIERS
2 µm	20 µm	50 µm	0.2 mm		2.0 mm

The last table holds good for France (F) and Switzerland (S).

Representation of particle-size distribution
Particle-size distribution is the set of values of the proportions of *sand*, *silt* and *clay* in a sample. The distribution in a sample can be represented on the plane of the paper in a two-dimensional diagram, although three values have to be reported. This is possible because the three values are not independent. Actually, 'sand + silt + clay = 100%', at least if we limit ourselves to sampling and analysis of the fine earth (the <2 mm fraction). There are two ways of proceeding with this representation. The first consists of putting silt on the abscissa and clay on the ordinate in an orthonormal diagram, sand representing the difference from 100. It is impossible to cross the line defined by the points (0, 100) and (100, 0). This delineates a *triangular texture diagram*. The second method of representation consists of working in the plane 'sand + silt + clay = 100', obviously restricting oneself to positive values for the three variables. The texture triangle is equilateral. Locating a point exactly is a little complicated (Fig. 4.2). The content of an element (say *silt*) is read by drawing a line parallel to the 'preceding' side, counting clockwise (see arrows in the drawing).

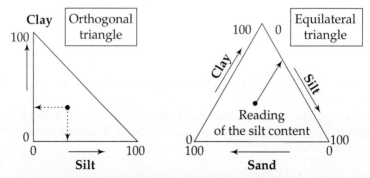

Fig. 4.2 Representation of particle-size distribution in triangular diagrams.

Concept of texture
The texture triangles can be divided into small areas of geometric shape so that all the samples the particle-size distribution of which fall in the same area have almost the same mechanical properties. The concept of texture covers two very different things:
- the name of the area in the triangle corresponding to the particle-size distribution obtained; this is the *texture class*; it is, therefore, a sort of grouping of the sample on the basis of its particle-size distribution; for example, such or such sample falls in the *silty clay loam* class;
- the name of the area in which the particle-size distribution will probably be found if it is determined; this is the *field texture*.

The two approaches are different. In the first case, we are representing the results of an analysis. In the second, a soil sample is kneaded between thumb and index finger to make an estimate of its particle size distribution. There are three reasons justifying qualitative estimates in the field.

First of all, mapping yields many observations of the surface soils with direct estimation of the texture. To follow this up by systematic particle-size analysis would be too costly and will require enormous time. So the qualitative test is quite useful. It is obvious, however, that checking is necessary so that the mapper could verify or improve the quality of the diagnosis. This is done when the profiles are characterized, since the horizons are first described, then sampled for analysis. The field texture is thus periodically checked by the particle-size analysis done later.

Secondly, identification of field textures for the horizons of a profile is a safety measure: anyone can mix up samples or make a mistake in copying down numbers. In particle-size distribution, as in many analyses, it is good to know the range of values the expected result should fall in before sending the sample to the laboratory... .

Lastly, the concepts of field texture and particle-size distribution are not the same as regards near-errors in diagnosis. Particle-size distribution, as said earlier, gives the proportion of sand, silt and clay. Texture indicates the *behaviour* of a sandy soil, a loamy soil or a clayey soil, which is slightly different. In particular, very fine particles of clay size may be bound strongly into assemblages so firm that they form small aggregates of sand size and behave as sand particles (light, porous soils). This, for example, is the case in some Ferralsols of the tropical zone. Pseudosands occur in them in the form of associations of iron and clay particles (kaolinite). The texture is sandy to the touch at first. After a minute's rubbing between thumb and forefinger, the aggregates are destroyed and it will be realized that clay predominates. Particle-size analysis, done after removing the ferruginous cement, reveals that the material contains more than 60 per cent clay! In temperate or dry zone, calcium carbonate often binds the particles.

Particle-size analysis can be done with or without destruction of cements. The choice depends on the soils and what is sought to be characterized: size and proportion of different particles, or their behaviour. Thus, it is usual to find quite different particle-size distributions from what the field texture would indicate.

Various systems of defining texture
The physical and chemical properties of a sample do not change abruptly when there is a slight change in particle-size distribution. Consequently, there is no international accord on texture classes. The triangles are divided in as many ways as there are countries! Some examples are given in Fig. 4.3.

Assessment of texture
For making a correct assessment of texture, one must know how to identify sand, silt and clay by feel. For this, at first a little moist soil is kneaded until

Checking the assessment of texture
To check the quality of an assessment, it will suffice to mark in the texture triangle the point given by the particle-size analysis data of the sample to be tested. In the ideal case, this point will fall in the *texture class* predicted in the field. The examples in Figure 4.4 have been obtained by this method. The 102 horizons predicted to be *limono-argilo-sableux (LAS)* and 83 identified as *limono-sablo-argileux (LSA)* were extracted from a soil survey report (Bonfils, 1993). The matching texture classes (in principle) are enclosed by thicker lines. In the former case, the error is almost *random* (a cloud of points centred around the true texture class). In the latter, a *bias* is seen, that is, a systematic error (cloud not centred around the true texture class).

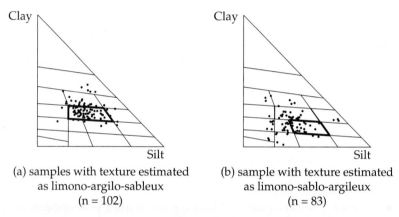

(a) samples with texture estimated as limono-argilo-sableux (n = 102)

(b) sample with texture estimated as limono-sablo-argileux (n = 83)

Fig. 4.4 Comparison of textures predicted in the field (*cases within thick boundary*) and particle-size analyses done in the laboratory on the same samples (Bonfils, 1993).

It will be wrong to deduce from this that the surveyors have done poor work. Who is right after all is said and done? The designers (not many) of the French texture triangle or the users, practically all, who have a different assessment of the feel of *limono-sablo-argileux?* Moreover, it can be verified that the points considered as *limono-sablo-argileux* by the French mappers are actually *limono-sablo-argileux* in other texture triangles! This shows that it is necessary to review the triangles and simplify them. The instances are too many if we refer to what a mapper in the field is on average capable of finding. But the fact is that a trained scientist, always working in the same environment, can pick a given texture triangle and then find quite well the slightly arbitrary names that it suggests.

New method for assessing texture
We must try to make progress. Actually, the manipulation of a few grams of soil with the fingers enables us perceive only two things:

- first, the *presence of sand* (*cf.* granularity mentioned above);
- second, the *plasticity* if the sample is wet enough or the *friability* if the sample is just moist or dry. Of course, nuances of detail can be introduced in the assessment conditions, but friability is roughly *rupture resistance* of the Soil Survey Manual, *strength* of the Australian Handbook and *dry consistence* of the Canadian manual.

The proportion of sand, silt and clay is more or less consciously deduced from just these elements. By examining the thousands of data stored in the STIPA soil bank of the INRA, Montpellier, France, we were able to determine which particle-size distributions (and which textures in the French GEPPA triangle) corresponded to samples identified in the field as more or less friable and more or less plastic. We suggest that this information be used in the field in future. The protocol proposed is as follows. An attempt is made first to assess the friability (dry sample), then the plasticity (wetted sample). The definition of plasticity has been given elsewhere (Fig. 4.7). Friability is assessed by trying to crush a small soil clod between thumb and forefinger:
- *not friable:* the sample cannot be crushed unless it is placed on the palm of the hand;
- *slightly friable:* crushing is possible between two fingers with force;
- *friable:* crushing requires only slight force;
- *very friable:* the sample is very fragile and it is difficult to hold it without crushing.

Then the granularity (sand predominant, present, absent) is determined. Table 4.3 can be used to assess the texture as precisely as possible.

Table 4.3 Aid for identification of texture in the French texture triangle (GEPPA)

		Plasticity estimated on a wet sample			
		Very low	**Low**	**Moderate**	**High**
Friability estimated on a dry or moist sample	**Very small**		AS, ALS LA, AL	AS, ALS AL, A	ALS A, AA
	Small	SA, SAL	AS LAS, ALS LA, AL	AS LAS, As, ALS AL	ALs
	Moderate	SL, SA, SAL LSA L	SA, SAL LSa, LAS LA, AL	LAS AL	LAS
	High	S, SL, SA, SAL LSA L, LL	Sa, SAL LSA LA	Almost impossible combinations	

Instructions. Friability and plasticity of a sample are determined on the dry sample then on the moist sample. The square to be used is then identified in the table. Lastly the granularity is taken into account. If there is much sand, the textures are taken from the top line, for more or less sand the middle lines and for scarcely any sand the bottom line. For example, a friable and plastic sample without sand corresponds to AL.

It should be noted that a table constructed in the same manner has been published for the Swiss texture triangle (Kaufmann, 1990). But such tables also show that a given physical behaviour may correspond to very different proportions of sand, silt and clay. Hope of textural expertise of very high precision should be given up. It is a pity, especially when texture is used in very large scale mapping for defining soil units as homogeneous as possible.

One last remark is obvious. In what was said above, we did not use all the observations and data that could help to reach the best diagnosis possible of the texture. For example, presence of shrinkage cracks is a sign of probably high clay content. But observations should not be replaced by interpretations and one should not introduce in the soil description subjective correlations between factors that will rig subsequent statistics if needed on the subject!

4.2.3 Coarse fragments

Relevance of study of coarse fragments
It is necessary in a soil description to define the *abundance, resistance, nature* and *degree of weathering* of the coarse fragments. Certain crops tolerate stones poorly (beet), others are impossible to harvest mechanically in stony soil (potato). All plants suffer the effect of dilution of the fine earth by the coarse fragments present. For example, a soil containing 50 per cent by volume of elements larger than 2 mm is almost equivalent in reserves of water and easily available nutrients to a non-stony soil half as deep! However, coarse fragments can be useful. When they are abundant on the soil surface, they restrict evaporation, an advantage in dry climate. Moreover, a soil covered with stones need fear much less erosion by water. In certain cases, the chemical action has to be considered. Thus in mountains, coarse limestone fragments resist dissolution and prevent soil acidification (Legros et al., 1987). Lastly, their shape and orientation can play a part. For example, in micaceous schists they consist of elongated laths that get arranged flat in the soil and impede penetration of roots to a greater degree than round stones would. A book in French is specifically devoted to these questions (Gras, 1994).

Total content of coarse fragments
This is a matter of calculating the percentage by volume or by weight of the soil occupied by coarse fragments. The estimation is difficult and there are many ways of going about it.
- *Visual estimation* by using boxes showing relative occupancy similar to those used for mottles linked to excess water (§ 4.2.5).
- *Point-count method* (Hodgson, 1978). A piece of square-mesh grid is applied on the soil surface or on the vertical face of the profile. It is

sufficient to count the number of times the wires cross on a coarse fragment and compare this number to the total number of observations made of the nodes of the grid to get the proportion of these coarse fragments by volume. The grid should have a mesh size larger than the size of 90 per cent of the coarse fragments. A hundred points should be counted. The coarse fragments observed should be dry enough and clean for being easily distinguished from the fine earth. Various papers have explained the statistics underlying this kind of determination (the binomial law).
- *Direct determination.* Obviously this is the most accurate method but it needs time and so cannot be used routinely. Necessary sieves and balances are taken to the field for separating the coarse fragments, weighing them and comparing their weight to the total weight of soil they came from. Warning: in this case the determination is not by volume. To always use the same units, the data can be converted by considering the bulk densities. Here again, there is the problem of sample size. For a precise estimation, the larger the stones the more soil has to be weighed (*cf.* Table 4.11).

Size of coarse fragments
The total quantity of coarse fragments having been defined, it is useful to define which sizes are dominant:

G = *gravel*   0.2-7.5 cm
S = *stones*   7.5-25 cm (ISO); 7.5-60 cm (USDA)
B = *boulders*   >25 cm (ISO); >60 cm (USDA).

Subdivisions can be introduced for more detailed characterization. Where many sizes are present, one can say in which order they dominate. For example, 'SG' would indicate a mixture of stones and gravel, the stones being more abundant. However, only the coarse fragments with >5 per cent abundance should be mentioned.

Nature of coarse fragments
We have seen that the nature of these elements has to be considered in certain cases at least. It is generally useful to compare the nature of the coarse fragments of the soil and of the parent material. Often the difference is only due to the degree of weathering, which increases from the bottom to the top of the profile. But at times certain coarse fragments present in the parent material might have totally disappeared from the horizons by dissolution or weathering (in humid climate, limestone pebbles are not present in the upper part or are highly weathered in fluviatile or glacial deposits).

Reference can be made to specialized publications for the description of other properties of coarse fragments (shape and orientation).

4.2.4 Colour

Importance of observation of colour

The accurate identification of soil colour is something of significance. Actually, colour is linked to the presence of certain principal components, in particular organic matter, iron oxides and hydroxides and calcium carbonate. Characterization of the colour sometimes enables us guess the very approximate proportion of one or other of these elements and to predict the agronomic behaviour they determine. Also, if one of them is very abundant, it is the result of specific pedogenetic conditions. Furthermore, certain colours are characteristic of Gleysols (WRB), that is, soils having an aquic moisture regime (Soil Taxonomy). Thus colour is an important criterion among those that allow identification of a given soil and understanding its functioning and behaviour. An entire book has been devoted to this question (Schultz et al., 1993). Authors of old were not mistaken. They distinguished, for example, *Red Mediterranean soils, Black Cotton soils, Yellow Tropical soils,* etc. Lastly, colour helps to distinguish at first glance the principal horizons of the profile.

Identification of colours

The tool used for identification of colours is a booklet in which are pasted small chips of different colours. The problem is to find out which resembles most the soil sample considered. The *Munsell Soil Color Charts*, manufactured in U.S.A. since 1954 by the *Munsell Color Company,* is known the world over and has become a de facto standard. Another soil colour chart published in Japan under the name *Revised Standard Soil Colour Charts* can be considered equivalent, though it does not match the Munsell perfectly. These books are costly. But then so would be any published book containing more than 200 colours! In the Munsell, the colours are classified on the basis of three criteria (see extract, Fig. 4.5).

- *Hue* indicates the dominant colour, that is the chromatic composition of the light that reaches the eye:

    Y   *yellow*
    R   *red*
    YR   *yellowish red*
    G   *grey.*

    These basic colours are supplemented by intermediate colours by a numerical prefix. For example: 10R, 7.5YR, etc. A colour corresponds to one page in the Munsell. It is rather easy to identify the colour of the soil sample studied. When aligned between any page and the eye, the sample immediately appears redder, more yellow, less green, etc. The appropriate page is quickly found by turning the pages.
- *Value* is the degree of lightness or darkness of the colour from zero (very dark) to 10 (very light). It is obtained by mixing the colour with

DESCRIPTION OF SOILS IN THE FIELD

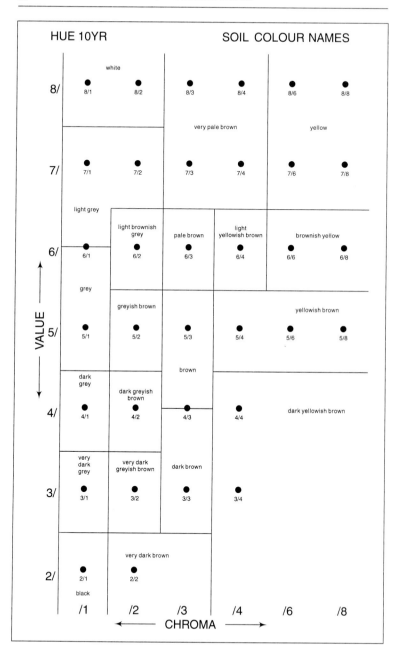

Fig. 4.5 Arrangement of colours on a page of the 'Munsell' chart.

black (low values) or with white (high values). Grey corresponds to 5. The value is, we will see, often related to the organic matter content.
- *Chroma*, the relative purity or strength of the spectral colour, indicates the vividness of the colour on a scale of 0 to 10. When we come down the scale, we pass from the maximum vividness to a colour progressively mixed with grey. At level 0, the colour disappears and a neutral grey intermediate between black and white is obtained (*cf.* value).

The result is given in the form of a code. The *hue* is given first, then the *value* and lastly the *chroma*. For example, 10YR 4/2. But the Munsell system also gives the translation of the code in clear:

10YR 4/2 = dark greyish brown.

Precision of the estimation of colours
The precision of this kind of determination is not high for soils with very dark colours. We can no longer see the basic colours and hence may not be able to choose between different hues. One is generally less in error for lighter-coloured soils. We did the following test: the colour of the same sample of soil was determined with the same copy of the Munsell chart by nine persons, teachers and students at the l'École Polytechnique Fédérale de Lausanne, Switzerland. The results of the test are presented in Table 4.4.

Table 4.4 Test of definition of a colour by 9 observers

2.5Y 5/3	2.5Y 5/3	10YR 5/3
10YR 4/3	10YR 4/4	2.5Y 4/3
10YR 4/4	2.5YR 4/4	10YR 5/4

Determination of the Munsell colour thus gave a close reading of hue, a close reading of value and a close reading of chroma. But one could not choose among very close colours! This is still better than giving a colour without precise references.

Among all the factors defining precision, the most important is obviously selection of a representative soil sample. This is no problem in the case of a horizon with very uniform colour. One should however be warned that the matrix of peds often has a colour different from that of the surface of those peds. If the difference is large, both colours should be characterized.

If mottles are seen on the faces or in the matrix, it is desirable to describe them and define their colour (see § 4.2.5).

For estimating colour in a proper manner, it is recommended that one work with natural light at 45 degrees. Determination should be avoided in very poorly lit areas (forests), at midday if the light is too strong or even in the evening when the rays piercing the earth's atmosphere are reddish. The reading of the value depends on moisture content: the colour may be described at the moisture state when the sample is described (the moisture content should then be determined and reported), it may be moistened for examining its colour when moist, or a sample may be taken for drying in the laboratory and the colour determined when it is dry. In certain soil classifications the change of colour from the dry state to moist state is a criterion taken into account for recognition.

4.2.5 Mottles

Nature and significance of iron mottles
Mottles are related to the phenomena of oxidation and reduction of iron and manganese in the absence of oxygen but with excess of water present.

Transformation of the oxidized form of iron to the reduced form is schematically written as follows:

$$Fe^{++} \Leftrightarrow Fe^{+++} + e^-$$

At the usual pH of soil, ferric iron in solution is most commonly found in the forms $Fe(OH)^{2+}$, $Fe(OH)_2^+$ and $Fe(OH)_3$. The concentration is very low, even negligible: between 10^{-6} and 10^{-9} mole L^{-1} depending on the pH (Chamayou and Legros, 1989). Ferric iron is thus mainly precipitated as *mottles, soft segregations* (also called *soft concentrations*), *nodules* and *concretions* with dominantly yellow or reddish-brown colours. On the contrary, ferrous iron, which occurs as Fe^{2+} and $Fe(OH)^+$, is likely to be in much greater proportion in solution, from 0.6 mole L^{-1} to 10^{-4} mole L^{-1}. In other words, ferrous iron is up to 100,000 times as soluble as ferric iron. When precipitated, it appears in grey, greenish or bluish colours.

At a partial pressure of oxygen equivalent to that in air (0.2 bar), ferrous iron is unstable and is transformed to ferric iron. On the contrary, in soils in which the water is poor in oxygen and so much in excess as to prevent air from entering the pores, the iron can become ferrous. The iron then migrates and disappears from the milieu or even remains in place in solution to be precipitated later in the ferric form at the same place or close by when oxidation takes place. Various studies have been conducted to link these manifestations to the hydric functioning of the soil and in particular to the duration of saturation by the water table if it exists. The most sensitive criterion is decolorization. A whitish colour is often caused by a water table present for several months. In some cases, reduction of iron may be very local and related to the consumption of oxygen by combustion, that is mineralization of organic matter. Such, for example, is the case of ploughed gley linked to burying in the wet season of vegetation or manure that was earlier on the soil surface.

Mottles are accumulations that can be seen by their colour but are not hardened enough to constitute elements with mechanical properties distinct from those of the matrix. They are conventionally described by their colour, size, contrast, distinctness of boundaries, shape and position relative to other features and constituents of the soil.

Abundance of mottles

Abundance of mottles is estimated by comparison with 'boxes of relative occupancy' (Fig. 4.6) the object of which is to visualize what actually represents a certain proportion of mottles: 10 per cent, 20 per cent, 30 per cent, etc.

   20%   25%   35%   40%   50%

Fig. 4.6 Boxes of relative abundance; extracted from those published in the Canadian glossaries.

Colour of mottles

It is usually sufficient to describe the colour of mottles using simple terms. For example, white, blue, red, etc. If needed, the Munsell code can be used. But are these very precise data entered in the data bank for later use?

Size of mottles

The systems foresee more or less detailed descriptions. Here, for example, is that of the FAO.

Very fine	<2 mm
Fine	2-6 mm
Medium	6-20 mm
Coarse	20-60 mm
Very coarse	>60 mm

For description of other properties of mottles (contrast, distinctness, shape, orientation and distribution), and for characterization of soft accumulations, nodules and concretions, one should refer to specialized books.

4.2.6 Organic matter

Significance of organic matter

Organic matter plays an essential role in the soil. It forms an important reserve of nutrients that are slowly supplied to the plant by mineralization and absorption. Organic matter acts in stabilization of structure and in water retention. It develops a higher cation exchange capacity, weight for weight, than clay. However, very high accumulation (peat, thick humus horizon) indicates very low biological activity.

Assessment of organic matter in the field

Field estimation of organic matter content addresses concerns similar to those evoked by texture (§ 4.2.2). Besides, it is traditionally accepted that the quantity of organic matter (OM%) is related to the carbon determined in the laboratory (C%) by the empirical equation

$$(OM\%) = 1.72 C\%$$

The organic matter in some surface horizons is distinct from the mineral material. There is superposition of organic layers and mineral layers or juxtaposition of organic debris and mineral particles. It is then possible to recognize this organic material, describe it and estimate its abundance. To achieve this in proper conditions, reference should be made to the suggestions in the Référentiel Pédologique (RP, 1995), the most detailed in the literature. But in most cases, the organic matter and mineral material are intimately associated and plant debris is too much transformed and fine to be observed in the field. Under these conditions, estimation of the abundance of organic matter is done chiefly based on colour, darker soils being supposed to be richer in it. This is more or less true. Using the data stored in the STIPA-Montpellier data bank, we verified the two statements below:

- a light-coloured soil does not contain large amounts of organic matter;
- a dark soil may be rich or poor in organic matter: some mineral skeleton grains (from schists, gabbros, etc.) are dark and may lead us to think that carbon is present in the milieu.

But if this trap is avoided, one is right to assign to colour the role of indicator of organic matter content. With a series of 1477 'colour/carbon content' pairs pertaining to cultivated soils of France, we could demonstrate the following relationship, very simple to memorize and use:

$$C\% = 9\text{-}value - 0.5 chroma$$

For example, a soil with colour 10YR 4/6 has on average $9 - 4 - 3 = 2$ per cent carbon and a soil with 5YR 7/4 colour is more or less devoid of organic matter. Unfortunately, this expression is associated with a multiple-correlation coefficient of just 0.43. One can find in the literature numerous correlations relating to the same subject (Schulze et al., 1993). In all cases it is difficult to demonstrate anything more than an approximate relationship between carbon and colour. When the sample belongs to uncultivated soils, the organic matter content is higher for a given *value*. Fortunately, the mapper does not base his diagnosis solely on colour. He also draws on his experience of the milieu and knowledge of the soils. For example, a specialist working in the South of France knows that the soils of vineyards rarely contain more than 1 or 2 per cent organic matter.

Scale of approximate organic matter contents

The assessment scale of organic matter content that we recommend and has been adopted by the ISO is:

0 None (0%) 4 Moderately high (4-10%)
1 Unknown 5 High (10-20%)
2 Low (< 1%) 6 Very high (20-30%)
3 Medium (1-4%) 7 Extremely high (> 30%)

The limits were chosen to match the thresholds used in the RP for characterization of humus and to harmonize with the limits of the *diagnostic horizons* of the FAO-UNESCO legend of world soils.

4.2.7 Effervescence

Relevance of assessment of effervescence
It is important to know if a soil contains calcium carbonate in view of the important properties that depend on presence of the Ca^{++} ion. We may mention in particular the pH (higher in calcareous soil), humus type (certain humus types are linked to presence of calcium), structure (generally more developed and more stable in calcareous soils and in soils with high amounts of Ca^{++} on the exchange complex. We also know that some plants, called calcicole, prefer calcareous soils whereas others, called calcifuge, tolerate them poorly or not at all. But the reaction of vegetation also depends on the hardness and degree of fragmentation of the rock present, which would determine the ease of diffusion of Ca^{++} in the milieu (Legros et al., 1987).

Estimation of effervescence
When a drop of dilute HCl is placed on calcium carbonate, the following reactions take place:

$$CaCO_3 + 2HCl \rightarrow CaCl_2 + H_2CO_3$$
$$H_2CO_3 \rightarrow CO_2 + H_2O$$

Thus CO_2 is released as indicated by bubbles within the drop of liquid. Therefore, for examining if a soil sample contains calcium carbonate, it is sufficient to drop hydrochloric acid diluted with water (one volume of acid mixed with two of water) on it. The carbonate content of the soil is then very roughly estimated as shown in Table 4.5.

It must be remembered that magnesium carbonate, $MgCO_3$, is decomposed with difficulty by dilute hydrochloric acid in the cold. Actually, effervescence is observable if the $MgCO_3$ is finely divided (fine earth) or if the surface of the rock one wishes to examine has been reduced to powder by repeated blows with a hammer or again if the bottle of HCl is gently warmed in the hands.

Table 4.5 Relationship between effervescence and estimated carbonate content

Reaction of soil with HCl	Diagnosis
If no effervescence is visible or audible	Noncalcareous
If effervescence is audible and if some bubbles of CO_2 are seen only after a few seconds (not to be confused with expulsion of air by introduction of liquid into the pores)	Slightly calcareous
If effervescence is visible and bubble are quite numerous	Moderately calcareous
If the bubbles form a thin layer but this layer is more or less continuous	Highly calcareous
If the reaction is vigorous with bubbles rapidly forming a thick froth	Very highly calcareous

4.2.8 Identification of horizons

The habit of identifying the soil horizons by the first letters of the alphabet (A, B, C, D,...) was picked up in the nineteenth century. Presently, the basic principles of using letters of the alphabet are the following:
- O: surface horizons chiefly containing organic matter,
- A: surface horizons chiefly containing mineral material,
- E: impoverished subsoil horizons, often light-coloured,
- B: horizons of weathering and/or concentration of clay, iron, aluminium, humus, carbonates, gypsum,
- C: horizons having some characters linked to soil development and most part of others pertaining to the original geological material,
- R: hard bedrock.

Further, the transition horizons are codified in the following manner. Taking for example a transition between A and B:
- AB if the transition horizon resembles A a little more than it does B,
- BA if the transition horizon resembles B a little more than it does A,
- A/B if the transition horizon is actually irregular and contains recognizable and separate pockets of A and pockets of B.

The system is more complex in its details. We can use prefixes and suffixes to define things exactly and identify, for example, a 'IIBs'.

4.3 DESIGN OF A DESCRIPTION SYSTEM FOR COMPUTER PROCESSING

Too many databases are created without sufficient thinking in advance regarding the type and form of the information to be stored. However, the preparatory analytical effort suggested here is very worthwhile because it avoids much trial and error in the setting up of the system. As in everything, design should precede execution.

4.3.1 Mass of information

To build up a soil data bank, it must first be decided which part of the 'universe' is going to be the object of a characterization for computer storage. This is more difficult than it appears. For example, one is going to record the data concerning the parent materials of the soils. But this is not a matter of building a geological database. We must restrict ourselves to the data of this discipline having pedological interest. The boundary is not easy to locate. Will the petrological composition of the rock be recorded, if it is known? At present, the DONESOL system, developed by INRA for managing cartographic data (Gaultier, 1990; Gaultier et al., 1992b) is able to allow entry of considerable information as summarized in Table 4.6. The definition of layers is given in Chapter 6, § 6.3.3.

Table 4.6 Amount of information that can be stored in the DONESOL system (Gaultier, 1990)

General references	38 variables
Description of map units and the environment	109 variables
Description of surface features of the soil units	49 variables
Summarized description of layers with their variability	25 variables per layer
Detailed description of profiles	
Environment	55 variables
Description of horizons	94 variables
Analysis of horizons	77 variables
Total	447 variables

Under these conditions a map comprising 80 map units, 100 soil units and 200 profiles each having 4 horizons could theoretically represent 167,000 data. Actually the volume of information is much smaller for three reasons:
- First, the DONESOL system avoids, by its relational structure, duplication of information. For example, the values of climatic parameters are not repeated for each map unit if it clear that many units have the same climate.

- Then, many variables are not used because, in a given case, they would be pointless. Thus the forms of calcium carbonate are not described for soils not containing lime and all that pertains to relief (slope, aspect) is ignored in level areas....
- Lastly, the characterization is almost never complete in the group of variables that could be recorded. Such is the case with analyses. In general, only the most common determinations are done for each profile.

4.3.2 Form of the data

The form the data will take in the database must be defined. How will the reality seen in the field be expressed, in words or in numerals that can be stored? A task resembling an exercise in translation and modelling as well must be accomplished. Many possibilities are open in each case. The following, mainly extracted from an earlier publication (Legros and Nortcliff, 1990), presents the list of major questions for which answers must be obtained. This concerns every case, that is every variable (in the mathematical sense) that will be in charge of carrying a small amount of information from the field to the computer. Such a review is not indispensable for using the field sheets, but it helps us understand on what bases the field sheets are drawn up. This also shows that the design of a database is necessarily an extensive job demanding much thought.

Synthetic or analytical approach?
For describing an object, a tree for example, we can directly give its name and say 'this is an oak'. We can also as well set out the detailed list of all its characters and thus describe it completely. The former approach, here termed synthetic, has the advantage of brevity, but it is rarely very precise (that is, there are all sorts of oak, different in species, height, etc.). Also, it may be inapplicable if the non-specialist observer does not know how to fine directly the term that has overall diagnostic value. The latter approach, that is the analytical approach, is more precise and easy to handle. In our example, it will consist of characterizing the trunk of the tree, shape of its leaves, etc. But it will be translated by a long description within which the main point is often drowned in details.

In the domain of concern to us, characterization of the soil is essentially analytical. On the other hand, characterization of its environment (geology, climate, ...) is often synthetic because of the anxiety to be concise. In some cases, both approaches are used simultaneously. For example, we are not restricted to describing the soil, we also give its name in different classification systems.

Advanced or primitive language?
For explanation, let us take the example of colour. Wee can expect to see various coloured elements in the soil, such as organic matter, carbonate accumulations, ferruginous mottles, etc. But experience shows that all the possible coloured components are not present at one time in the same profile. Theoretically, it is enough to have two or three 'colour' variables and use them according to need by detailing each time to what they apply. For example, we may set up a variable 'colour of coloured objects of Type 1' followed by 'nature of coloured objects of Type 1' (here 'Type 1' has no other use than to ensure the relationship between colour and nature). This solution is appealing because it is compact. It is closer to an advanced language in which many words have their meaning defined by the context. Unfortunately, serious problems arise in using it. Suppose we are searching in the database for the soils having ferruginous mottles. They could be described before or after the other coloured objects in the profile. To find them, therefore, we have to explore the content of all the variables of the type 'nature of coloured objects'. Thus we do not know where exactly the information sought will be found. Still more serious, the same soil may be described in many different ways since the objects of Type 1 may correspond to organic matter and objects of Type 2 to ferruginous mottles or vice versa.

What a lot of complications in prospect when we use the data bank! So we must discard 'service variables' created in small number, which will be used for defining abundance, size, shape and colour of indifferent objects. We should then agree to keep to a sufficiently primitive language in which each word (each variable) has just one meaning and possible content. For example, 'colour of oxidation mottles', then 'colour of reduction mottles' and so on.

Master or subordinate variable?
Certain variables are subordinate in the sense that they can be described only if others have already been recorded. For example, it is not possible to give the size of concretions if it was earlier indicated that this type of accumulation is absent! This is an obvious fact that does not require that one stop there when the soil description is being entered on sheets. Actually, the observer manages by instantly skipping with eye and pencil all the portions that are shown to be inoperative on the sheets. The difficulty arises from using a portable microcomputer in the field. To avoid the machine scrolling through all questions pertaining to useless categories, jumps must be provided for. This means that we cannot economize on analysis of the hierarchy of different terms used to describe the soils or the environment. This is not necessarily simple. There can be many levels. For example:

(1) Are organic elements present?
(2) If yes, is it litter?
(3) If yes, does this litter contain needles?
(4) If yes, what is the degree of transformation of these needles?

Level of perception?
Let us take the case of a steeply sloping sideslope, organized into terraces or cultivated berms. If we ask an observer located on one of these terraces to estimate the slope, we could as much get the reply 'steep slope' as the response 'no slope'. All depends on the scale of observation at which we work. This problem is general for many variables: for example, definition of type of vegetation or crop, definition of soil type (local or dominating the environment), etc.

Experience has shown that different specialists of the natural environment (climatologists, soil scientists, agronomists, geomorphologists,...) work by considering geographic systems fitted together but whose size and boundaries are not the same from one discipline to another: parcel, biotope, microclimate, soil series, etc. So the scale of observation should be given, by imposing it on the observer or by requiring the observer to define it exactly. At a pinch, the slope or vegetation can be described at various successive levels, which obviously leads to increase in the volume of data to be recorded.

Quantitative or ordered qualitative variable?
For describing, say, the size (class) of structure we have a choice of two systems:
- give a numerical value such as 10 mm or 15 mm;
- choose and tick against the appropriate term in a list showing, for example, *fine structure, medium structure, coarse structure*, etc.

Each of these has advantages and disadvantages. The numerical approach involves difficult measurements in the field, but results in storing information that is easy to process. When published, it could manifest itself as a displeasing presentation. Thus, a description that accumulates numerals (colour, 7.5Y 4/2; sand, 20%; silt, 60%; stones, 30%; mottles, 15%; roots, 15 dm^{-2}) will rapidly become unreadable!

As for the qualitative approach, it enables definition of classes the limits of which are not regularly spaced but, on the contrary, are selected for their significance in a particular context.

The principal disadvantage in the qualitative approach is that it leads to use of adjectives whose meaning is not easy to define precisely. For example, coarse granular structure is smaller than fine prismatic structure. In plain language, the adjectives only take their meaning according to the substantives they are applied to. So we are obliged to develop different tables on the model of Table 4.7.

Table 4.7 Conversion between numerical values and qualifiers

Size of structure	<2 mm	between 2 mm and 20 mm	between 20 mm and 200 mm
GRANULAR	fine	coarse	
PRISMATIC		fine	coarse

These tables should then be known both to the author and to the user of the description. Obviously this is hardly practicable!

Type and degree of codification
For all sorts of variables, for example for characterizing a rock, it is possible to provide an answer in clear such as 'granite' or in code such as '124'. From the viewpoint of data entry, experience shows that the better solution is a mixture of both systems. Actually systematic recourse to codes is annoying because the observer must refer to tables giving the codes. In the same way, generalized use of text in clear is also unsatisfactory because it involves writing a large number of words by hand. This leads to loss of time.

We should therefore appropriately choose the variables that will be recorded in clear and those that will be coded. Codes are practical for variables having few well-known states (example: types of structure). Text in clear is appropriate for variables that have very many states not listed in advance (example: type of vegetation).

Single variable, double variable or variable with dual content?
We saw above that it is at times useful to describe many kinds of slope, many rock types, etc. Then, for each variable the problem arises of a possible doubling or tripling. This multiplication of number of variables is often asked for by mappers because it seems to them that it would help in facing all situations including the most complex. But experience proves that these variables asked for are later scarcely or not at all used. Under these conditions, it is better they are abandoned for reducing clutter on the sheets and input screens, at the same time leaving the mappers with the option of mentioning the existence of such characters in the 'comments' field provided for this purpose. One could be tempted by the technique of doubling not the variables but their states. For example, the two solutions presented in Table 4.8 are, in principle, equivalent.

But in many cases the second solution turns out to be a rather unpleasant usage: it is at times necessary to read a long list of terms before spotting the one we are seeking, to select and tick it. Furthermore, it becomes impossible to express the fact that fine roots are many, while medium roots are rare. In our view, it is still preferable to duplicate the variable 'size of roots' and all those for which we have the same kind of hesitation.

Table 4.8 Structure of the coding for two kinds of roots

First method of coding		Second method
First variable:	Second variable:	A single variable presenting all the combinations:
fine	fine	fine, medium, etc.
medium	medium	fine and medium
coarse	coarse	fine and coarse, etc.

Example: One tick: *fine*   One tick: *medium*   One tick: *fine and medium*

Implied zero values or not?
The question here is of deciding if the observer who has established the absence of a given character, for example absence of coarse fragments, has to mention it explicitly or not. In other words, is it necessary to provide at the beginning of several variables the case 'absence of …'? It is clear that in certain situations this may be important. Also, ticking against 'absence of …' signifies that the observer has not forgotten to examine the case in question. Therefore, this represents a safety measure. But one will not know how far to go without overloading the soil description with a series of negative statements of the type: no stones, no mottles, no earthworms, no shrinkage cracks, etc.

In current thinking, it seems that the wisest solution might be to provide for the case in which the character is absent or zero … just by asking the observer to use the corresponding column only if it really conveys information considering the local or regional context. For example, 'no shrinkage cracks' is useless information if we are dealing with a Histosol, but desirable information if it is a Vertisol.

Table 4.9 summarizes all the questions we have just posed and which crop up for almost every variable all along the development of a vocabulary for characterization of the soil or the natural environment.

4.3.3 Compilation of glossaries

Choices have been made on the basis of the possibilities that have been reviewed. They have resulted in publication of glossaries (Field Handbooks) whose existence was mentioned at the beginning of the chapter. We are now in a position to give these documents a more precise definition.
- A glossary is first of all an explanatory dictionary because it gives the definition of the terms proposed.
- A glossary is also a translation dictionary helping to compare the descriptive terms proposed and their machine translation in the form of numeric or alphanumeric codes.
- A glossary is also an instruction manual in that it defines in some cases how to proceed with characterization of soil. For example, definition of the conditions for assessment of plasticity.

Table 4.9 Options for developing a descriptive system pertaining to the natural environment

	QUESTIONS	ALTERNATIVE RESPONSES	
1	Type of approach	Analytical Ex: to describe the characters of the object	Synthetic Ex: to give a name to the object
2	Type of language?	Primitive Ex: the variable 122 represents the colour of oxidation mottles	Developed Ex: the variable 122 is the colour of any element to be defined
3	Position of the variable in the hierarchy?	Master Ex: presence or absence of roots	Subordinate Ex: size of roots (has meaning only if there are roots)
4	Level of perception	Preset Ex: 'to give the slope at the hm scale'	Left to choice Ex: 'to give the slope and to define the scale of observation'
5	Type of variable?	Quantitative Ex: structure of 3 mm size	Qualitative Ex: fine structure
6	Type of coding?	Code Ex: 124	In clear Ex: granite
7	Type of doubling?	Two variables Ex: roots (1) fine and (2) coarse	One variable with complex content Ex: 'fine and coarse roots'
8	Treatment of zero values	Explicit null field Ex: NO roots	No 'zero' to be ticked

- Lastly, some glossaries, as that of the ISO in particular, often are multilingual technical dictionaries.

Figure 4.7 attempts to illustrate the various functions of glossaries by presenting a page extracted from one of them.

4.4 TECHNIQUE OF DESCRIBING SOILS

In the present state of techniques there are many ways of describing soils in the field and then recording the descriptions in the database. We shall now review them. This is an important question because digitization of data is a delicate step, tedious but essential to ensure their protection. Data entry

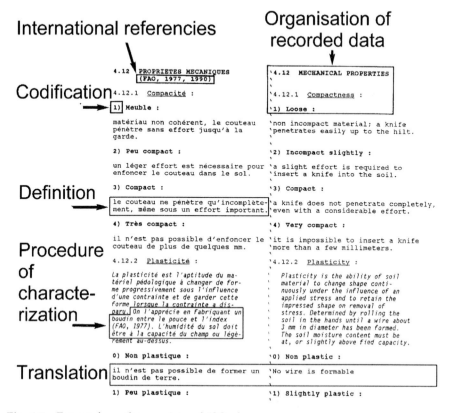

Fig. 4.7 Extract from the provisional ISO glossary.

remains, if not manual, at least through human intervention. Experience shows that there is no ideal system at present. Many methods of data entry have been devised. Each one has its own properties and, in a given case, can turn out to be the best. Sometimes, for lack of precise analysis of his problem, the user selects a poor method and loses time.

4.4.1 Review of the field sheets

Conventional sheets of paper certainly do not constitute the quickest method of entry. But their ease of use and their durability under difficult conditions (rain, heat) explain their being retained and even their preferential use by many mappers. We shall examine the different types of data sheets possible. Actually, this can help those who want to build up a data bank and are forced to compose the data entry forms. We will show that all the models of data sheets come down to a few principal types. This is understandable because we have two constraints that limit design possibilities. First, we reject the data sheets that cannot directly serve as entry sheets and require recopying and restructuring from paper to paper before reaching the stage of keyboarding. These data sheets do not contain the information necessary for using the computer. Thus they are technically outmoded. Their only use is in structuring a minimal description of the soil and to serve as a checklist for avoiding important gaps in the description. Second, for the sheets not to be too voluminous compared to the information they are allowed to carry, the same columns (texture, structure, etc.) must not appear several times. This involves bracketing together the descriptions of different horizons.

Fixed-format sheets

The translation of the principle stated above is the construction of grids defining the cases to be ticked at the intersection of a horizon number and a state liable to be found; for example: dry, slightly moist, moist, wet. Horizons can appear as lines on the sheet of paper; they are superposed as in nature. This is what we call the **horizontal mode**. The text is then written vertically, which makes reading it a little difficult (example on the right in Fig. 4.8). But we can also rotate the blocks 90° and present the text in rows and the horizons in columns. This is the most commonly used system. We call it the **vertical mode** (example on the left in Fig. 4.8).

In both cases, the successive grids corresponding to different variables are bracketed together to represent a total length of 80 boxes. In the early days of computers, this corresponded to a punch card. In data sheets of this

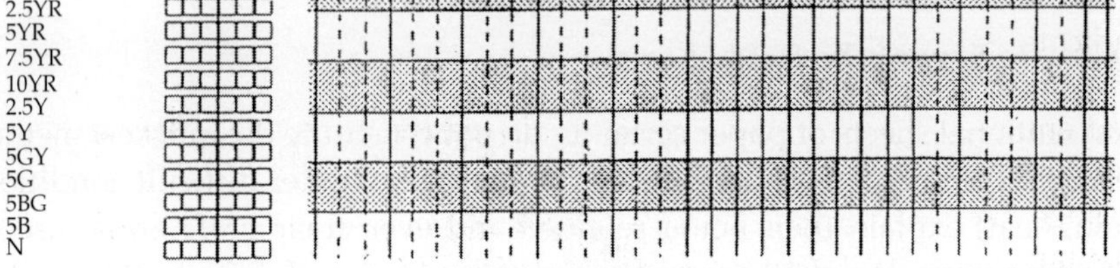

Fig. 4.8 Vertical arrangement of the fixed-format sheets proposed in 1970 by the Compagnie National d'Aménagement du Bas-Rhône-Languedoc (France) and horizontal structure proposed at the same time by the Soil Survey of England and Wales.

type an item of information is identified by its position in the record. These data sheets are compact, but have a rigid structure. So it is very difficult to make up for omissions or to introduce an additional category later. It is also difficult to simplify them to retain only those categories necessary for the description of auger holes. This solution is now virtually abandoned.

Sheets with ten-character codes
The principle of these sheets was developed by the Canadian scientists of the CanSIS system. It was copied out by a working group coordinated by the author of the present book to develop, in 1979, the 'STIPA' sheets (Bertrand et al., 1984; Bonneric et al., 1985). The system has been particularly used in France and French-speaking Africa. It has served to enter information pertaining to 10,000 or 12,000 profiles. It was completely revised and modified in 2002. Figure 4.9 gives an example showing how one can in the STIPA-1979 system indicate by crosses that the second and fourth horizons of the profile are wet.

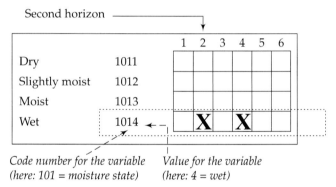

Fig. 4.9 Principles of coding in the STIPA-1979 system; example of the data block used to characterize the moisture state of the different horizons.

Thus a group of ten characters including blank spaces (b), here '1014bXbXbb' (see box enclosed by broken line in Fig. 4.9), completely codifies the data. The first three numerals represent the number of the variable and the fourth numeral the state of the variable. The boxes marked **(X)** indicate that the second and fourth horizons of the profile are concerned. The information thus organized may be presented anywhere on the sheet and entered before or after any other record whatever. The data blocks are independent. This allows development of simplified sheets suitable for such or such milieu or suited to description of auger holes. Addition of a supplementary category poses no problem, either in the sheets or in computer processing. The fact remains that the sheets with ten-character codes impose constraints.

- First, the information is entered in some disorder or rather in the order in which it is presented in the sheet! Thus, for a variable such as *plasticity*, we can find a datum pertaining to the last horizon before finding similar information concerning the first horizon. For example:

	1	2	3	4	5	6
1 1 9 2					X	

before

	1	2	3	4	5	6
1 1 9 4	X					

This can sometimes be upsetting to the user.
- Then, the relative complexity of the presentation results in reluctance of the staff of the data entry centres. The ten-character-code sheets are designed to be filled up by the mappers themselves or their assistants.
- Third, the states of the variables cannot exceed 9. This is annoying in many cases: types of structure, signs of biological activity... . In principle, nothing prevents provision of a 5-character code, the last two serving to mark 99 possible states for each variable. We then have eleven-character-code sheets. But in most cases the numeric code with five positions will seem quite pointless and will make data entry tedious.
- Lastly, unlike in fixed-format sheets, one is greatly restricted in this system as regards the number of horizons that can be described. The maximum number is generally six. To go beyond this, two sheets must be used in succession and the 'continuation' sheet suitably processed. It is also possible to use a compaction system. We shall see this later.

Coded sheets

The ten-character-code sheets are not very compact. Actually there must be a 'horizon-state' data block for every variable in the description. Thus the STIPA-1979 sheet takes five pages to describe a profile. To obtain more brevity, we discard the principle of box to be ticked. Boxes are provided for just entering the codes or the appropriate numerals. Coding tables allow going from a state (ploughed, for example) to the appropriate code (1, for example).

	1	2	3	4	5	6
Tillage	1	2	0			

0 none, 1 ploughed, 2 subsoiled

For the system to be efficient, the coding tables should not be in a different document. They should be printed on each sheet at the exact place where they are needed, such as in the example above taken from a sheet produced around 1983 by the French organization GERSAR. The STIPA-2000 system, which will be presented later, applies the same principles. In other models, the coding tables are printed on the back of the sheet or on the next page. This is convenient when the sheet is presented in the form of two leaves joined at one edge. The codes for page 1 are on page 4 and those for page 2 on page 3. Thus it is possible to fill in pages 1 and 2 with the necessary codes in front of the worker. It is enough if the sheet were opened out. This arrangement of coded sheet is as suitable in vertical mode (*cf.* the example above) as in the horizontal mode.

Compaction methods

There are other ingenious ways to gain space, but they make errors easy. We return to the tick boxes in horizontal or vertical format, then take advantage of the possibility of providing more data in the same box. For this, one can play with the number of horizons or number of variables.

The number of horizons is doubled by stating that the boxes are ticked with the symbol '+' if the horizon number is 1 to 6 and by the symbol 'X' if the horizon number is 7 to 12. For example:

```
                       1  2  3  4  5  6
   slightly moist     |  | +| X| *|  |  |
                       7  8  9 10 11 12
```

(*signifies that the cross and the plus sign are superposed). Thus, in the example, the second, fourth, ninth and tenth horizons are slightly moist. The ingenuity of the procedure consists of finding signs that can be superposed without ambiguity in the boxes. The most popular systems are:

```
    +     X     *
    |     —     +
    1     2     3
```

In the case of numerals, the sum (3) replaces superposition of values 1 and 2.

The same system can be used for doubling the number of variables instead of doubling the number of horizons. In a soil sample it is often necessary to describe two types of structures, two types of coarse fragments, etc. On compact sheets the writing system used allows us to indicate if the first or second variable is referred to.

Thus,

```
                   1  2  3  4  5  6
   prismatic      |  | +|  |  | X| *|
```

signifies that the first (perhaps the only) type of structure of horizon 2 is *prismatic*, the second type of structure of horizon 5 is *prismatic*, and horizon 6 exhibits two types of *prismatic* structure, undoubtedly differing in size. This second compaction procedure is better than the first. It is easier to handle and more useful because doubling of variables is often needed. On the contrary, description of more than 6 horizons is a rare case that can be processed otherwise. As in the case of doubling of horizons, other signs can be used:

```
    |     —     +
    1     2     3
```

Use of numerals seems preferable, permitting keyboard entry of exclusively numeric data and thus to gain a few seconds in recording a profile.

The procedures for coding and compaction reviewed above might appear complex. However, it must not be forgotten that many factors make their use easy. Firstly, the users are experienced people describing very many profiles. They end up juggling with all the systems they are accustomed to. Secondly, the filling up of sheets in the field is done slowly, horizon by horizon; the appropriate box and symbol are quite easily found. Lastly, it should be remembered that only entry is required of the user. Deciphering is done by the computer.

Other sheets
For our review to be complete, a few other types must be mentioned:

Sheets with instruction to cross out. In this very strange type of sheet, all the useful terminology is copied out (Boulaine and Girard, 1970). The user is asked to cross out all that is not applicable, category by category and horizon by horizon. For example:

~~No effervescence,~~ slight effervescence, ~~strong effervescence~~

It is obvious that sheets of this type are not compact at all. They can form a useful system when the terminology handled is very limited (description of auger hole?).

Sheets not structured by horizons. All the sheets presented till now accept as fundamental basis the existence of horizons in which different characters were described one after another. At l'École Polytechnique Fédéral de Zurich (ETHZ), Switzerland, sheets based on the reverse principle were devised in 1977: one considers feature by feature and then defines graphically for each the zone of occurrence in the profile. For example, the appropriate column is blackened between 0 and 45 cm to indicate the depth to which organic matter is observed. This procedure, very picturesque, can help as an introduction for schoolchildren.

Sheets for optical reading. It is theoretically possible to use the optical reader, that is, to fill up the sheets in the field and then to read them by an automatic system. Indeed, many trials have been done in USA (Fig. 4.10).

But one comes up against a major difficulty. As there are very many data to be entered and as the reading systems recognize only simple signs (bars, numerals written in a manner that will not lead to confusion), the field sheets are transformed into pages of hieroglyphs. It becomes very difficult to use, even more difficult to consult them for rereading. Further, this way of doing things is less appropriate than use of the portable microcomputer because it does not allow introduction of immediate controls in the field for rejecting wrong codes. In the laboratory, it is often too late to verify an observation. The optical reader has practically been abandoned at present.

DESCRIPTION OF SOILS IN THE FIELD 137

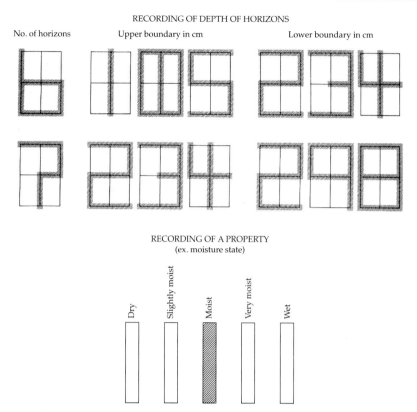

Fig. 4.10 Presentation of data for optical reading.

4.4.2 Presentation of the STIPA-2000 system (Falipou and Legros, 2002)

General organization

We tried to modernize our system in 2000. It works best with two persons. One will be in the pit taking observations. The other records the information. This second person can use field sheets or a portable microcomputer, or even be in the office and be linked to the observer by telephone (Fig. 4.11).

In all cases, the person who enters the data does not have a passive role. Eye on screen or on field sheet, the operator asks the questions: texture?, structure?, abundance of earthworms?, etc. The person describing the profile only has to know how to answer. This procedure ensures that no column is missed. An hour must be allowed for properly describing a routine profile, if the team is trained.

The STIPA-2000 system is distributed free with its manual. But it should be remembered that the vocabulary and instructions are in French.

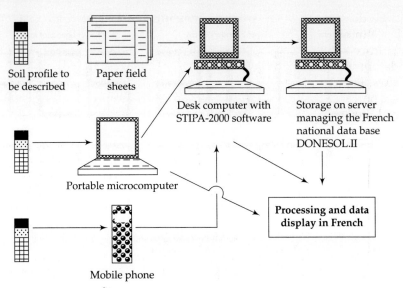

Fig. 4.11 Organization of data capture with the STIPA-2000 system (Falipou and Legros, 2002).

Paper field sheets
Paper field sheets are of various types as indicated in Fig. 4.12.

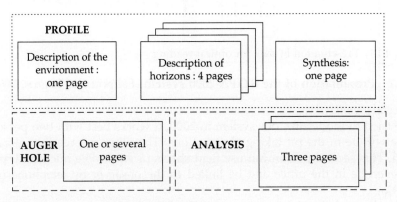

Fig. 4.12 Sheets of the STIPA-2000 system.

The synthesis suggested at the end of the description pertains to the following elements:
- designation of the horizon if it is organic (O, H, etc.),
- letter designation of the horizon (ex: $2B_{ca}$),
- profile diagram,
- summary description of the profile emphasizing the characters of homogeneity (ex: weakly structured profile) and characters that lead to

differentiation into contrasting horizons (ex: sandy texture on clay texture).

In coding we have two types of variables, *textual variables* that have a datum indicated in clear, such as 'Montpellier' or '10 cm', and *coded variables*. In STIPA-2000, for each variable six boxes are available representing six horizons (Fig. 4.13). The user should report in the appropriate box the code corresponding to what is observed.

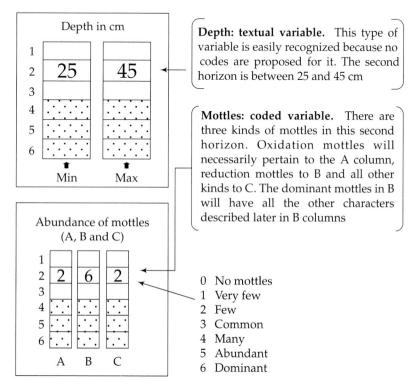

Fig. 4.13 Principles of coding for the variables in the STIPA-2000 system. Examples of horizon depths and quantity and nature of mottles.

When a variable is doubled, for example shape of coarse fragments, to be able to state 'rounded AND flat', or when a variable is tripled to state, for example, 'reddish-brown, green and white mottles', the entry boxes are doubled or tripled as shown in the bottom portion of Fig. 4.13. Compaction systems presented earlier are not used. The case of structure is a little peculiar when a hierarchy is introduced between different types (ex: prismatic structure parting to fine granular). The field sheet then indicates how to proceed.

The dots in Fig. 4.13 are to aid the eye in distinguishing between horizons 3 and 4. No variable is obligatory, except the profile number and the number of the project. We can create, according to demand, all sorts of more or less simplified sheets. Figure 4.14 shows the first of four pages of the description.

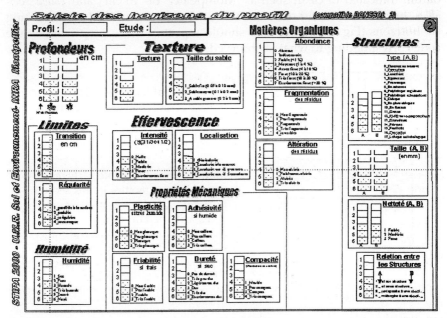

Fig. 4.14 The first of four pages of the description of profiles in the STIPA-2000 system.

A glossary comes along with the sheets to give the exact definition of the terms used. But the use of sheets no longer involves reference to external coding tables because all the codes needed appear near the boxes to be filled in.

If the profile has more than 6 horizons, two sets of sheets are used, with the horizons in the second set renumbered from 7 to 12.

In the laboratory, these sheets are copied out on the input screens, the organization of (order of columns in) which is close to that in the field sheets so that the operation is easy.

This way of proceeding is not ideal inasmuch as copying out data is an additional intermediate operation. It can be avoided if a portable computer is used. Still, the sheets are quite suitable for work in difficult conditions. This is why some users do not want to abandon the procedure.

Portable microcomputer
Data can be directly entered on a portable microcomputer in the field. To facilitate data entry under these conditions, the input screens are noticeably different from those used for copying out descriptions already put down on

paper sheets. Use of a portable microcomputer does not slow down description of soils. Actually, the speed of profile description is not restricted by the speed of data entry (paper or keyboard) but rather by the difficulty felt by the scientist in the pit in determining the characteristics of the horizons: texture, structure, Munsell colour, etc.

However, desktop computers are not meant to function outdoors in the wet, dust and heat. They are too fragile. Great precaution has to be taken, for example placing the machine on a clean sheet on the ground and sheltering it if needed with a sunshade or an umbrella. Also, they are not independent enough and provision must be made for spare accumulators as well as the appliance for recharging them overnight. Lastly, backlit screens do not allow proper work in bright light. Anyway, the market represented by mappers and other scientists working outdoors is very limited. This is why manufacturers do not appear pressed to distribute equipment that is eminently suitable for us.

Mobile phone

The mapper uses a portable 'hands-free' telephone for dictating the soil description to the secretary sitting in the office. This method, which might appear revolutionary or even hare-brained, is sometimes the best. The intermediate paper field sheet is avoided; no fragile equipment is transported to the field; a second person is called only for those periods strictly reserved for describing soils. This method is very suitable for studies carried out in difficultly accessible areas where the stages of delimitation of soils and sampling are combined (ex: mapping in the mountains). The cost of telephonic communication is very little compared to the total cost (travel expenses, salaries, digging of pits, analysis).

4.4.3 Auger holes

Various efforts deserve to be pointed out regarding direct data entry from auger holes in the field.

In Switzerland, the Service de l'Aménagement du Territoire du canton du Vaud in Lausanne (Gratier and Kissling, 1994; Gratier, 1995) uses a PSION, a very small, powerful, shockproof machine that is carried at the waist or against the thigh like a revolver. The auger holes are 'made to formula' according to a system comparable to that described in Chapter 2, § 2.3.3. The data, including the coordinates, should be downloaded in the evening to a larger machine at the base camp. They are later retrieved and recorded in the ARGIS system, the geographical database of the canton. The whole system is characterized by its simplicity (small amount of data stored) and efficiency.

In the USA, the CALLISTO system has been developed, which functions in the field on a *pen-based computer* compatible with *Microsoft Windows* (Teachman et al., 1995).

In France, attempts at recording from auger holes in the field have been made by the Service d'Étude des Sols et de la Carte Pédologique de France (Delvaux-Galerneau, 1987). A small (21.6 cm × 15.6 cm × 3.2 cm) portable microcomputer was used. It was the Hunter from HUSKY. This is a shockproof, waterproof machine. It was earlier not MS-DOS-compatible and its liquid crystal display was too small. This greatly restricted its use. Later versions were free of these disadvantages.

Simplified data-entry sheets for recording from auger holes in the field are available in STIPA-2000.

But, above all, objectives suited to these tools must be drawn up. Indeed, although computer storage of profile data is naturally a must for backup of very expensively obtained information (*cf.* cost of the analyses), the same is not true of storage of auger-hole data. Such storage is justified only if it facilitates use of that information every day in the field. For example, we can think of retrieving the representative auger hole of a soil unit or expressing the heterogeneity of a map area by comparative analysis of the auger holes in it. Much progress remains to be made in this domain. Moreover, it is an essential thing inasmuch as the objective is, for the first time, of going beyond use of the data generated by an already completed mapping to the application of data that can aid in preparing the map.

4.5 QUALITY CHECKS RELATED TO DESCRIPTION OF SOILS

The considerable pains taken to describe the soils in detail will be worthwhile only if the quality of description is assured and checked. Various quality-control methods are used. They are not all at the same level. Some can be routinely used for checking the batches of data entering the data bank. Others serve to occasionally check the quality of the description system or the certainty of the interpretations of the observers working in the field.

4.5.1 Occasional checks

Comparative test
This consists of asking several mappers to separately describe the same soil and then comparing the descriptions obtained. The test was conducted at Montpellier, France, where 18 persons successively described the same *sol brun calcaire* of the Lavalette domain (Legros and Argeles, 1973). Let us examine, as an illustration, the opinion of the observers regarding size of pores and size of roots in the (B) horizon of this soil (Table 4.10).

Table 4.10 Opinions of 18 observers who described the same horizon

Pore size		Size of roots	
Type of answer	Number of answers	Type of answer	Number of answers
Very fine	1	Fine	3
Fine	3	Medium	3
Medium	8	Coarse	3
Large	2	Fine and medium	3
		Fine and coarse	4
No answer	4	Medium and coarse	2

Very heterogeneous answers (size of roots) result when the glossary does not provide precise guides to estimation. Thus, in the description experiment, the observers did not have numbers available enabling them to precisely define the adjectives 'coarse', 'medium' and 'fine' for characterizing the roots. This shows that the observers are not always primarily responsible for lack of quality. For working correctly, they need to have a fully defined vocabulary for description. The vocabulary should have templates for measuring the abundance of mottles, tests for determining plasticity, techniques for calculating abundance of coarse fragments, etc. Elsewhere, we have stated that dispersion of the answers is increased when observation is difficult. Obviously it is easier to state the absence of coarse fragments than to choose between 'many' and 'very many' when the need arises!

A posteriori checks
Certain observations can be later subjected to quality check by specific determinations:
- field texture is cross-checked by particle-size analysis (Fig. 4.4);
- the estimated organic matter content is verified in the laboratory by determination of carbon content;
- effervescence with hydrochloric acid is taken over by determination of carbonates;
- estimated porosity can be compared with bulk density measurements and checked against values of permeability.

Checks of this type are very useful. They enable mappers to learn to sharpen their sense of observations and also to better judge the limits of all qualitative assessments field work makes them do. So one should not take offence at some redundancy in information that enables introduction of occasional checks and even systematic checks as we shall see below.

4.5.2 Systematic checks

Different types of systematic checks (Legros et al., 1992)

Validation checks are also called direct checks. They consist of verifying if the data entered satisfy a certain set of known and well-defined rules. For example, some references are compulsory, in particular the number enabling identification of the profile considered. Also, some data should be numeric, such or such code cannot exceed such or such value, etc. The corresponding checks are done as soon as possible, for example in the field when the information is entered through the computer keyboard. Any attempt to break the rule results in an error message or a sound alarm: the system (software) refuses the suggested data or refuses to continue without a datum defined as obligatory or appropriate. In other words, the system eliminates right from the start all causes liable to result in its getting locked up later when such or such type of data manipulation is done. At this stage, the data are 'correct', at least from the computer point of view.

Coherence checks. The data entered pertaining to a soil are examined to see if they are coherent and compatible with other data. This method is used by all soil scientists. For example, they would be worried about a very low pH in a calcareous soil. We have seen that this coherence check is the easier the more the data entered are redundant.

Checks of plausibility. Plausibility checks are also called indirect checks. They consist of evaluating if the entered data have acceptable values, in the light of available general information on the environment. For example: is it normal to describe soils at altitude greater than 322 m in the Netherlands? In the present state of the art, these checks are not entrusted to the machine. In the future we may think of introducing, in the databases, expert systems equipped with some knowledge and hence capable of detecting certain improbabilities. However one should not have too many illusions. These expert systems will remain rudimentary and many incongruities will continue to slip through the checks. What expert system could know better than a specialist if baobab, basalt, salt lake, Vertisols and peat bog are acceptable terms for describing things observed in specific areas in the world?

Implementation of coherence checks

Software has been developed for automatically conducting coherence checks (Dunand-Divol, 1988; Legros et al., 1992). For compiling this program called CHECKUP, all the data already stored in the data bank were reviewed and the relationship existing between data taken two by two (for example, the relationship between colour and organic matter content) or three by three (for example, the value of CEC in relation to organic matter and clay contents) were examined. For two quantitative variables x and y, the scatter (more or less elongated) of points showing the relationship was first drawn

in an orthonormal system. This was done for thousands of pairs (x, y). It was then assumed that, for the first time, all possible situations were effectively observed. From this three possibilities were to be considered for a pair x_i, y_j that enters the data bank and pertains to a new soil:
- if the point (x_i, y_j) clearly falls outside the scatter of points, an error is very likely and an error message is sent to the screen or to a paper output for the user;
- if the point (x_i, y_j) falls clearly in the middle of the scatter of points, the data are in all probability correct and no warning is given;
- if the point (x_i, y_j) falls near the edge of the scatter, error is possible; the user must be careful; a warning message is sent.

For facilitating implementation of tests, the scatter of points is divided parallel to the graphic axes into small rectangular compartments that are definitely approved or are prohibited, or give rise to warnings. The method is also applicable to qualitative variables: such or such value of one, for example 'prismatic structure', is approved or forbidden according to such or such value of another, for example 'pure sand'. In total, 15 tests were introduced for judging the coherence of 17 properties pertaining to each horizon of the profile (Fig. 4.15).

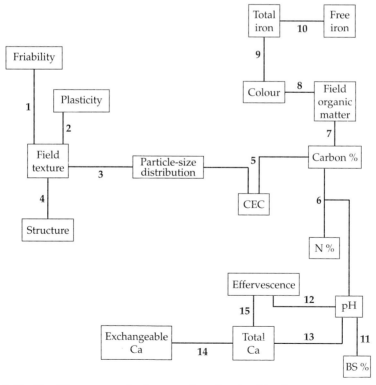

Fig. 4.15 The fifteen tests of coherence done by the CHECKUP program.

But the data already recorded and used to define the permitted zone are not perfect. All possible situations are certainly not represented in the scatter. Furthermore, the edges of the scatter surely have doubtful points that are already from erroneous data and do not have to be taken into account. It comes to mind that a more rigorous test (smaller zone of acceptance) should be designed so that no risk is taken. But this amounts to loading the data-management system with very many warning messages, which run the risk of being ignored completely. It seems that one must think in terms of probability and navigate between two difficulties: an error that exists may not be detected; an error that does not exist may be signalled (*cf.* risks of the first and second kinds in statistics). In practical terms, it is arranged that the user does not, on average, get more than 2 or 3 warnings per profile. But this is an average. If the data are very good or greatly doubtful, these numbers can change. Lastly, it should be noted that to detect an incoherence does not mean that the error has been found. In the example taken earlier (low pH and calcareous soil), one may ask if the pH had been underestimated or the carbonate content overestimated! A complex and not always efficient method has been devised to attempt to resolve this sort of indecision.

In all cases, it is the responsibility of the mapper, the defendant, to maintain or, on the contrary, invalidate his original diagnosis. The machine is not programmed to modify the data from the outset!

Tools for plausibility checks

Since we do not know how to automate plausibility checks, they are left to the observer. However, a suitable display of the recorded data may help in finding out errors.

Editing in clear. The data input, soil descriptions and soil analyses are printed out on sheets that the mapper is invited to reread carefully.

Drawings and graphs. Another important way of visualizing the data is to draw the profiles by computer on the basis of the data contained in the bank. Originally providing for illustrating soil survey reports, these drawings are seen to be an excellent means of checking (Fig. 4.16). For example, if a horizon has been recorded with a thickness of 300 cm instead of 30 cm, this is glaringly obvious on the diagram created by the machine!

The profile drawings provided as examples show the following variables:
- *Horizon thickness;*
- *Kind of boundary* (continuous line for clear boundary, broken line for others, wavy if necessary);
- *Thickness of boundary* (numbers to the right of the profile);
- *Texture* (dots = sand; vertical lines = silt; horizontal lines = clay);

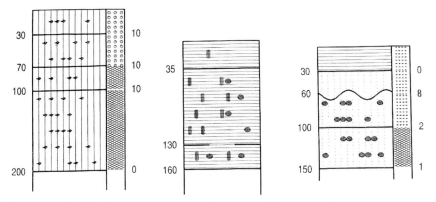

Fig. 4.16 Profile diagrams automatically created from the data contained in the STIPA-Montpellier data bank.

- *Stoniness:* the coarse fragments are displayed taking account of their abundance as written down in the field description and also their shape (they are drawn rounded or angular as the case may be);
- *Mottles* (reddish-brown mottles are shown, for example in the profile in the centre of Fig. 4.16, by close vertical hatching in clusters);
- *Structure* (displayed in the small profile diagrams drawn close on the right of the principal profile: circles = *granular*, lattice pattern = *blocky*, etc.).

Of course improvements could be made in the display modes and the quality of the drawings.

The data entered are not directly dumped in the data bank, at least in the STIPA system (Chap. 6, § 6.4). They are accumulated in an intermediate scratch file. They are copied out into the bank after they have undergone all the checking operations presented above.

4.6 ORGANIZATION OF A PROFILE-STUDY PROGRAMME

4.6.1 Principle of organization of projects

In everyday life, organization of task comprises many phases linked one after another often done without specific thinking in advance. We proceed by a sequence of blockage and release of the type 'I see I am missing a map for use as a base for my third draft soil map, so I am telephoning the Institute of Geography to order this document urgently'. This way of doing things is unacceptable when the work in question concerns many persons. Carelessness of just one could slow down all the teams. It is necessary then

to be organized and develop what some call a *process model* that we more readily call **organizational scheme**. In practical terms, this means to present on paper all the intermediate steps of the work and to indicate what conditions should be satisfied so that each one can be achieved. For example, soil survey will be started if simultaneously (Fig. 4.17):
- the contract has been signed with the sponsor;
- the necessary documents (photographs, base maps) have come to the cartography centre;
- the vehicles essential for the survey have been hired, bought or overhauled;
- the travel orders have been issued (for government employees);
- hotel accommodation has been reserved.

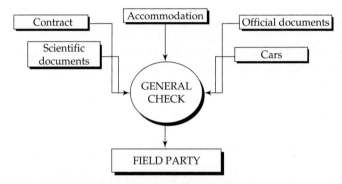

Fig. 4.17 Presentation of part of an organizational scheme.

It is obvious that the scheme shown is only an extract. The chart extends upwards (for example, the list of procedures necessary for obtaining travel orders and permissions) and downwards (for example, details of the phases of the survey).

Experience has shown that by formulating a scheme of this type the mapper, aided by an expert in organization methods, mentally goes through two stages. First, he is somewhat reluctant, having the feeling of handling obvious facts and therefore of losing his time. Then comes the moment when he realizes that the effort in organization is not pointless and that certain phases of the work are in a tricky sequence. He then finds out how to find a way to avoid loss of time, stoppages and pointless duplication of certain efforts.

4.6.2 Methods of layout of profiles

Stratified sampling
The most widely used method rests on free selection of sites for digging profiles. There are two purely technical constraints:

- the reasonable limit for total number of profiles to be opened and described;
- location in the field because it is desirable to reach the chosen sites without special effort.

For this, we proceed with *stratified sampling,* that is the profiles are laid out taking into account knowledge of the milieu; this means (Legros, 1978):
- area of the units to be characterized, a very large unit generally giving rise to more numerous observations;
- homogeneity of the milieu inasmuch as a unit identified as very homogeneous during auger hole survey could theoretically be properly characterized by a single profile;
- scientific and applied interest in each type of soil.

This procedure is generally satisfactory and economical. But it has disadvantages. In particular, it introduces a sort of more or less conscious cheating. Thus it is tempting to chose profiles at the exact location of auger holes done earlier. Noteworthy auger holes often leave a mark on the memory of the mapper by showing up exceptional situations. For example, the most clayey pedon or that having the brightest red colour.

Stratified random sampling

The layout is organized as follows (Walter, 1990). First of all, a square-mesh grid is superposed on the draft soil map in the office. Then an intersection point (x, y) of the grid is picked at random. If it does not fall on a unit studied, the process is repeated. In the opposite case, one of the points of the set is chosen. The process is continued until one-fourth of the necessary points are obtained. Later, in the field, one gets into position at each of the points selected earlier. Then with the help of a pocket calculator with a random-number generator, a value between 0 and 360 is picked to represent the direction in degrees. This defines the second point of observation with the help of a compass x metres from the first, x being chosen. From this second point the operation is started over again and the mapper moves $x/2$ metres in the direction found. The last point is finally obtained in a similar way $x/4$ or $x/5$ metres from the preceding one (Fig. 4.18).

The only condition is that the mapper does not go out of the unit studied. If this happens, the selection of direction is considered worthless and the procedure started all over again. In total, therefore, a pattern is obtained of points selected at random with spatial intervals that fully satisfy the requirements of geostatistics. Actually, contrary to what would be obtained had the points been at the nodes of a square-mesh grid, there are as many pairs of points separated by a short distance as points separated by a large distance. Thus, variograms can be prepared under appropriate conditions. But this procedure takes a lot of time. So it is reserved for certain research projects.

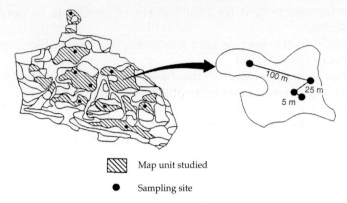

Fig. 4.18 Method of sampling at controlled distance and random direction (Walter, 1990).

Other methods

The first method outlined above has the advantage of suggesting a rational choice of sampling sites. The second claims, legitimately, greater scientific rigour. A compromise is possible by having the sites chosen by a scientist who has not studied the area and who has the draft soil map. Under these conditions, it is easy to respect certain constraints (ease of access in particular), then to locate observation sites at random in the remaining areas without previous knowledge of the area distorting the choice.

Thus, 1650 profiles were located in the pedological investigations in Pays d'Ouche (Favrot and Bouzigues, 1975). This enabled checking of the agreement between the predicted soil (draft soil map) and that actually found. Between 60 and 80 per cent agreement was obtained depending on three factors:
- environment in question,
- mapper who has worked in the area,
- rigorousness of the checking (is a small variation from the reference soil acceptable?).

This method is a means of checking the accuracy of the soil map with no additional work and therefore no additional cost.

An algorithm for automating the selection of observation sites or sampling sites has been proposed (Warren et al., 1990). The system considers the total number of samples permitted, then the nature of the units to be inventoried and lastly their area. This amounts to first selecting areas in which the sites are then randomly located.

4.6.3 Pitfalls to be avoided

The mapper of course has available some measuring instruments: field pH meter, Munsell colour charts, etc. But generally the main thing in the quality of the work depends on the mapper's observation ability, thinking, synthesis and also eyesight and sense of touch. The mapper is thus the main instrument of mapping. One can state very seriously that he must be in good physical and mental condition. Mapping activity presupposes preventive steps against occupational hazards, in particular excessively hasty diagnosis, routine and isolation. It is worth examining all these.

Hasty diagnosis. The most surprising thing for an inexperienced person is that the soil does not allow itself to be easily and directly understood, even by a specialist. Many soils are brown, without very clear distinguishing characteristics and without marked horizons. The mapper who comes to a setting that is new for him should not claim to identify the soil, name it unequivocally and begin to discourse on its agronomic properties. On the contrary, the observer must take his time, describe many profiles and move around the region several days, even several months, before understanding what type of soil mantle he is faced with or knowing how to determine what part of that soil mantle a given site is representative of. Thus, experience about a given soil mantle is acquired slowly and times with difficulty, most of all when contrasts between various soils of the region are slight. After some time our mapper becomes a veritable expert on the region. Like a wine taster, he can recognize with certainty the subtleties of the regional soil vintages. In his best form he knows how to establish thin, precise boundaries between soils. But this experience gained in such a hard way is not retained if it is not regularly kept active. A year later, memories become hazy whether they concern colour of soils, their sand content, plasticity or any other property tricky to estimate. This is why reopening of surveys to modify them is not advisable. Except for the odd case, the modified boundary will be less correct than the original one….

Routine. The mapper who continues his activity in a given environment for several years is faced with another problem: the danger of routine. A classic symptom of this sort of thing is seen in soil descriptions that get progressively briefer until they vanish completely to be replaced by statements of the type 'same as above'. This must be fought. Before getting down into the profile pit to study a type of soil already described many times, he must give himself the following obligation: not to leave the pit without having inventoried the characteristics showing in what way this profile is different from what was expected. If one is not in this state of mind, if all curiosity has disappeared, why still pretend to be observing soils!

Isolation. For remaining alert, it is anyway necessary not to spend many long years mapping a single pedogeographic environment. Otherwise the scientist loses his general references a little. He ends up exaggerating small variations in the soils and takes into account details with no practical relevance. He will almost pretend to have identified in silty soils the range of textures from pure sand to true clays! This might seem exaggerated. However, we have come across situations of this kind in regions where mappers had worked alone for a very long time.

4.6.4 Opening of profile pits and technique of description and sampling

Opening of profile pits

The most efficient way of opening pits is to use, when possible, a mechanical digger worked by a specialist, who must be asked to avoid smoothing the face to be examined (working the mechanical teeth forward) and to make steps for the observer to go down into the pit. The topsoil (A horizon, *cf.* Fig. 4.19) must be separated from the rest of the earth to be put back on the surface when the pit will be filled up. A scoop at least 50 cm wide is used so that a pit wide enough to admit the observer's shoulders is obtained. A 60-cm scoop is ideal.

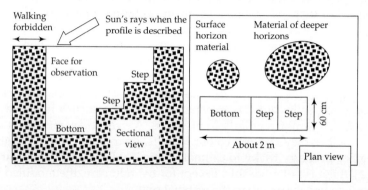

Fig. 4.19 Profile seen in section and from above.

The depth is generously calculated because for clearing the face for study and removing the machine's tooth marks with the spade a large amount of soil has be thrown on the bottom and will hide the base of the pit. The orientation of the pit should be such that the face to be studied is illuminated by the sun. But it is sometimes hard to predict where the sun will be at the time the profile will be described when many pits have been opened in advance. In such a case, it is efficient to take along a compass and, wherever possible, orient the pit north-south.

Some elementary precautions should be taken because hitches are not rare: for example, the mechanical digger might cut a drain or a pipe. This happened once to the author. For a few hours, a field was converted to a pond. At the same time the five thousand residents of the town of Langeac (Haut-Loire, France) were deprived of drinking water. *Thus it is essential to be insured* for this sort of mapping activity and to check that the firm engaged for opening pits is also insured. Further, it is necessary to provide the address of the mapping laboratory to the owners of the farms affected by the profiles. This will enable the laboratory to be informed of any difficulty (for example, a pit had been forgotten and was not filled up).

Use of a mechanical digger is restricted to large-scale work (up to 1:50,000). At smaller scales, travel times between two sites will lead to prohibitive cost. Also the machine must be able to reach the sites. This is why in many cases we are obliged to dig the pits manually, even though the actual cost is higher and the quality of work is lower (generally a shallower pit gets dug).

Studies in mountainous areas represent a particular case. To avoid having to do the same approach twice, the survey phase is often merged with the soil description phase. All cuts, natural or manmade, identified as useful during delimitation of soils, are described on the spot and samples are drawn. The technique is not very practical. Actually a profile-study programme requires efficient organization of equipment and assembling some tools. It is also necessary to rediscover automated methods and a certain bent of mind. Describing a profile now and then is neither very efficient nor very pleasant. There is also the problem of optimization of selection of sites: sampling from the first cut come across is not very good of course! But sometimes kilometres of cuts are seen along forest roads. In that case, representative sites can be pinpointed.

The particular case of large projects
We consider it useful to detail the organization of a field programme when at least 60 profiles are to be exposed and described. The following are the steps:
- To begin with, the approximate locations predicted for the profiles are determined.
- Then in the field, the owners are asked for the necessary permissions; if the owners refuse, the location of the pit in question is changed.
- The finally decided sites are then numbered and marked on the detailed map. For finding them in the field, it is convenient to drive in stakes (at least two per site). One is in the field or meadow and marks the profile, the other is on the approach road in such a way that the line made by the two stakes is perpendicular to the road. This way of working is very useful when there are hedges and the stake planted in the field is not visible from the road or highway.
- A highway route for going from one site to the other without detours or unproductive loss of time is drawn up.
- Each description team gets the list of profiles to be described and a copy of the corresponding location map showing the route to go easily from one site to the next.
- The pits are then opened with a mechanical digger pulled along by a car. This car is driven by a mapper having the location map and who knows the location of the pits to be dug. It is actually unpleasant to be

lost on country roads with such equipment and to spend valuable time in rediscovering the route. The worst obviously is to make a mistake and open up a profile where the description teams will be unable to locate it.

To avoid loss of time and minimize costs, the profile-description team should comprise seven persons, not counting the operator of the mechanical digger. One surveyor is assigned the digging of profile pits and guiding the mechanical digger. He carefully follows the digging to stop it at the depth according to the levels found (the attempt is to reach the C horizon). He has a 'digging card' on which he records the water table level so that it is seen when the pit is dug. He can also record other notes such as resistance to digging, particular abundance of earthworms, etc.

The six other persons are formed into teams of two. The first team describes profiles 1, 4, 7, 10, etc., the second 2, 5, 8, 11, etc. and the third 3, 6, 9, 12, etc. If we proceed in this fashion the mechanical digger still goes twice as fast as the three teams together. This is what is looked for. Actually, when the mapper would have described half the number of pits, all pits would have been opened. The mechanical digger makes the circuit a second time for filling up the pits. In principle all the work will be finished at the same time. The digger will not be slowed down by the teams and will hardly fall behind them. This method has two advantages:
- the pits are left open for the shortest time;
- the mechanical digger and its operator work only once without break, which pleases the contractors.

But when it is possible, the leader of the study should visit all the profiles in order, including the ones whose description does not depend on him. By spending 5 to 10 minutes at sites pertaining to other mappers, he can meet his collaborators at work, help the teams to properly identify the soils (grouping, classification) and thus ensure coherence in the ensemble of names. He is advised to take notes. His partner follows him because there is no alternative. Therefore, the team leader will have a smaller daily output in number of descriptions actually done. But the advantages of this way of working outweigh the disadvantages.

Time devoted to describing profiles
The profile-description teams, as stated earlier, are ideally composed of two persons. One studies the profile and describes it, the other enters the information on paper sheets or in a portable computer (and tries to keep his fingers clean). The speed of description depends on the depth of the soil, distance between profile pits and the way they were dug. If the mappers have not opened and filled up the pits, the average time needed for each description may be estimated at a little more than an hour. To describe six profiles a day is reasonable. Reaching eight as routine is possible only during the long days in summer. This requires considerable speed and very smooth-running mappers for the work to be of high quality.

Sample collection in bags
Most of the time, soil samples are destined for chemical and not physical analysis. These samples are put in sturdy plastic bags, suitably labelled.

Several notations are made on the label. Firstly, one marking gives the reference of the study in progress. Then the profile number profile is indicated. Lastly the horizon number, symbol (A, Bt, etc.) or, even better, its depths.

Sampling must be done from the bottom upwards, that is the soil extracted from the subsoil horizons is bagged first. This avoids the first sample contaminating the entire profile face, which would involve a clean-up before proceeding to the next sample. Also, always going from the bottom up, strictly in that order, limits the risk of error in labelling. This danger exists above all at the end of the day when the mappers are tired, in particular because the one who takes the sample (and so has soiled hands) is not always the one who writes the labels and puts them on the bags.

In theory a few grams of soil are enough for most laboratory analyses. Thus a sample representing about a hundred grams should allow all the analyses desired. However, the main problem is to ensure the representativeness of the volume drawn compared to the pedon. This is obviously easier if this volume is large. In general, one to two kg soil per horizon is sent to the laboratory. But for an overall characterization of the material (including the fraction larger than 2 mm), it is necessary to sample all the fractions in the proportion they are found in, without sorting. The sample should be the larger the bigger the coarse fragments because the probability of selecting a small representative sample diminishes. The ISO recommends the following sample sizes according to the size of the most abundant coarse fragments (Table 4.11).

Table 4.11 Weight of sample (kg) as function of the size of the most abundant coarse fragments

Size of the most abundant fragments [mm]	2	5	10	20	30	40	50	60
ISO standard [kg]	0.5	1	1	2	4	7	12	18
Author's suggestions [kg]	1	1	1	2	7	16	32	54

The experts were optimistic. An experiment was conducted in a soil in which 50-mm particles predominated (Buchter et al., 1994). Samples of increasingly larger size were drawn and the coarse-fragment content in them calculated. The values obtained, at first very variable, converged when samples of about 27 dm^3 (or 37 kg) soil were drawn. Under these conditions we recommend drawing the weight equivalent to that of 200 particles of predominant size (presumably spherical) without going below one kg of soil (last line in Table 4.11). In some cases, the cost of transport of samples must also be considered. Sometimes an aeroplane is necessary!

Later on, mixing operations in the laboratory will ensure homogeneity and representativeness of the test sample.

Taking undisturbed samples
In some cases it is useful to draw samples of soil without destroying its physical organization. This may be necessary for studying the soil with micromorphological techniques and also for doing porosity measurements, or determinations of permeability or water-retention at different tensions (water-retention curves). Soil clods are very fragile most of the time. To avoid destroying them or even breaking them, there are three ways of proceeding:
- The *first method* is to forcibly push a **metal cylinder** with a sharp cutting edge into the soil. Special equipment is available for this. The soil is then levelled off at both ends of the cylinder. Thus, trimmed with a pastry-cutter and protected by the cylinder, the block is firm enough to be transported.
- The *second method*, close to the preceding, consists of extracting with a spade a block of soil then dipping it in *plaster*.
- The *third method* (Fig. 4.20) consists of cutting a horizontal step in the profile (do not put your foot on it!), then cutting with a knife a cubical block of soil without separating it from the step at the bottom (1). A metal box with detachable base and lid is positioned around the block as a collar and encloses it (2). Generally, the size of the block is calculated for a 1-cm gap to be left between the soil and the box. This gap is then filled with paraffin wax previously heated on a camping stove (3). When the wax sets it ensures the whole thing is firm. Then the lid is placed on the box (4). The base of the cube can now be separated from the rest of the soil with a knife (5) and the bottom of the box is placed in position. This method is known by the name *Vergière cube* method.

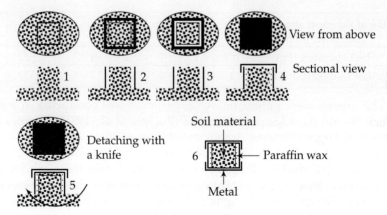

Fig. 4.20 Making a Vergière cube, sectional and plan views.

Using smaller boxes of the same type, and even without the need for paraffin wax, we can take samples for being dried, then impregnatthem with resin for micromorphological study. In all cases, the laboratory must take considerable care while separating the samples from various wrappers and coatings.

4.6.5 Sampling in special cases

Taking composite samples for agronomic purposes
For characterizing agricultural parcels and obtaining an overall diagnosis at that level, about twenty small samples are taken from the plough layer more or less throughout the parcel. These samples are then mixed thoroughly and a small portion of the mixture is sent to the laboratory. According to the recommendations of the ISO (TC 190, WG 2, *sampling*), there are two ways of obtaining a proper composite sample using an oilcloth or any other cloth that can be properly cleaned between uses.

- *Simplified method.* Mix the soil of 20 samples by forming it into a cone with the hands. Then raise two opposite corners of the cloth on which the soil is placed so that the soil is distributed in the form of a long pile. Remove half of the pile from the middle and repeat the operation from the beginning until the quantity of soil left represents about 1 kg, which will then be taken to the laboratory.
- *Optimal method.* Mix and spread the soil in a thin layer on the cloth, draw two lines at right angles passing through the centre of the heap, discard the soil in two opposite quadrants and repeat the operation till a kilogram of soil is left, which will be saved.

The hypotheses below have to be checked properly when working in this manner. Let us take the example of an ideally uniform field and a barely precise analytical method. It is obvious that by grouping 20 samples, time is lost and there are risks in doing just one determination. In such a case, it is better to draw 5 samples and analyse them separately. In other words, the sampling strategy must be carefully thought out in relation to the ratio between the sampling variance and the analytical variance (Gomez et al., 1986).

Let us lay down that
$$\Delta = N(1/nk - 1/N)$$
where
N = total number of samples
n = number of composite samples submitted for analysis
k = number of determinations per sample.

Let SS^2 be the sampling variance and SA^2 the analytical variance.
If $SS^2/SA^2 > \Delta$, the composite sampling is more accurate.
If $SS^2/SA^2 < \Delta$, the single sampling is more accurate.

However, we are not going to give the 'magic formula' that will enable all people to substitute themselves for a specialist to draw soil samples and so characterize the soils. This will deny the suitability of every procedure presented in this book that greatly values the survey of the organization of the soil mantle and the study of its place in the environment. It is clear anyway that we will not be able to make a soil map on the basis of 20 samplings from each field!

Sampling for monitoring development of soil pollution
There is fear now in industrialized countries of increase in contamination of soils by several substances foreign to the system: heavy metals related to use of composts of poor quality, atmospheric dust of industrial origin, accidental pollution like that caused by the Tchernobyl power station, etc. This is why soil-monitoring networks have been established. These networks comprise parcels in which the contamination level is examined every five or ten years. We can cite, for example: *l'Observatoire Quebecois de la Qualité des Sols* (Rompre, 1994), the *Réseau National d'Observation des Sols* in Switzerland (NABO), *l'Observatoire Français de la Qualité des Sols* (O.Q.S., 1988) now replaced by the *Réseau National de Mesure de la Qualité des Sols* (RMQS) and *l'Observatoire Français des Écosystèmes Forestiers* (RENECOFOR). But in Europe, the countries most advanced in setting up a soil-monitoring network are Austria, Belgium, Norway, Sweden, United Kingdom (King and Montanarella, 2002).

Actually, monitoring pollution is very difficult technically. The substances that one wishes to monitor are present in extremely small amounts and their increase with time is slight. For drawing valid conclusions, several conditions must be satisfied:
- Use high-performance analytical techniques.
- Take into account the drift in their readings over long periods of time; for this, samples drawn and analysed should be preserved for fresh analysis along with those of the next sampling programme (Legros et al., 2001).
- Avoid artificial contamination of samples; beware of metal tools if it is desired to determine heavy metals later.
- Take many samples according to an appropriate sampling plan that makes it possible to return to the same sites, give or take a few decimetres, five or ten years later; for this the skills of a land surveyor, who installs markers after taking alignments, are at times used.

An interesting study was done at the European level and has yielded a large and pertinent report after synthesis (Wagner et al., 2000). A site for collaborative study was selected in Switzerland. It measured 0.61 ha. This site was known to be polluted by airborne heavy metals. It was first the object of a very thorough characterization based on 301 samples. Then the 15 participants who represented Switzerland and 13 other countries in Europe (two laboratories in Spain) were invited to work in the field, one after another. They were allowed a choice in number of samples, choice of sampling techniques and choice of methods of drawing samples. The objective was to determine the pollution level individually. But all the analyses were entrusted to the same laboratory. Among the many results from the work, we may retain these:

- The error related to differences in field methods and in sample preparation was as large as that linked to analyses when different laboratories did the work.
- Europe lacks a common methodology for this type of study and also has no common standards for interpreting the results obtained.
- The best plan of sampling depends on the objectives: is the average content of the metal wanted or the extreme values occurring in the parcel?

Conclusion and perspectives

Profiles constitute the very small amount of information that will represent in a special way the soil in the reports with the soil maps as well as in the soil data bank. Consequently, their location should be chosen very judiciously and the algorithms aimed at optimizing their positioning must be followed. Furthermore, since the publication of the *Soil Survey Manual* more than forty years ago, very great efforts have been made to ensure precise and exhaustive characterization of these profiles. We have now reached a stage where it seems difficult to do much better, even though we already foresee the systematic digitization and recording of profile photographs. But this development has resulted in more and more lengthy and detailed soil descriptions. Research with great objectivity has led to preferring a strictly analytical approach; each characteristic of the soil, even minor, is described individually. In the future, simplifications must be introduced to reduce costs and also because it is not worth describing, recording and storing data in the computer when they will practically never be used later. So we should evolve towards a more synthesising approach, involving attempts at interpretation in the field. For example, we will seek to directly characterize the degree and duration of excess water instead of recording all the corresponding properties (colour, number and size of mottles, etc.). But such development will not be possible without a better understanding of the functioning of the soils in place. Although mappers are still naturalists, attached to the qualitative description of objects, it is fine because soil chemists, physicists and biologists have not yet achieved all the expected breakthroughs. The latter would be very wrong to criticize the empiricism of the former.

CHAPTER 5

MAP PREPARATION AND QUALITY CHECKS

In this chapter will be presented some of the principal steps in mapping, in particular those pertaining to delimitation and characterization of soils in the field. As indicated earlier, we shall essentially place ourselves in the context of conduct of a free survey. Description of auger holes and of profiles, already covered in Chapter 4, will not be gone into again.

5.1 METHODOLOGICAL BASIS OF ZONING

We saw in Chapter 3, § 3.3.4 that the first step in mapping is to demarcate on aerial photographs the different milieus that are to be studied. Then we go to the field. Each natural milieu should then be examined. There is no question of rightaway digging the soil and observing nothing else. It is necessary, on the contrary, to analyse the environment in which the soil occurs. In particular, we seek to understand how the geological strata are organized in space, how the local morphology is constituted and how the vegetation is organized and distributed (Legros, 1986). For this it is essential to have maps available, especially geological maps. All the time needed for doing this study of the environment should be taken. The mapping will then be of better quality and will be done very speedily. *'A mapper who does not know to sit down to read the geological map and reflect is a poor mapper'* said my old supervisor, Professor Servat.

5.1.1 Prediction of the soil composition of landscape units

Auger holes represent very limited information: only a few dm^3 of material is directly studied each time. We have stated in Chapter 1, § 1.4.2 that all the auger holes of a free survey taken together constitute a soil area not exceeding $1/10^8$ of the soil mantle to be characterized. No direct interpolation is possible on such a weak foundation. For meaningful interpolation it is necessary to conduct grid mapping, that is, to bring the observation points closer so much so the time required and costs increase considerably. In the case of free survey, for generalizing the information provided by the auger holes we proceed by a classic scientific reasoning that involves observation, establishing a hypothesis on the distribution of soils, formulating a rule (law) for this distribution and, lastly, verifying it. All this can be summarized as a kind of algorithm (Fig. 5.1).

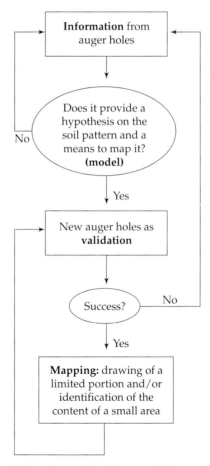

Fig. 5.1 Method of utilizing the soil-landscape relation, *a step-by-step procedure.*

A first set of auger holes informs the observer and possibly helps to establish a hypothesis on the local distribution of soils as a function of the environmental factors (vegetation, topography...). For example, it might be established that the soils of the depressions are deep with redoximorphic features. This is a kind of model that, in principle, allows prediction. Other depressions, visually observed or identified on the aerial photograph, must also contain deep soils with redoximorphic features. A second set of auger holes enables us verify if this is so. If the hypothesis is sound, that is, always verifiable, one can later be allowed to space out the checks. But it is necessary to be careful; after some time the model used always ends up unproductive by not being valid, probably because the soil landscape has changed. Then the mapping has to be based on something else, that is, reasoning all over again!

To sum up, in free survey auger holes have two absolutely local roles of interest: they are established either *for study* in a sector that appeared interesting or *for verifying* a hypothesis regarding the nature of the soil at such or such place. In the former case, one gathers data, in the latter, checking is done. But in neither case is extrapolation or direct interpolation allowed!

Mapping proceeds by simultaneously using several rules similar to that given above as example. These rules have a limited life span and disappear to be replaced by others. Ideally these rules should be quantified by using probabilities and Baye's theorem (Cook et al., 1996). For example, what is the value between 0 and 1 representing the probability that a sector in a given region contains soils with excess of water if this sector is known to be in a basin and covered with meadows? Quantification can be achieved in certain kinds of mapping with the aim of extrapolating around a Reference Area. But generally it is difficult because by definition the soils of the region are still unknown. Thus one proceeds step by step and updates this kind of expert system as and when knowledge advances.

The method described above enables us attribute portions of the study area to such or such mapping unit; it does not propose a process of accurate separation between two contiguous units. We shall now see how one proceeds to that level.

5.1.2 Delimitation of soils

Drawing of boundaries is, to some extent, the culmination of mapping work and is thus an essential component of the work.

Types of boundary
We have seen that there are, in principle, two broad types of boundary, the drawing of each presenting specific characters. Some boundaries are *crisp boundaries*, observable at least locally. One can follow them in the field or

recognize them on aerial photographs. Most of the time the reason for appearance of these boundaries can easily be identified, for example change in geomorphic position or of rock type. Other mapping boundaries are *fuzzy boundaries*, which are not easily observable because they are very gradual. In the Massif Central of France for example, the soils of moors become richer in organic matter as altitude increases. Rise in carbon content is regular between 500 and 1600 m. Thus there is no observable natural boundary that can be relied upon for distinguishing humus-rich soils from humus-poor soils. It is then necessary to introduce an artificial separation in a continuum.

But a fuzzy boundary can result in accurate positioning. For example, in the Massif Central, it was decided that soils rich in organic matter are those that contain more than 8.5 percent carbon, then the sampling procedure and analytical methods necessary for drawing the line on this basis were given. This line should not be considered a priori less relevant than a crisp boundary and can be used for pedogenetic or applied reasons.

Crisp boundaries can be very difficult to find if, for example, they only affect features of the deeper horizons.

In sum, the natural *fuzziness* of a boundary is an intrinsic property. The difficulty in observing it is extrinsic (*uncertainty*) and the two notions do not always match (Hadzilacos, 1996; Lagacherie et al., 1996).

Detection of the boundary
In a small-scale study, the boundaries are often based on the environment. Some of them are marked directly on the aerial photographs or on satellite images; others are indirectly found out by using the geological map, for example. In any case, their exact location is rarely indispensable.

On the other hand, in large-scale mapping we are often forced to follow the boundary in the field and on the aerial photograph for tracing it with great precision. The study we did in the plain of the Orbe (Vaud canton, Switzerland) will enable us illustrate the reasoning used (Legros et al., 1997). In the plain is seen a drained marsh characterized by very black, highly humiferous soils that form a large darker map delineation on the panchromatic (black-and-white) aerial photograph. But accurate delimitation of the boundaries of this delineation of humiferous soils is difficult because the patchwork of cultivated fields complicates the observations. Various elements are to be considered on the photograph taken in spring:

- Parcels have dark colour on one side and light colour on the other. These parcels correspond to ploughings or to young seedlings that do not hinder viewing the soil. In this case, the boundary can be directly perceived on the photograph, but in a broken manner because the concerned parcels are few. Tracing the boundary on this basis alone relies mostly on interpolation. But it is possible even if the proportion seen of the boundary represents only one-third or one-fourth of the total length. The objective of Figure 5.2 is to persuade the reader who also should be able to do it quite well and without much error.

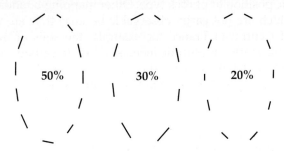

Fig. 5.2 Reconstruction of a soil boundary based on observation of a small proportion of it (see numbers indicated).

- On the same photograph, parcels have black colour, which is explained by a bare soil highly humiferous in the surface layer or by a field with thick plant growth (maize or wheat in particular). Delimiting the marsh by assimilating it to the black zone is thus locally a very wrong gamble. But the probability of making a correct diagnosis can be improved if the dotted boundary presented above is considered along with the colour of the fields.
- On the same photograph, white parcels correspond to ploughing in zones where the bare soil is light-coloured, that is, non-peaty. It cannot be a matter of mature wheat or of stubble because the photograph was taken early in the season. This time, unlike in the preceding case, we have a certainty: the zones concerned are outside the map delineation to be demarcated. Unfortunately, fields of this type are very few, so much so the information carried by them represents very little of the area.
- Other indicators are further added to the list presented above. Topography is used. It is obvious for example that the marsh is confined to the lower portion of the plain. Thus, a black field located on a slope does not correspond to peat but more surely will to maize or wheat. Besides, experience and commonsense show that two additional rules can still be applied. On the one hand, there is no marsh in the immediate vicinity if orchards, houses or villages are seen. On the other hand, presence of marsh is probable if areas of free water (old peat-exploitation areas), poplars, drainage channels or vestiges of old outflow channels are seen.

In sum, demarcation of the old marsh is done by simultaneously using all the data that have been reviewed. Also, every time there is doubt, the mapper uses the direct method, that is, he goes to the field concerned and takes an auger sample or at least a surface observation.

Drawing the boundary
It will be seen in Chapter 7, § 7.1 that we can push the reasoning further and estimate the risk of error when the method described above is applied. But in general the experienced mapper does not embark on statistical calculations. To draw the boundary between the marsh and its borders, he works very fast in an almost intuitive fashion without having to examine every detail of the procedure he is following. The beginner, or the observer, who accompanies such a specialist in his survey finds it hard to integrate all the data and, not having understood everything, may find the demarcation random or arbitrary. Yet the volume of data taken into account is considerable. But it is regrettable that the *mental model* used exists only in the mind of the one who has developed it (Hewitt, 1992). Soil reports that explain how the map was constructed are rare. Only block diagrams, if they exist, convey pictorially the soil-landscape model used. This is not sufficient, particularly when a map has to be generated or integrated in a study done in the same area but at a different scale. A good part of the work of analysis of the environment is to do it all over again to fully use all the information already available. We may add that there are methods for disaggregating the information presented in the maps and thus to find the environmental factors that were considered for making the delimitations (Bui and Moran, 2001).

The major difficulty in mapping for a beginner is to draw his first boundaries in the field. It is essential not to wait till he returns to the office. On the contrary, when three or four good observations are available, drawing a line is desirable. The hand and pencil will no doubt hesitate in the beginning, but this is of no importance. The field observation that follows will enable improvement in the line or even its complete revision. Not only will it bring additional elements pertaining to the soil but, since there will be displacement in the field, it will also enable observation from a different angle of the small sector in question, which is very valuable. Figure 5.3 shows how one proceeds through successive corrections. It is perhaps the most important in the entire book!

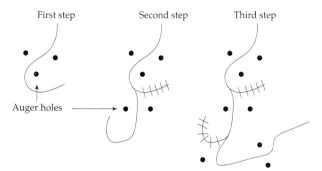

Fig. 5.3 Drawing a boundary through successive approximations by field observation and taking some auger observations for checking.

For the beginner, the best method is to draw the lines with black pencil directly on aerial photographs printed on matt paper. Erasers and pencil sharpeners should not be forgotten, several sets of them, because these precious tools are easily lost in the field and in woods. The expert can use glossy aerial photographs and an appropriate felt-tip pen that can write on all surfaces and uses an ink soluble in alcohol (also to be brought along). If alcohol was forgotten, eau de Cologne or even a deodorant will do the job.

5.1.3 Establishment and study of auger holes

Layout (Lo and Watson, 1998)
Let us first examine some generalities regarding methods of sampling. The observation sites can be chosen at random, chosen arbitrarily or systematically established in the field, in particular at the nodes of a regular grid. In some cases the sampling is stratified, that is, organized on the basis of pre-existing boundaries. For example, 10 auger holes may be done in each unit drawn on the geological map. Figure 5.4 summarizes the situation, which is very complex. Actually these different methods may be combined to some extent. Also, instead of establishing each time an observation point according to the methods we have just seen, the mapper can establish a cluster of

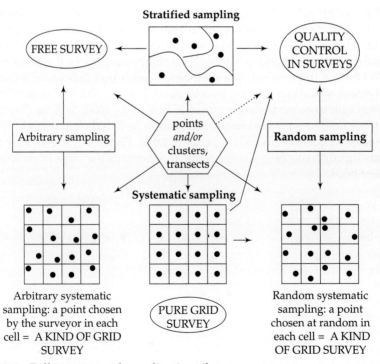

Fig. 5.4 Different types of sampling in soil survey.

several auger holes or a transect (several aligned auger holes) or even a mixture of points, clusters and transects. Furthermore, the sampling schemes for networks devoted to monitoring soil pollution can be very complex, for example simultaneously systematic, circular and in clusters (Martin et al., 1999).

An expert system has been proposed for defining in a given case the best sampling plan simultaneously with prediction of cost and the error made (Root Mean Square Error, RMSE) (Domburg et al., 1997). The exercise is a bit complex and it is certain that respect for the standards presented earlier (Chap. 3, § 3.2.1) permits escaping from it in common cases. In special field conditions, it is a tool for pointing out the various facets of the problem and their links.

In a free survey, which concerns us most here, it is advisable to follow certain rules:
- Large areas must not be left without observations. This is to limit error when the cartographic reasoning used is not correct, and is the compromise made in systematic sampling.
- One is well advised in certain cases to align the auger holes along the slope (concept of *transect*) to discover the possible existence of a *toposequence* of soils. It has been shown (Legros, 1978) that proceeding in the field along a straight line is an efficient method for intersecting the largest number of soil units in a given milieu.
- It is also necessary to vary the distance between successive auger holes. Specialists would refer to geostatistics in this connection. For example, ten auger holes done in one square metre can bring out soil variability over a very short distance. This is the concept of *cluster sampling*. On the other hand, if the mapping concerns a limited area, it will no doubt be useful to make some observations broadly outside the area to better understand the context in which the study is done.

Density

The elements for calculating the number of auger holes per unit area appropriate for different kinds of mapping were presented in Chapter 3, § 3.2. We shall also see later in this chapter how the number of samples useful in a specific situation is determined.

But let us take the simplest case, which is also the most common. If we are trying to obtain information per cm^2 of map on average (coefficient of mapping efficiency $K = 4$, see Chap. 2, § 2.4.3) and if surveys are done at a scale double the final scale, one observation must be done every 4 cm^2 of the photograph or base map used in the field. This corresponds to *one observation on average every two centimetres* on the same document. This is a very simple and efficient rule for achieving appropriate accuracy without difficulty. Some of these observations are not auger holes and require very little time as, for example, study of the surface soil of a ploughed field.

Entry of auger-hole descriptions
There are three ways of recording in the field the data relating to auger-hole descriptions:
- *The first method,* the best, consists of using a small portable microcomputer resistant in principle to shocks, water and dust. One can refer to Chapter 4, § 4.4.3 for relevant technical details.
- *The second method* of handling auger-hole descriptions consists of entering them on a special sheet such as what the STIPA system enables us to design by simplifying the profile sheets. This way of proceeding is quite satisfactory, but it is not often used because it has the disadvantages of the first method (rigidity of the menu) without the advantages (direct computer entry of the data).
- *The third method,* the least suitable, is also the most generally used. It consists of taking notes in a notebook originally without any indication or guidance. If the mapper persists in using this obsolete system, at least two precautions must be taken. First, one should indicate on the first page of the notebook the name of the user, address (in case of loss), date of use and the area in question. Second, it is necessary to number the auger holes for use by all mapping teams so that references are not duplicated. For example, one team uses the numbers 1 to 200, another 201 to 400, etc.

In addition, the location of the auger holes should be recorded. This can easily be done by marking a cross on the aerial photograph or, if need be, on a topographic base map. In view of the low cost of a pocket GPS, all mapping departments are now in a position to pinpoint the location of observations by means of this system (Chap. 3, § 3.5.4). But time is required for attaining good accuracy in the location.

5.1.4 Daily and weekly syntheses

Work in the evening
It is imperative to make a fair copy of the day's work. This can be trying after a long day in the field. But it is absolutely indispensable. At least an hour to hour and a half must be available each evening after dinner to finish the necessary tasks:
- refining under the stereoscope the boundaries drawn or sketched out in the field; if they were done in pencil, they must be redone with a felt-tip pen or in India ink;
- verifying that each soil described is geographically pinpointed (cross and numeral on the photograph or base map);
- reading again the day's notes to be certain that they will be comprehensible a few months later (watch out for abbreviations, illegible writing, faulty magnetic recordings, etc.);

- filling out as soon as possible the appropriate sheets for characterizing the different map units and soil units found; this ensures that essential elements are not forgotten and enables the mapper to progress with compiling the report in the field; in USA one does not work in any other way: *'the soil survey manuscript is developed concurrently with mapping'* (Hartung et al., 1991); a check-list inspired by the DONESOL system (Gaultier, 1990) is presented in Table 5.1; it will help those who do not envisage computer processing and who do not, therefore, have to use the entire range of sheets proposed by that system (Chap. 6, § 6.3.3).

Table 5.1 Check-list for the description of map units (MU) and the soil units (SU) included

NAME OF THE MAPPING UNIT:	
No.:	Type: Consociation
Estimated area (ha)	Association
Symbol colour	Complex
PRINCIPAL CHARACTERISTICS	Undifferentiated soils
General relief/geomorphology:	
Land use (vegetation, crops, parcel characteristics):	
Altitude (minimum, mean, maximum):	

SOIL UNITS (SU) INCLUDED, SPATIAL ORGANIZATION

	A	B	C
SU NAMES: No: Relative area (%):			
Location in relation to the environment			
Type of pattern: polygons, bands, etc. **Intricacy of the pattern:** m, hm, km? **Spatial integrity:** number of soil areas			

	A	B	C
Reference number of the profiles			

	A/B	A/C	B/C
Boundaries, contrasts			

SOIL UNIT DESCRIPTION

	A	B	C
ENVIRONMENT *Geology* *Slope (minimum, maximum)* *Vegetation belt or type* *Specific land use*			
SOIL PROPERTIES *Physical:* depth… *Chemical:* acidity, salinity *Biological and humus* *Hydrological* (regime, water table…)			
MANAGMENT/ PROTECTION *Land-use limitations, degradation risks, dynamics of phenomena*			

- in addition, it is advisable to accurately write down in a notebook reserved for this purpose observations of all kinds that will be worth mentioning in the explanatory report of the soils and the environment; the mapper should not hesitate to write down in this notebook questions or thoughts because all of them will be valuable later on;
- it is necessary to identify those areas mapped that day for which a supplementary field check is required to ensure quality of the map fragment in question; except in particular cases, mapping will continue the next day by study of the immediately neighbouring area; thus the least time is lost in starting work in the morning with a rapid retraverse of the sectors that have a problem, at least if they are well-connected by road;
- lastly, the portion of the provisional map corresponding to the day's work will possibly be drawn (see below).

Some mappers who have the feel of the land and experience on the job still remain poor professionals because they do not know how to force themselves to these nocturnal efforts. In such a context, very rainy days are not days lost; they serve for fair-copying of notes and records to make up for the delay caused.

Weekly correlations

In the presentation of free survey (Chap. 2, § 2.1.5) we had indicated the necessity of organizing weekly correlation trips. The objective of these is to verify that the various teams work in the same way and also that they work in a way that does not change with time, especially when field work takes many months or several years. It is not considered as time lost. Moreover, it is often possible and even advisable to take advantage of it to further the work. Efficiency is lower since a single team brings together all the mappers.

5.1.5 Consideration of the experience of farmers

Farmers have great knowledge of their soils. But the farmer-mapper dialogue is difficult: the vocabularies and concerns are not the same. Credit for proposing a method for introducing farmers' knowledge in a large-scale soil map goes to the SOL-CONSEIL Society (Strasbourg) of France (Party et al., 1986). The procedure is based on four points:

- Mappers conduct a *reconnaissance* in the field at the rate of half-a-day for 1000 hectares. This enables them find out the mapping criteria and envisage the questions that possibly could be asked of the farmers. For example, if the soils are seen to be non-uniformly stony, one could ask 'where are the stones?'.
- This reconnaissance mission thus helps in developing a *questionnaire* for the farmers' use, if possible taking into account the local terminology of soils and their properties.
- A *meeting* with the farmers is organized. They are invited to show on the aerial photographs or on a cadastral base the information they have: stoniness, zones of excess water, sectors where the soils are very shallow, etc.
- A *final phase of systematic survey* then enables the mappers to complete the work, that is, to verify the data provided by the farmers and to delimit the soils in the zones left without indications.

This method of mapping with the help of farmers has various advantages. First, it has the merit of making farmers participate in preparation of the soil map, which is diplomatic. Secondly, mappers and their assistants do not forget to integrate in their work soil properties that appear important to the local people. Thirdly, the method is efficient. According to SOL-CONSEIL, the mappers proceed at least twice faster than in traditional mapping. More generally, it is surprising and instructive to see on the draft map at the end of step 3 the considerable volume of things known from the farmers! This can help, if necessary, to restore a little modesty in the mapper. The difficulty in such work is obviously the mobilization of farmers. This can be done if some conditions are respected: the date chosen for the meeting should not interfere with the agricultural calendar, and the press and local officials should participate. Success is assured if, in mapping a *commune* (the smallest administrative division in France), at least six participants representing more than 20 per cent of the cultivated area are brought together. This is, in general, sufficient to get indications on 60 to 80 per cent of the area because everyone knows the soils of his neighbourhood. The ideal thing would be to assemble farmers of very different ages. Under such conditions, a dialogue is established and it becomes possible to locate 'problem soils'.

Besides, one can relate precisely, through statistical approach, on the one hand the behaviour of the soils as perceived by the farmers and recorded in a questionnaire and, on the other hand, the same behaviour appreciated by the mapper on the basis of field observations (Arrouays, 1989).

5.1.6 Number of mapping teams

We have already mentioned that the work of mapping should be organized for several teams of two persons. But the project leader, who will be in charge of the synthesis, will see his work complicated by the presence of too many teams. With a single team (of which he will be a member), he inspects the entire area himself. With two teams the leader inspects one half and, with three, one third, even if the companion is changed regularly to facilitate standardization of methods and diagnostic criteria. Beyond three teams, it is difficult to ensure uniformity in work, even if several correlation tours are organized. Thus, it appears that a configuration of two teams of two persons each is ideal: correlations are easy and the work proceeds at an appropriate pace. For profile-study programmes, however, one can ask for reinforcements if necessary and go to seven or eight mappers, which is the optimum for that phase of the work (Chap. 4, § 4.6.4). As the techniques for describing soils are well codified, a scientist who has not participated in demarcation of soils in the region can nevertheless describe the soils perfectly.

5.2 PREPARATION OF THE MAP

5.2.1 Construction of draft maps

Practical arrangements
Construction of the first draft map is done in the field, if not on a day to day basis, at least once a week, taking advantage of rainy days. It is possible to make corrections and deletions on this draft. It is often prudent to leave it behind in the hotel because it is the only copy of a major part of the synthesis of all the efforts. But sometimes it has to be taken to the field for

verification. In spite of all precautions being taken, this draft might get soiled and crumpled. In other words it is a rough sketch that necessarily should be redone. It is, however, indispensable because it will be very risky to do without this first draft and attempt much later in the laboratory a direct synthesis of all the elements dispersed on aerial photographs, field sheets and in the cartographers' minds. It is very convenient to make the first draft at a scale twice that of the final map. For the second draft, there usually is a choice: the scale of the first map may be retained or that of the final map adopted. Everything actually depends on the methods that will be used to reproduce and publish the reports. If the study is a big project, a draft map may be in several sheets.

Management of the list of profiles
It is very useful to draw up the list of profiles in the form presented in Table 5.2, in chronological order, as and when the profiles are described. It will be very easy later to obtain an automatic grouping in order of increasing profile numbers.

Table 5.2 Structure of the list of profiles

Profile number	Draft	x-coordinate	y-coordinate	SU	MU
325 (R)	X	−2°34′45″	44°34′30″	4d	(14)

A profile is first of all marked on the draft by a cross and the corresponding number. It is also advisable to report the same number in the blank margin of the map, as close as possible to the actual position of the pit. When this is done, the column headed 'Draft' in the table is ticked. These precautions are essential to avoid forgetting things, which will involve loss of time in correction later. For example, to have described 300 profiles and to find only 298 on the draft guarantees that hours of research will be spent in identifying and locating the two that are missing! A profile is characterized by its belonging to a soil unit (SU). Three cases should be considered: it is typical and serves as *Reference* (R) for defining this unit; it shows some specific characters giving a *local variant* (V), or it is only *related* or *atypical* (A). If it represents a soil unit still not described, it is marked 'N' for *new*. The soil unit name is indicated, at least provisionally. When possible it is also necessary to give the map unit number. This is not simple. Actually, a profile can be characteristic of one SU-MU pair and still occur as an *inclusion* or *impurity* in another SU-MU pair. In this case, the number of the MU that actually contains the profile considered should be mentioned in the table. To call attention to the anomaly this number is placed in parentheses. On the base map, the profile number is also placed between parentheses to indicate that it is not characteristic of the MU in which it occurs. We will see later that the geographical relations between profile and MU can be established and analysed automatically (Chap. 6, end of § 6.3.3).

During field survey, the numbers or names of the MUs and SUs are not final. Table 5.2 is then corrected and updated as many times as necessary.

Rules for drawing boundaries
Suppose we observe a polygon of *Histosols* (peat) including two small hills of *Cambisols* that form islands of the same nature as the soils outside the polygon of Histosols. This is a situation we have seen in the morainic deposits of Switzerland. We want to give an account of this on a map the scale of which necessitates generalization. Figure 5.5 summarizes the methods that can be used. According to Shea and McMaster (1989) there are seven ways of doing it.

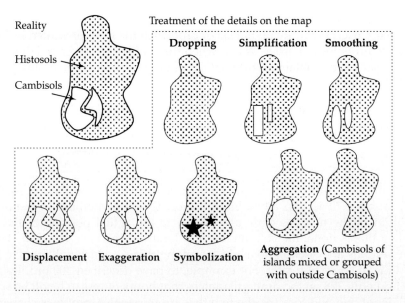

Fig. 5.5 How to treat details and increase map legibility; example of a marsh with two islands of Cambisols.

These seven operations are not, in fact, strictly independent, so their presentation is somewhat distorted.

Rules for graphically presenting the content of units
For presenting the drafts and the map, the specialist can play on *colour* and its intensity, or with the *shape, orientation, size* and *density* of symbols used. The choices are made more or less rationally:
- For a quantitative property, the size or density of symbols will be varied. For example, a design with varying density could be introduced to indicate soils with varying stoniness.

- The number of cases that must be shown should be counted. Thus, while changing the orientation of a hatching, more than four different situations hardly can be drawn, which restricts the use of this kind of graphic to simple things.
- Above all, it must not be forgotten that colours are the most directly visible elements. They should therefore be used for essential data one wants to highlight.
- Logic is also considered. For example, red soils will be represented in red and brown soils in brown, rather than the other way round! There are also accepted conventions to be followed in this regard.
- Lastly, one should think of aesthetics and work in a spirit of synthesis for attaining high legibility. Some mappers have indisputable talents. Others only know to produce unsightly maps. Mapping is a bit like the art of painting….

5.2.2 Construction of the legend

The legend, as far as possible, *must be simple but comprehensive.* Some mappers, in their anxiety to show everything and justify the method whereby they reach synthesis, draw up complex two-way legends that have to be studied a long time before the way to read them is understood. Of course the mapper who has spent many months and sometimes several years on the work is theoretically right to demand some effort from the reader. But is this demand quite reasonable? There is risk of pure and simple rejection, even if the work is basically excellent.

The legend must be structured vertically and horizontally. In general, it is not desirable to arrange from top to bottom the index 'boxes', that is the small coloured rectangles, without classing them into larger groups. We have already mentioned this point (Chap. 2, § 2.3.2). The legend should be structured horizontally as well. The data furnished for each box should be presented in strict order for the reader to find them.

The legend must be coherent. For example, it is inadequate to distinguish the soils as follows:
- deep soils,
- shallow soils,
- calcareous soils,

when calcareous soils can obviously be deep or shallow. One may wish to treat them separately, but this must be properly explained.

The legend must be explanatory. Some years ago the tendency was to draw up very brief legends distinguishing, for example, without other statements:
- Podzols,
- Cambisols,
- Luvisols.

This is understandable to the extent the map is accompanied by an explanatory report giving all details. In addition, an experienced specialist knows the concepts and realities hidden behind terms such as *Podzols* and *Luvisols*. But this procedure makes the map incomprehensible to the non-specialist, who has early on associated lack of transparency with the idea that the work is theoretical and why not useless? In addition, it requires readers to refer to the explanatory report, find the correct page and often to sort out themselves the important points from the non-essentials.

We think, therefore, that the legend should synthesize in a clear manner the major points of the pedological information (Fig. 5.6). Reference to the report should only be necessary for deeper understanding of the soils.

5.2.3 Ordering for and use of the analyses

Pretreatment of samples
It is impossible to give general rules on this topic. Everything depends on what is desired to be determined. For the non-volatile and almost non-reactive components such as quartz, no particular precautions are required. For volatile components or those entering organic compounds, on the other hand, appropriate pretreatments should be done immediately. Depending on the case, this may consist of storing the samples in a cool dark place, of strong drying in a hot-air oven, of freezing, of sterilization, etc.

Choice of analyses
Standard soil analyses are required for some major objectives: refining the physical characterization of the profile, doing a chemical characterization, and resolving problems in pedogenesis, classification, agronomic improvement. More can be understood from a specialized book (Baize, 2000) and by referring to a discussion on the subject (Bock, 1994). Table 5.3, inspired by the book by Baize (2000), does not pretend to exhaustively justify the usefulness of any specific type of analysis. It is only given as an indication.

But two basic limitations in soil analyses must be borne in mind:
- they only give an average composition whereas the precise location of the elements is very important; for example, in a soil analysed as slightly calcareous, so much $CaCO_3$ might be concentrated around the roots that they react as if the soil were extremely calcareous; thus nothing can take the place of field observation;
- for each chemical element it is possible to propose a full series of increasingly efficient extraction methods, from cold water to mixtures of hot strong acids; to claim that a particular reagent extracts the soluble, mobile, amorphous or available fraction is always an approximation verified by indirect evidence; at least unless we proceed the other way and define, for example, the amorphous fraction as that extracted by a particular reagent!

Unit number	Functional units Comprehensive description-dominant descriptive characteristics	Topographic position. Relationships between functional units	Major feature influencing evolution of the unit	French CPCS classification
	SOILS OF THE RHÔNE PLAIN AND TRIBUTARY VALLEYS **RECENT ALLUVIAL PLAINS**			
	Topographic position lower than the other landscape units: drainage and water circulation axes. Recently deposited materials with frequent rejuvenation by floods and overflows; very variable lithological and particle-size composition. Shallow ground water present; soil-evolution stages directly related to the ground-water regime.			
1 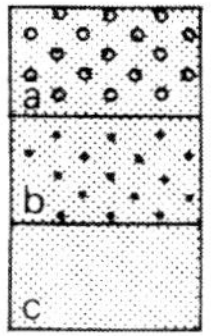	Calcareous soils: (a) thin or discontinuous (sand, gravel and pebbles deposited by the Rhône river) (b) generally deep with light texture, single-grain structure, rapid infiltration, permeable (c) generally deep with medium texture, strong massive structure, often with pseudogley at depth when fine-textured	More or less submersible gravel benches of the Rhône and its tributaries (a) Alluvial plain formed by exceptional floods of the Rhône and its tributaries (b and c)		IMMATURE ALLUVIAL SOILS
2 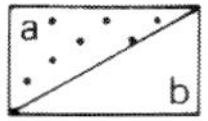	Noncalcareous soils, variable depth: (a) light texture, often with gravel (tributaries on right bank) (b) medium texture, locally hydromorphic	Submersible sectors of the tributary valleys of the Rhône	Deep, circulating ground water, aerated	
3 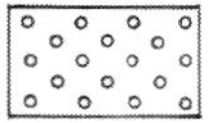	Loamy soils, noncalcareous, weakly humiferous surface layer on the silty clay with pseudogley (60 cm) limited by a more or less indurated stony layer	Recent alluvial deposits in the Würmian plain of Bièvre-Valloire	Moderately deep ground water (100 cm)	
4 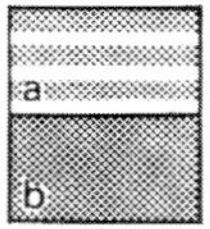	Soils with pseudogley throughout and gley at depth, weakly humiferous surface layer, medium to heavy texture, calcareous (a) or noncalcareous (b)	Medium longitudinal section of the tributary valleys of the Rhône and edges of piedmonts, terraces	Temporary shallow water table	HYDROMORPHIC ALLUVIAL SOILS

Fig. 5.6 Example of a legend: soil map of Saint-Étienne on 1/100,000 (Legros and Bornand, 1985).

Table 5.3 Examples of certain types of laboratory analyses

ANALYSIS	Use in pedogenesis and classification	Use in characterization of soil functioning	Use in soil protection or utilization
pH	Identification of soil type, weathering mechanisms	Solubility of elements, biological conditions	Choice of fertilizers
Organic matter	Identification of Histosols and humic soils	Calculation of C balance at the soil/atmosphere interface	Choice of crop rotations, agronomic potential of soils
Particle-size analysis	Examination of allochthony of surface horizons	Study of aggregate formation or destruction	Estimation of erosion hazards
$CaCO_3$	Study of carbonate leaching	Ca-ion budget, risk of acidification in certain soils	Choice of fertilizers
Cation exchange capacity	Aid in clay type identification	Study of adsorption phenomena	Appropriate levels of fertilizers
Soluble salts	Identification of salts and of corresponding soils	Mechanism of salinization	Choice of suitable salt-tolerant plants

Also, analyses are time-consuming and necessarily costly. So we must economize, for which there are several strategies:

- The first consists of rationalizing the requirements bit by bit. Only those analyses necessary for resolving just the persistent questions are ordered for. For example, a granulometric analysis will be done for a particular horizon for which there is doubt about the texture identified in the field.
- The second procedure consists of completely analysing the apparently most typical profiles and applying to the others a standard menu of routine analyses (such as determination of pH, organic matter and particle-size distribution). The standard menu is variable and established according to the aims of the study. This is the method most generally used.
- The third strategy is to decide after the field phase the profiles that will be analysed and which others will be left with just a description. If necessary, only one profile in two is sent to the analytical laboratory.

The third method is a bit frustrating because it leads to discarding profiles that are otherwise relevant and representative. We however feel that it is best if the descriptions and analytical data are then entered in a data bank so that they can be processed later. Indeed, let us suppose that in the data bank the places reserved for storing the results of each analysis are 50 per cent filled (half-full database, which is quite good). Let us also suppose that we are seeking to highlight the correlations among four variables x_1, x_2, x_3 and x_4 characterizing the horizons. Lastly assume that the bank contains 10,000 soil horizons. If either of the first two strategies was used to fill the bank, the probability that a horizon has been analysed for all the four variables is $(0.5)^4$. Thus we will have available about $10,000 \cdot (0.5)^4$ or 625 useful sets of four values, which is very few and certainly does not justify the data-storage efforts agreed to. On the contrary, with the third strategy there will be 5,000 useful sets of values! It is better to have a detailed characterization of a few hundreds of profiles than an 'analytical sprinkling'.

Ordering for analyses
Soil-analysis laboratories, on request, provide very well structured forms for ordering an analysis, for example, the form given by the INRA laboratory at Arras in France. But these forms are suitable for farmers' needs often applicable to a single parcel characterized by just one composite sample. Soil mappers have different requirements. They only send samples to the analytical laboratories once or twice a year. But the samples are quite numerous because they belong to a whole mapping project. It must therefore be possible to order for analysis of samples not on paper but directly from computer to computer. This saves time, enables validation and allows calculation of the cost in advance. Communication of analytical data should also be computerized. The laboratories have their data on the computer and so do the mappers. It would then be a shame to send the information on paper, which requires entry afresh on magnetic media with risk of error as a bonus. Unfortunately in many countries, the mappers are small clients of the laboratories because the number of analyses they request is small compared to the total. In addition, these same mappers do not always have identical or even compatible computer systems. In short, many laboratories are not urged to create all the interfaces that would be useful. However, it seems to us that such interfaces will serve at least to enhance the prestige of these laboratories.

5.2.4 Compiling the report

Compilation of the report can only be done after the mapper receives the results of analytical determinations. Considerable latitude is generally given to the mapper in compiling the text accompanying the map. In certain cases standardization may regrettably be lacking. The typical design of such a publication, extracted from tests and experiments, is described below.

Part one: structure, context, description of the milieu
The structure of the report is presented in the beginning: identification and location in the report of mapping units considered, organization of appendices.

The general features of the study are then indicated: *objectives of the study* (amelioration, systematic inventory, etc.), *methods* (types of survey, scale, etc.), *terms for carrying out the work* (organizations and persons involved, contribution of the mappers, types of existing studies incorporated, etc.) and *results obtained* (area covered, number of profiles and auger holes studied, data bank established, documents prepared).

The natural regions of the area are then reviewed: geology, geomorphology and land use (natural vegetation, crops).

Part two: description of soil units

For each map unit (MU) the list of soil units (SU), relative spatial arrangement of these units, their area, their characteristics and the corresponding profiles are defined (Table 5.1). Then, for each SU are indicated:

- characteristics of the *type profile*: this type profile generally does not exist in reality; it is reconstructed, feature by feature, by studying the most frequent case on the basis of all the data known; local variations in the type profile that can occur in the unit are indicated;
- generally, as illustration, the description and analytical data of a *reference profile* dug in the SU and characteristic of it is added in a box; if possible a few photographs and diagrams showing this soil and its location in the milieu are introduced;
- interpretations pertaining to pedogenesis could follow.

Part three: protection and amelioration of the area

The situation is not the same if the study is done with an applied objective or within the framework of a systematic inventory. In the former case, an answer is given for a practical question posed. In the latter, the regional constraints for utilization of the land are presented. For example, in the Arles region (Bouteyre and Duclos, 1994), the problems of salinity were dealt with and the irrigation network was presented. In the Privas region (Bornand et al., 1977), the problems of giving up agriculture were brought out: abandoning farming terraces, soil erosion, decline of chestnut groves and pasture, growth of forest.

Appendices

In the appendix are presented some of the characteristic profiles that have been fully analysed (R and V types) but could not be accommodated in the main report.

5.3 QUALITY CHECKS RELATED TO GRAPHIC INFORMATION

The checks envisaged below essentially pertain to map drawing, that is, to the soil boundaries. The checks pertaining to semantic information have been covered in Chapter 4, § 4.5. Many tests exist for examining item by item the quality of different mapping steps. We shall review the principal ones. Review articles have periodically given updates on the question (Beckett and Webster, 1971; Legros, 1978; Wilding and Drees, 1983; Burrough, 1993).

The errors can be grouped into three broad types:
- *Mapping errors* directly linked to field work. They will be discussed in detail later.
- *Transfer errors*. Transfer errors are linked to the difficulty in translating reality, presumed to have been correctly interpreted, by a precisely located line to be drawn on a map showing roads, rivers and other geographical features (base map). Most of the time the corresponding error is much smaller than one mm, except when details are considered. This has already been mentioned (Fig. 5.5).
- *Locational error*. Even if no error has been introduced during the preceding two steps, inaccuracy remains because the map on which the soil boundaries have been drawn is not perfect. For example, a road intersection is not exactly where it is shown, if its coordinates that can be picked up from the map are compared to its real position on the Earth determined with a high-quality GPS. Specialists are examining the question extensively (Maling, 1989; Goodchild and Gopal, 1989). But there are many reasons to neglect this inaccuracy in the purview of this book. This error is usually small. In Australia for example, the roads on maps at 1/50,000 scale are traced within 25 m of the actual position over 78 per cent of the route (van Niel and McVicar, 2002). Secondly, this error keeps reducing rapidly and can be reduced to a few metres if necessary using GPS and orthophotographs (Barrette et al., 2000)! Lastly, except for operations that require using several maps at the same time, we can be satisfied in soil mapping with proper positioning relative to the base map. For example, if it is observed that Podzols go up to the road, it is so shown on the map. It doesn't matter in the end that the position of the road is not correctly picked out on the Earth!

Some scientists use another vocabulary. They distinguish *absolute positional error*, corresponding to our locational error and *relative positional error*, which is our mapping error.

5.3.1 Validity of pre-zoning

It was mentioned in Chapter 3, § 3.3.4 that the preparation for field work could be facilitated by a pre-zoning executed through photointerpretation, above all in physiographic mapping. To estimate the accuracy of this step in the work, we developed the following test (Legros, 1997). Forty students of l'École Polytechnique Fédérale de Lausanne (EPFL, third-year students of the Rural Environmental-Survey Engineering section, 1992-1993 admissions) were asked to do a pre-zoning on a monochrome aerial photograph on 1/30,000, pertaining to a high-mountains sector. The relief and vegetation types were varied and contrasting. The students had good stereoscopes, lateral photographs enabling a relief view and the topographic map. They were grouped into pairs, so that 20 pre-zoning interpretations were obtained for the same sector.

Superposition of these interpretations was done by means of a vector-type GIS. Figure 5.7 presents the product obtained.

Various comments can be made now. One, boundaries identified by all exist on the photograph studied, but for small error in positioning them. The landscape concept, therefore, is not totally subjective and it is possible

Fig. 5.7 Superposition of pre-zoning interpretations done on the same aerial photograph by 20 pairs of observers working separately (Legros et al., 1997).

to use it many cases. Two, sectors also exist in which the boundaries are perceived completely differently by the observers. Then the lines cross each other and no clear zoning can be seen. One can understand why this is so. Mainly, the observers did not hesitate with a boundary that pertained both to land cover (the grassland/forest boundary for example) and to topography (for example, the plain/slope boundary). On the other hand, when such a boundary becomes double, that is, when the meadow/forest boundary is different from the plain/slope boundary, the opinions of the observers differed on what is appropriate to be done from amongst four possibilities: discard both boundaries, choose one or the other for tracing, or even let both appear.

This exercise leads to practical conclusions. First, in this pre-zoning stage when the boundaries have no recognized pedological significance, it is necessary to first draw the major boundaries having two or more determining factors. Second, to avoid introduction of too many boundaries of doubtful significance, that is, having just one determining factor, we recommend that the mapper draw altogether no more than a length $2L$ of boundaries of all kinds if L represents the length of the reliable boundaries revealed by several determining factors. Last, and most important, one should not forget that the principal objective is not to find discontinuities on the photograph but to isolate the areas with some specificity—not always the same thing. In the end we must insist on the fact that, at this stage, it is not a question of soil mapping but just a rough sketch that must be validated and refined or even abandoned.

5.3.2 Relevance of mapping criteria

The relevance of mapping criteria can be checked through simulation of accurate detailed mapping by sectons conducted as an experiment. It should be remembered that observations made at the nodes of a regular grid are later classified, that is, grouped by an automatic process without considering their relative location on the land. Similar observations close to each other on the map are then included in the same mapping unit. Under these conditions, the map can be altered by modifying the classification procedure, in particular by playing with the weight assigned to each soil property. In other words, one can visualize the consequences of the choices made in mapping criteria and in the weight assigned to each one.

Norris (1971), Van den Driessche et al. (1975) and, later, other authors (Girard, 1983; King, 1986) developed algorithms for making such classifications based on soil horizons (the DIMITRI program) or even on entire profiles (the VLADIMIR program). These efforts are particularly interesting. They enable us to calculate:
- the degree of instability of the map as function of the criteria used to draw it (up to 30% variation in the area of delimited units);

- the dangers of delimiting the soils on the basis of surface features, whereas it is the features at depth that are judged interesting (for example, presence of limestone fragments on the surface for estimating the depth to the rock stratum);
- loss of precision linked to the fact that mapping is often conducted on the basis of auger holes that do not allow us to study all soil features; in such a case, simulation is restricted to using the observations corresponding to profile pits. The work is done twice: one map is drawn using all the useful characteristics, another with just those features that would be observed in auger holes; comparison of the maps will show how hard it is when profile pits cannot be dug all over.

5.3.3 Verification of accuracy of the boundaries

Suppose the true position is known of a boundary that the map only shows approximately. In practice, this implies that the mapper returns to the field in a chosen sector for establishing the true or 'ideal' boundary as perfectly as possible by putting in unusual effort. We did this with third-year students of the Agricultural Engineering Department of l'École Polytechnique Fédérale de Lausanne in Switzerland. These students were asked to delimit with great accuracy the boundary of the reclaimed marsh of the Orbe plain (Vaud canton). This required separation of a black soil very rich in organic matter from a lighter-coloured clayey soil. Altogether, the efforts involved three successive classes and represented 110 to 115 field days for identifying the boundary of the marsh, a map delineation of perimeter about 15 km.

Vector-mode GIS's allow drawing a buffer of chosen width around this ideal boundary. This is the concept of *epsilon band* or *Perkal's epsilon error band*; it represents in principle the tolerated error (Chrisman, 1982). This band may or may not contain the boundary drawn under usual conditions by the mapper (Fig. 5.8).

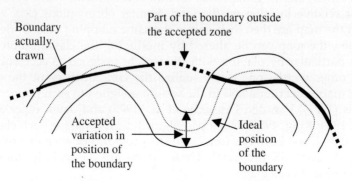

Fig. 5.8 Position of a normal boundary compared to its ideal position.

It is possible, by using the so-called intersection functions in the GIS to calculate the proportions of the drawn boundary located within the buffer and outside it. The curve given in Fig. 5.9 can be drawn by giving the buffer 5 or 6 successively increasing widths.

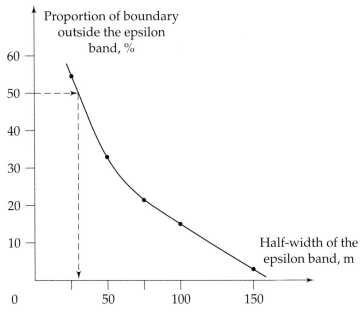

Fig. 5.9 Relationship between width of the buffer and the proportion of the normal boundary within it.

From this, by interpolation, one can deduce the buffer width such that one-half the length of the boundary drawn by the mapper is within the buffer (proportion = 50%). The median (not the mean) accuracy was thus found out by this method. In the case of the marsh of the Orbe, this means an error of about 30 m for mapping at scale 1/10,000.

In regard to accuracy of boundaries, the first French edition of this book (1996) had recommended that an error of 4 mm measured on the map should not be exceeded for more than 10 per cent of the length of the lines. But so much progress has been made in the use of GIS, orthophotographs and remote sensing that soil mappers should follow and try to observe the new norms recommended in Table 5.4. These values are applicable to conventional mapping and not to methodological studies that scientists tend to conduct on a few hectares with very great accuracy.

Table 5.4 Standards for accuracy of map boundaries

Scale	Accuracy in the field	Accuracy on the map
1/250,000	500 m	Error < 2 mm, but with greater error accepted on 10% of the length of the boundary
1/100,000	200 m	
1/50,000	100 m	
1/25,000	50 m	
1/10,000	20 m	

5.3.4 Accuracy of a *raster map divided into pixels or sectons*

If required, the definition of the concepts of *raster map* and *pixel* will be found in Chapter 6, § 6.2.4. Let us consider two maps, the first being the one whose accuracy is to be evaluated and the second, the reference map for validating the first. This reference map is often obtained by returning to the field and making a very special effort to prepare a map as accurate, at least locally, as possible. This is the *Best Possible Map* or BPM (Beckett and Burrough, 1971).

Confusion matrix and overall accuracy
For easier comparison of the two maps, let us divide them into the corresponding pixels and examine the content of each pixel on the reference map and on the map to be evaluated. Let both the maps comprise three map units A, B and C. On this basis, the frequency table is drawn up as given in Table 5.5 (Story, 1986).

Table 5.5 Comparison of a map with the BPM (raster maps). In this table, the numerals correspond to the number of pixels

Confusion matrix	A on BPM	B on BPM	C on BPM	Row total
A on tested map	**280**	100	50	430
B on tested map	10	**190**	50	250
C on tested map	10	10	**300**	320
Column total	300	300	400	1000

For example, 10 pixels that belong to unit A have been mapped in unit B by error.

Table 5.5 has been variously named by remote-sensing specialists as *Error Matrix, Confusion Matrix, Misclassification Matrix* or *Contingency Table*. The *Overall Accuracy* of the evaluated map is measured by the ratio of the number of pixels in the diagonal of the table to the total number of pixels. In the present case, it is (280 + 190 + 300)/1000 = 77%.

Other quality-control parameters
We can also determine the area that has been attributed by error to such or such unit. This is the *Error of Commission*. For unit A, for example, it is (100 + 50)/430. Similarly we can calculate the area that, by error, has not been attributed to the unit it was normally observed in. This is the *Error of Omission*, which is (10 + 10)/300 for the same unit A.

The *user's accuracy* is also identified. It corresponds to the proportion in which the pixels grouped in a category actually do belong to that category. For example, it is 280/430 for unit A and 190/250 for unit B. So too, the *producer's accuracy* represents the efficiency of the mapper. For example, the mapper has correctly identified 280 out of the 300 pixels of unit A. Thus we have (Janssen and van der Wel, 1994):
- user's accuracy (%) = 100 − error of commission (%)
- producer's accuracy (%) = 100 − error of omission (%).

Some writers define the *correct percentage*, which is nothing but user's accuracy. If all the categories are considered at the same time, the correct percentage and user's accuracy are confused with overall accuracy.

In some cases we can allow a positional error more easily than an error in the areas involved. For example, it may be important for planning to know that 50 per cent of the area is suitable for wine-growing even if the areas are not located accurately. This is why some authors are interested in the *Areal Error* (Lowell, 1994), also called *quantification error* (Pontius, 2000). This error corresponds to relative under-estimation or over-estimation of the area of each unit. For example, 430 pixels were found for unit A instead of 300. The error then is (430 − 300)/300 = 43%. In some remote-sensing applications, the proportion that should be obtained in each category is known. For example, it may be known from cadastral data that the area studied comprises 30% forest and 70% grassland. In such a case the constraint is followed, which means that the Areal Error is arbitrarily reduced to zero. We can proceed in this manner in soil mapping, for example when automated generalization is done around a Reference Area (see Chap. 7). In such cases, the errors of commission exactly equal the errors of omission.

The error matrix can be normalized through an algorithm that works by iterative proportional fitting and reduces to unity the row totals and column totals (Congalton, 1991). Many matrices can be so compared. If the objective is, say, to compare the performances of two mappers who worked in parallel in the same region but in different areas, the BPM (best possible map) then serves as standard.

Size of quality control sample
Such quality checks are, for practical reasons, never done over the entire map but only on a fraction of it. Then the question of representativeness of the tests arises. Let us suppose that an accuracy of 80% is required for the whole of the map. Is an accuracy of 77% obtained in a sample restricted to

1000 pixels high enough to vouch for the hypothesis that the accuracy of the whole map is equal to or more than 80%? To debate this, the mapper (i.e., the producer faces the user who is a sort of consumer. The following are distinguished (Aronoff, 1982, 1985):
- *producer risk*, the probability that a map having the required accuracy is rejected by the accuracy test; in this case, the mapper risks being required to make improvements even if they are useless;
- *consumer risk*, the probability that a map of unacceptable accuracy successfully passes the quality check.

These two concepts are different from producer's accuracy and user's accuracy seen above. To go further, the following question must be posed: if the map to be studied has a proportion p of its area wrongly identified, what is the probability of finding 1, 2 or 3 to a maximum of n wrongly mapped pixels per N pixels checked on the map? One would have recognized in it a problem that comes within the purview of the binomial theorem and can be expressed as follows: what is the cumulative probability of finding 1, 2, 3 or n black balls in a draw of N balls from a box containing black and white balls in the proportions p and q (= $1 - p$)? We shall not give here the numerical solution, which will be found in the articles cited. Let us only note that to be sure of obtaining an accuracy of $100q$ (in %) or $100(1 - n/N)$% with a risk of error in diagnosis restricted to 1% or 5% of the number of tries, the user would have to have to demand an accuracy greater than this value of $100q$ in the check sample. Table 5.6 gives the acceptable number (n) of wrongly identified pixels for attaining a *consumer accuracy* of 80% with probability limit of 5% when the number of pixels checked is N.

Table 5.6 Number of pixels to be checked and the tolerated number of errors for the user to obtain confidently, that is at 5% probability limit, an accuracy of 80% (Aronoff, 1985)

Required accuracy (user)	80%	80%	80%	80%
Pixels checked (N)	30	50	100	200
Number of errors tolerated (n)	2 (instead of 6 = 20%)	5 (instead of 10)	13 (instead of 20)	30 (instead of 40)
Risk taken by mapper	0.18	0.04	0.00	0.00

If the verification is restricted to a very small number (say 30) of pixels, the user should be rather demanding and aim for an accuracy of 28/30 (= 93%) to be certain of getting 80%! At the same time the mapper risks seeing the map refused 18 times out of 100, while it is correct at the 80% level chosen. The practical consequence of all this is that it becomes necessary to check 50 to 100 pixels for the benefit for both parties. But three times as many must be checked if the concern is not overall accuracy but the accuracy with which each of the soils A, B and C is identified.

If the product $Npq \geq 30$, the distribution becomes less asymmetrical and the binomial distribution tends to a normal distribution. The problem is simplified. For applying the binomial law, the *standard error* (SE, %) is first calculated (Stehman, 1992). This error is that of the population, estimated on the basis of what is known of the sample studied:

$$SE = \sqrt{\frac{(T-N)pq}{(N-1)T}}$$

where q is the overall accuracy (calculated in % because the variable and standard error are both in that unit), p is equal to $100 - q$, N is the number of pixels checked and T the total number of pixels in the map.

When N tends towards T, the standard error tends to zero. Then the calculated values of p and q are indeed no longer vitiated by uncertainty, because all the pixels have been checked.

By applying the normal distribution law, one can easily calculate the range (+1.96 to −1.96 times the standard error), in which 95% of the values of the distribution will be found. For example, with a map of 1000 pixels, a sample of 200 pixels and a required accuracy of 80%, we have an application of this simplified procedure. First we get $SE = 2.53\%$, then a range of ±1.96 times the $SE = \pm 4.97$ at the 0.05 probability level. We then must aim for an accuracy of $80 + 4.97 = 85\%$ for getting 80% (indeed, if the tail end of the distribution is at a value greater than 85%, the mean will be greater than 80 because the standard error is defined). The more accurate method using only the binomial law (Table 5.6) indicated that for 200 pixels checked, it was necessary to have 170 correct pixels, which is also 85%.

Numerous tables have been published for avoiding all calculations. Some of them are very practical (Hord and Brooner, 1976). Caution: for the test to be valid, the pixels checked must not be arbitrarily selected, but taken at random or in a systematic manner (for example, one pixel in 50 is checked). It will be more useful to avoid checking contiguous pixels that thus form a portion of the map, geographically close pixels not being independent in composition. However, it is sometimes useful to choose pixels in a line (§ 5.3.6).

Kappa coefficient of agreement
Even if a map whose quality is desired to be tested had been made at random without any reference to reality and knowing only the soil types likely to be present, its accuracy will not be zero because pixels might be correctly identified by chance and will be found on the diagonal of the matrix of confusion! The Kappa coefficient was formulated to take this phenomenon into account and thus to reduce the accuracy assigned to the map tested. But to succeed at random is still to succeed all the same! Therefore the usefulness of this correction can be questioned. Be that as it may, the Kappa coefficient is (Cohen, 1960 given in Hudson, 1987):

$$\text{Kappa} = K = \frac{\theta_1 - \theta_2}{1 - \theta_2}, \text{ with}$$

$$\theta_1 = \sum_{i=1}^{r} x_{ii}/N = \text{observed matching}$$

$$\theta_2 = \sum_{i=1}^{r} x_i + x_{+i}/N^2 = \text{chance agreement}$$

where x is the number of one of these cases in the matrix (Table 5.5), i is the number of the row or column, $i+$ and $+i$ are the totals for the ith column and ith row, respectively.

The value of Kappa is zero or even lower when the degree of coincidence is solely because of random chance. It is 1 when the two maps match perfectly. It is seen rightaway that the *overall accuracy* corresponds to the portion θ_1 of K. It should be noted that many published papers have given erroneous formulations for the Kappa coefficient. The reader is advised to refer to the article cited above (Hudson, 1987) and not to introduce any other, above all if it is an earlier publication. This coefficient can be used for testing the quality of the delimitation of a single soil unit i. Then it is given by

$$K_i = \frac{Nx_{ii} - x_{i+}x_{+i}}{Nx_{i+} - x_{i+}x_{+i}}$$

There are other similar formulae. They too are useful for checking accuracy, such as the methods of Turk, of Hellden and of Short (Rosenfield and Fitzpatrick, 1986). The one proposed by Brennand and Prediger (Foody, 1992) would be more accurate for exact elimination of the influence of luck. It consists of replacing the quantity θ_2 in the formula by $1/n$, n being the number of categories taken into account in the matrix of confusion. But some of these formulae do not have the statistical qualities of Kappa whereby a variance can be determined and the probabilities be calculated.

The *Classification Success Index* (CSI) is a variant of the *Overall Accuracy* (Koukoulas and Blackburn, 2001).

5.3.5 Accuracy of a vector map

If necessary, the definition of the concept of *vector map* can be looked up in Chapter 6, § 6.2.1. In the case of a vector map, the problem is particularly complex because it is necessary to have a tool available for quantitatively measuring the similarity between two map delineations of whatever shape but presumed to represent the same thing (Fig. 5.10).

Many authors distinguish the error in position of the boundaries (termed *positional* or *cartographic error*) and error in the composition of delineations, which is the *thematic, attribute* or *descriptive error* (Veregin, 1989). But one

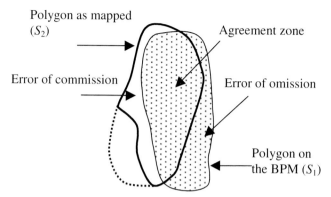

Fig. 5.10 Definition of the similarity between two map delineations.

must be careful when it is evident that an error in position generates an error in composition. Very exact rules must be given for permitting one to decide which error arises from a minor discrepancy in the soil of a delineation and which error actually constitutes one of identification of the composition (Chrisman and Lester, 1991).

Following the work that we did, we proposed the formula below, called *Similarity Pattern Index (SPI)*, for comparing two polygons:

$$SPI = (\text{area in agreement})/(\text{total area})$$

If the area of the first portion considered is S_1 and that of the second S_2, this amounts to writing

$$SPI = \frac{S_1 \cap S_2}{S_1 \cup S_2}$$

It can be seen that this index has three essential qualities. Firstly, its value is 1 if the areas perfectly superpose and 0 if they do not match at all. Secondly, it is symmetrical and gives the same importance to both areas, which is important when one has not a priori been considered better than the other (in comparing the work of two mappers, for example). Thirdly, the similarity between the two drawings decreases, although their area in common does not change (numerator constant) but the portion in one not covered by the other becomes larger (increasing denominator, see the dotted line in Fig. 5.10). This property is intellectually quite satisfying.

With this index we did various tests:
- Comparison of the same delineation made separately by many mappers; in this case, *SPI* is about 80 (80% similarity between the drawings).
- Comparison of several tracings done for the same zone by the same mapper by photo-interpretation, using if necessary different photographs from different flights. In such an exercise of measuring "fidelity', a few months is allowed between the different attempts to avoid

the memory effect of the drawing already done. The *SPI* is then close to 0.9.
- Comparison of the drawing done during a recent mapping and the drawing redone very carefully following a check. In such a case (measure of 'accuracy'), we found *SPI* close to 0.70.

The index we suggest a little later is generalized for comparing not only two polygons but also all the polygons of two maps. But the formulation is to be discussed.

5.3.6 Checking the composition of a unit

This verification applies important methods because they are used intensively, especially in USA, for verifying the quality of old maps and for updating them. Therefore, one should be sure of the quality and operational characteristics of these methods (Brannon and Hajek, 2000). For accurately checking the content of a unit the only means is to return to the field and take detailed observations. This is generally done through auger holes. There are three ways of selecting their location (§ 5.1.4):
- *systematic sampling* by locating the observation sites at the nodes of a grid;
- making transects, that is spacing the observations regularly along many straight lines; but to be able to usefully make statistical analysis of such an arrangement, these transects must be drawn at random; for example, the position of their middle point is calculated, then their direction; a variant of this approach consists of preparing a large number of transects located using arbitrary and conventional criteria (at a right angle to the slope for example), then to take at random a small sample of these transects for actual study (Young et al., 1997);
- using a random process for distributing all the auger holes at random (*random sampling*) (Dicks and Thomas, 1990).

The last two types of sampling can also be described as stratified if, for example, it is decided to work by Map Unit or Soil Delineation, putting in equal or unequal effort on these portions of the study area.

From the statistical point of view all these methods are almost equivalent (Stehman, 1992), but not so in practice. Location in the field of numerous points taken at random is tedious and difficult (Young et al., 1991). Positioning a grid also takes time. Therefore the transect technique should be recommended. The data thus obtained can be interpreted through various approaches. These have been examined and compared by Walter (1990) and Young et al. (1998).

Calculation of purity
This consists of examining in what measure the soils observed on a revisit to the area exhibit the characteristics the map and legend predicted for them.

Thus the purity of a unit will be said to be 80% if 80% of the area of that unit corresponds to what was expected. But everything depends on the taxonomic level taken as reference. For example, a map delineation in a study done in Texas (Nordt et al., 1991) had the following levels of purity; very poor for the *series* to which it was felt to belong (20% purity), but it matched the criteria for defining the *Great Group* of the American classification that this reference series exemplified (84% purity). Furthermore, it fell entirely within the *Order* that included this Great Group (100% purity). In other words, the less accurately the taxon is defined, the higher the purity, which is not intellectually satisfying.

Purity at the series level is calculated as follows: first an auger hole and the corresponding pedon are examined and compared with the recorded soil. We can define an *absolute purity* having a value 1 if no characteristic of the soil falls outside the range allowed and 0 in the opposite case. But this procedure is extremely selective. It is more useful to calculate the *average purity* of the pedon, represented by the ratio 'number of characteristics in the permitted range to the total number of characteristics'. From this the average purity of the map is easily deduced. This is the mean of purities of the units weighted by the corresponding areas.

Some authors refine the method a little by simultaneously taking into account the number and level of differences in the properties of the reference soils [Forbes et al., 1982 (in Nolin and Lamontagne, 1991)]. For example, the property which, without being perfectly representative, is nevertheless not very different from what is expected (one discrepancy class in the case of a variable having discrete values) only represents a *local variation* not counted as an impurity. However, when three characteristics of a pedon simultaneously come within 'local variations' that pedon is credited with an impurity that downgrades it totally or reduces its calculated purity. With the same thinking, Americans distinguish *named soils* (corresponding to the legend), *similar soils* (variations) and *dissimilar soils* (i.e., impurities).

Evaluation of the taxonomic composition
Beyond simple quantification of the purity, the method we have just seen enables analysis of the content of a map unit for evaluating its taxonomic composition. For example, is it not necessary to replace such or such *consociation* with an *association*? This type of question is often posed in USA, when old maps are updated.

Calculation of the heterogeneity of a map delineation
Conventional resources of descriptive statistics are used for this: calculation of variance, mean error and root mean square error (quantitative variables). Unlike in the preceding case, heterogeneity is judged without the need for referring to a legend or a classification.

When the data have been obtained from transects, the variance of the sample is calculated as follows (Young et al., 1991).
If
- Tr is the number of transects
- n_i is the number of observations in transect i
- N is the total number of observations
- x_{ij} is the value of one observation
- $\bar{y}_i$ is the mean of observations in the transect i,

$$\text{given by } \bar{y}_i = \sum_j x_{ij}/n_i$$

- m is the general average of all transects taken together,

$$\text{given by } m = \sum_i \sum_j x_{ij}/N$$

the variance is:

$$V_1 = \sum_i (\bar{y}_i - m)^2 / Tr(Tr-1)$$

But if the various transects do not have the same number of observations a weighted variance will be preferred:

$$V_2 = \sum_i \left[(\bar{y}_i - m)^2 (w_i)^2 \right] / Tr(Tr-1)$$

in which w_i is the weighting given to transect i and is equal to the number of observations in transect i divided by the average number of observations per transect:

$$w_i = \frac{n_i}{N/Tr}$$

Under these conditions, the general average m will come with a confidence interval

$$\pm t [V_1 \text{ or } V_2]^{1/2}$$

in which t can be read from Student's table at the desired level of confidence for a degree of freedom given by $Tr - 1$.

However, many statistical methods assume the populations from which samples are extracted to be normal. This is almost always untrue. Certain pedological variables have a distribution approaching log-normal. Above all, most of them have values strictly limited in the range 0 to 100, for example clay or organic matter content. Under these conditions, the calculated variances are often over-estimated (Young et al., 1998). That is why certain authors sometimes give expressions of the type: 'clay = 70% ± 35%. But it must be remembered that, in the case of large samples (>30), the sample means are well known to have a normal distribution even if the population they come from does not have this property.

Study of the spatial structure of heterogeneity
Two approaches have been explored. One of them is specifically applied to quantitative variables and falls within classic geostatistics. A variogram is prepared showing how the heterogeneity changes as a function of distance within the map delineation. The other method consists of examining the observation sites two by two and calculating the probability of occurrence or not of such or such property or horizon at a distance of, say, 50 or 100 metres.

The work cited earlier (Walter, 1990) gives some conclusions valid at least for the soils of Brittany, France. The average purity of the mapping units studied ranged from 61% to 77%. The variograms show that variation is large over short distances (<25 m). At the same time (but the two things are linked), the variability in a map delineation is almost as large as that across all the map delineations belonging to that same unit (this proves at least that the mapper has not made a mistake in the groupings). Lastly, the calculations pertaining to the lateral appearance or disappearance of horizons confirm the fact that beyond 100 m the soils are no longer spatially linked. This fixes an upper limit that should not be crossed when sampling schemes are drawn up.

5.3.7 Number of point data

Minimum sample size
It is necessary to calculate the number of samples that must be available to get a given accuracy in the estimation of an average value, considering the variability in the pedological environment of the property in question. The method has been in use for a very long time. The principle is as follows. In a soil unit, N samples have been drawn and a property has been determined on each sample. The mean value m and the standard error SE were calculated. If t is the value with probability P of being exceeded (to be obtained in each case from Student's table for $N < 30$ or the normal law for $N > 30$), the range on either side of the mean is

$$t \frac{SE}{\sqrt{N}}$$

Yet we look for the uncertainty to be less than e in relative terms, or $e = 5/100$, say, which leads to (if m is the mean)

$$t \frac{SE}{\sqrt{N}} \leq em \qquad \text{that is,} \qquad N \geq \frac{SE^2 t^2}{e^2 m^2}$$

The value of N thus determined is higher the greater the relative variation SE/m (coefficient of variation), the lower the risk of being exceeded is fixed (small P and large t) and the higher the accuracy demanded (small e). The method has disadvantages. The calculation might only be done a posteriori, after sampling and analytical determination. In other words, it is often too late to modify N if it is necessary. Also, if N is small, we have to presume a normal distribution for the variable, which is very erroneous. Let us keep in mind that a normal distribution corresponds to values that range from +∞ to −∞, whereas most soil properties have restricted range (for example, of percentages between 0 and 100).

However, calculation of N is often useful. Table 5.7 gives a summary of results drawn from the literature (Legros, 1978).

Table 5.7 Number of samples necessary for calculating a mean within 10% at the 0.05 probability level

Character studied	Size of sample		
	Minimum	Mean	Maximum
Organic matter	10	20	78
pH	1	1 or 2	16
Sand, silt or clay	1	2 or 3	6
Phosphorus	18	100	5100
Potassium	10	20 to 50	100

Accurate interpretation of these figures will need a detailed study of the conditions in which each of them was obtained. But they are sufficient to draw our attention to the fact that it is always necessary to do a few pH determinations, a few granulometric analyses, but many chemical analyses. Estimation of weight of carbon contained in a forest zone also involves a large number of samples, as high as 174 (Williot, 1995). In popular practice, it is difficult to attempt such a large sampling. The author last cited recommended that sampling points should be carefully chosen *after* many auger holes have been studied.

Number of profiles required
Am empirical method was earlier devised for calculating the number of profiles required, so that the accuracy is generally satisfactory for the specialists (Legros, 1973). It is based on the standards in the matter. We start from the principle that there should be more profiles the more the soils to be characterized, the larger the map area and the larger the scale chosen. The fineness of characterization F.C. is given by

$$F.C. = \frac{2P_R + P_V}{75E\sqrt{S_{tot}U}}$$

where P_R is the number of reference profiles, P_V the number of profiles corresponding to 'local variations', E is the scale (for example, 1/250,000), S_{tot} the mapped area in ha, U the number of Soil Units (not necessarily Map Units) and 2 and 75 are arbitrary coefficients.

Table 5.8 gives the values of F.C. for getting sufficient information at different scales.

Table 5.8 Reference values for the fineness of characterization

Scale	Recommended F.C.
1/10,000	60-90
1/25,000	70-120
1/50,000	90-160
1/100,000	100-200

Americans use the following certification rules for entering data in their MUIR database:
- At least three profiles described and analysed for each Map Unit (MU).
- One additional profile per block of 3000 acres mapped.
- Three transects of 10 observations each per MU.
- One additional transect per block of 3000 acres mapped.

5.3.8 Verification of representativeness

Representativeness of the reference profile
The reference profile presented in the explanatory report, which illustrates the characteristics of a unit, must be judiciously selected. To check if it is appropriate, we return to the area to do a series of samplings in the unit in question. Sufficient analysis is done later and lastly, it is examined if the profile is located in the middle of the scatter of points corresponding to the new samples. The following example (Fig. 5.11) pertains to the granulometric analysis of one soil unit in the middle valley of the Herault (Roques, 1990).

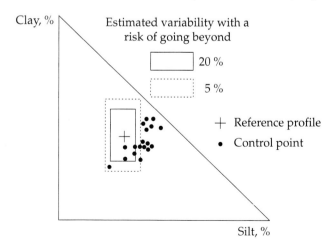

Fig. 5.11 Lack of representativeness of a typical profile in particle-size distribution (Roques, 1990).

Obviously the reference profile (marked by a cross) is not representative of the entire unit characterized by the other samples (shown by dots). In the diagram, the rectangles define, in theory, the internal variability of the unit. They will thus have to include all the dots with the risk, however, of crossing the boundary by 20% or 5% according to their dimensions. One can see that their size is quite correctly estimated, but they are poorly centred. In other words, the author of the map studied had properly appreciated the heterogeneity of the milieu but had not identified a representative pedon. There are two methods of avoiding such a situation:
- The mapper should first fight the natural tendency to select for reference profile not a common, that is, representative profile but a 'beautiful' profile, that is, an exceptional and rare profile. Who would not desire to draw attention to *the most* clayey soil found in the entire area to illustrate a unit of clayey soil? Actually this is not prohibited if the profile is presented as a variant and not as typical.
- Also, to avoid these difficulties it is necessary to exploit the possibilities of computer technology. Before writing the report and selecting the profiles to be described in it, the mapper can do statistics: calculation of the mean (quantitative variables) or the mode (qualitative variables) for all the properties of all of the horizons of all the soil units. The typical profile can thus be obtained. Some go further (Girard, 1983) and have formulated sophisticated algorithms for finding out the reference profile (true profile) that resembles most this calculated profile.

The representativeness of not only the reference profiles but also of the *Reference Area* will be discussed later (Chap. 7, § 7.3.4).

5.3.9 Discriminant character of the map

Verification of the discriminant character of a map is more often done by its utility than its accuracy. Be that as it may, the most traditional method is to use available information to compare the inter- and intra-unit variance for a certain property, clay content for example. It is a matter of demonstrating that mapping takes into account the variability in the milieu and that there is less variability in the characteristic within the units than between two units. This method may be useful, but it has several limitations:
- it is only valid for quantitative properties taken one by one,
- it is often impossible to verify the work of the mapper on the basis of so simple an approach. Actually the mapper is led to separate soil units that are nevertheless identical in such or such character, as in the case presented in Table 5.9.

Table 5.9 Logic of separation of three soil units on a map

	Soil Unit A	Soil Unit B	Soil Unit C
Depth	Large	Large	Medium
Texture	Clay	Sand	Sand

In such a case, proving by analysis of variance that the mapping does not have discriminatory value for depth or clay content is not really significant. It is necessary to reflect and group the units before starting the calculations.

Conclusion and perspectives

Mapping is a procedure for the expert. Some scientists are not happy with it and remark that it is very difficult to update old maps (Zhu et al., 2001). In fact, before modifying a boundary it is necessary to immerse oneself a long time in the work before being sure that the boundary can be altered without destroying the coherence of the whole. In the future, mappers must explain better the mental models they had used. On the other hand, great efforts have been made by specialists for checking the quality of their work. The map can be scientifically validated, which is a lot.

The portion of the mapper's experience going into the map is very limited. For example, it is quite misleading to represent all boundaries by the same kind of line, when some transitions are clear and abrupt but others are gradual and generally less accurately located. Even though the mapper has not provided precise indications in this regard, it will be possible to calculate the *contrast* between adjacent units by using a method codified by Villamayor and Huddleston (1991). This will, as a consequence, serve to represent the boundaries separating these units. We can also think of using the *Shannon entropy* as a measure of pedodiversity (Maclean et al., 1993) in spite of the difficulties expressed by Martín and Rey (2000) and Ibañez and Alba (2000). Also, even without coming back to an analytical presentation of the composition of Map Units (MU), it is usually not appropriate to represent all the delineations by a flat colour whether their contents are homogeneous or not (one or several SU).

But we should not get buried in useless details. Mapping admittedly has the objective of deciphering the variability in the soil mantle. We have presented the methods used for this. But it should be asked how it would serve Soil Science and practical agriculture to systematically work out soil variability at very large scale. Baveye (2000) very rightly posed this question when a review on geostatistics was published. The response of the authors of the review is also equally pertinent (Heuvelink and Webster, 2002)....

CHAPTER 6

COMPUTER PROCESSING OF DATA

In this chapter we set out to present the processing of graphic data pertaining to soil delineations, then processing of non-graphic data (composition of the units and profile data). Examples of publications will mostly be taken from France, but the latter part of the chapter presents a few of the most important international efforts.

Purely computer-related problems will not be touched upon. We wish to highlight here what in our opinion the main thing is: structuring the information in a logical manner for management in a computer system.

6.1 GENERALITIES

Soil mapping results in collection and use of a large volume of information. It is necessary, therefore, to ensure proper computer management of the data. This involves manipulation, quality checks, various treatments and protection operations. The three types of objects that need entry and processing in the computer are presented in Fig. 6.1.

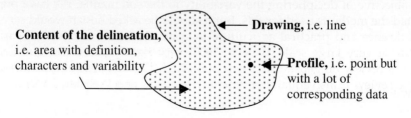

Fig. 6.1 Mapping objects that require computer processing.

It is thus necessary to manage through digital means: the map proper, the legend to the map and lastly the soil profiles collected together in an appended volume. Historically, these three objects have not been computerized simultaneously. The world over, systems for storing soil descriptions and analytical data were created first. This seemed urgent considering the volume of numerical data concerned. Later, when adequate tools made their appearance, computerization of soil boundaries was attempted. Lastly, interest grew in the treatment of map-delineation contents, that is, in processing of characteristics of Map Units (MU).

It should be asked rightaway if the management of areas is not redundant when profile data are being processed. Actually, profiles serve to define the content of the map units, at least when 'series-level' mapping is done (Chap. 1, § 1.4.3). In fact, many variables used to describe the profile are also found in the characterization of the corresponding MU: stoniness, depth, etc. But profile characteristics and MU characteristics are not completely equivalent for several reasons:

- the profiles correspond to accurate point data (say, 25% gravel); the MU's include the concept of spatial variability, for example, 5-35% stoniness, 60 to 100 cm depth, etc.;
- also, some soil properties with unknown variability have only local meaning (for example, size of roots); on the other hand, other characteristics only have meaning if a certain areal extent is considered (for example, type of relief).

When all is said and done, the non-graphic data characterizing the map units represent a synthesis limited to essential things. They integrate the idea of variability and summarize all the data known to the mapper: profiles as well as auger holes and surface observations. They should therefore be managed separately. In this context, soil profiles selectively bring a little accuracy and constitute local examples of the content of the units. One final argument could be advanced for the usefulness of a special computerization of the content of the map units as will be made clear later (§ 6.2.3). The mapping unit is no doubt the best computer object to tie up all the others: associated mapping boundaries, soil profiles included in them, map as an entity grouping together a series of contiguous map units.

6.2 TREATMENT OF GRAPHIC DATA

We will not study the treatment of graphic data here in detail. Actually, the tools created for this, *Geographical Information Systems* (GIS) deserve entire specialized books devoted to them. A good bibliography on the subject can easily be found through a reference book (Burrough and McDonnell, 1998).

There is also an excellent book in French (Collet, 1992) published by Presses Polytechniques et Universitaires Romandes. Here we shall limit our intention to some useful elements for presenting mapping applications.

6.2.1 Basic principles of digitization of maps

There are two ways of organizing graphic data on the computer.

The first consists of tracing manually with a cursor the boundaries of the map delineations. These are then entered in the form of a series of straight lines as short as desired. These lines actually are vectors, whence the term *vector data model* given to the procedure. There is a certain logic in proceeding this way in soil mapping because the lines introduced by the mapper remain approximately the lines in the database.

The second system consists of presenting the map to be digitized in front of a camera that will scan it by moving across it following a series of parallel lines (*scanner*). This enables us enter the soil type found at each point or, more correctly, each elementary area corresponding to a regular grid superposed on the map. The elementary areas thus created are in some sort of *pixels*. In the database, the information exists in matrix form, at least as a first approximation. This system corresponds to the *raster model*, which Collet proposes to term '*mode image*' in French. In theory, the digitization is automatic. But in reality, there are all sorts of precautions to be taken and corrections to be made, so much so that this second procedure is not quicker than the first in regard to data entry.

Links between the two approaches can be established, and a map originally digitized in the *vector mode* can be transformed to the *raster mode*. The reverse is also possible, but is more difficult. All this enables us to use the soil map in combination with available data in vector form (for example, geological boundaries) or in raster form (pixels indicating the type of vegetation, say). The possibilities are presented in Fig. 6.2.

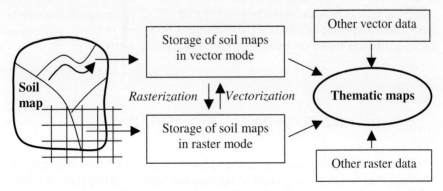

Fig. 6.2 Organization of storage and use of graphic data.

Conversion naturally distorts the original map a little, whether in raster mode or in vector mode, introducing an inaccuracy that in the second case becomes slighter as the pixels become smaller (Congalton, 1997).

6.2.2 The vector system

Logical structure of digitization
Digitization assumes that the objects to be stored in the computer would be correctly identified on the soil map (Fig. 6.3).

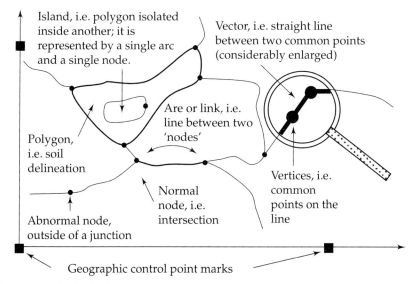

Fig. 6.3 Principal elements appearing in vector mode in GIS. Directions of the vectors and arcs have not been shown for the sake of simplicity in presentation.

Polygon. A polygon is an elementary area enclosed by a series of vectors. An elementary area or soil delineation digitized in vector mode thus becomes a polygon.

Nodes. Nodes normally correspond to intersections appearing on the map. Actually the systems tolerate, at least provisionally during digitization, unusable nodes placed outside the intersections by error.

Arc. An arc is the portion of the polygon boundary found between two nodes that normally correspond to two intersections. Also, if the polygon is isolated in the interior of another, larger one, its perimeter is then formed of a single arc with arbitrary origin (which is also its end point). This is somewhat like a snake swallowing its tail! In all cases the direction of travel is arbitrarily selected. But starting from there, the arc finally becomes directional and so too the vectors constituting it. This enables us define the origin of the arc and its end point to tell which polygon is on the left and which on the right.

Vertices. The end points of all the vectors included in an arc are vertices.

During digitization the *cursor* placed on a *digitizer* and, guided manually, traces the arcs. The vertices are digitized by a single 'click'. The nodes are marked with special clicks done with a different button to indicate that they are special points. The procedure can be more or less automated.

Topological model
In the graphic database, the information is automatically structured in such a way that it can easily be retrieved and used. For example, let us see how the small map below (Fig. 6.4) is described in the system by four tables called *arc table, nodes table, vertices table* and *polygon attribute table* (Table 6.1).

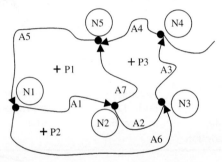

Fig. 6.4 Representation of a map in a vector GIS (*P* = polygon, *N* = node, *A* = arc). The '+' symbols are points marking the centroids of the soil delineations and correspond to polygon identifiers.

Let us take an example. Suppose one wishes to know if the soils of Map Unit 45 are locally neighbouring those of Map Unit 13 and, if so, the length of common boundaries. For this, each arc is considered (see Arc table). If the arc separates polygons belonging to the Map Units in question (see Polygon Attribute table), the corresponding length is taken (see Vertices table). These searches are considerably automated. Mainly, the user only has to fill up and query the Polygon Attribute table.

Such a structure is called the *topological model*. What has been described above is not the only one possible. There are others, but the principles remain similar. Besides, software developers are not much pressed to explain in detail which structures they have found to be more efficient than their competitors.

Table 6.1 Organization of the information within a GIS

ARC TABLE: relationships between polygons and nodes			
No.	Left Polyg.	Right Polyg.	Beginning Node	Ending Node
A1	P1	P2	N1	N2
A2	P3	P2	N2	N3
A3	P3	—	N3	N4
A4	P3	—	N4	N5
A5	P1	—	N5	N1
A6	P2	—	N1	N3

NODES TABLE: x and y coordinates of all the nodes		
N1	—	—
N2	—	—
N3	—	—
N4	—	—
N5	—	—

VERTICES TABLE: x and y coordinates for all the points, corresponding arcs			
V1	—	—	A1
V2	—	—	A1
V3	—	—	A1
Vn	—	—	A2

POLYGON ATTRIBUTE TABLE (information on soils)			
Map Unit	Ex: pH values	others
P1	—	—	—
P2	—	—	—
P3	—	—	—

This table is much longer than the others.

Map overlay

Several maps pertaining to the same area, corresponding to, say, the topography, vegetation, soils, etc., often have to be used at the same time. These maps are then considered as data *layers*. They can be superposed automatically. This is *map overlay*. As for computer processing, the superposition corresponds to creation of a new topological model and a new map. The new nodes are automatically detected and identified (Fig. 6.5). It is seen that the superposition of two polygons can give five polygons and sometimes many more. These new polygons are described with all their features as

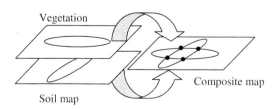

Fig. 6.5 Principle of map overlay.

indicated in Table 6.1, including their composition as had appeared in the original layers, in this case soil type and vegetation.

We can repeat the operation and combine the map obtained with a third, then yet another and so on. But the number of polygons tends to increase rapidly, even when the polygons exactly match from one map to the other. Actually, GIS are too accurate in a way. They do not recognize that two map delineations seemingly out of step by a few fractions of a millimetre are, in reality, identical. They generate polygons that are too small to be visible on the composite map but still encumber the internal tables of the system unproductively. These are *sliver polygons* (Fig. 6.6).

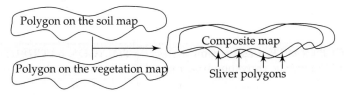

Fig. 6.6 Explosion in the number of polygons in map overlay.

The equation of McAlpine and Cook (Veregin, 1989) enables us predict approximately the number of polygons that will be obtained by superposing two maps:

$$N = \left[\sum_{i=1}^{c} P_i^{1/2}\right]^2$$

where

N = number of polygons expected
C = number of maps superposed
P_i = number of polygons in map i

For example, in the case of two maps, one containing 150 polygons and the other 50, we can expect a superposition to give, after reducing sliver polygons, $[(150)^{1/2} + (50)^{1/2}]^2 = 373$ polygons. This is nearly twice what simple 'addition' would give. Sliver polygons are suppressed by asking the system to eliminate every delineation smaller than a chosen value. This results in certain lines being erased. Still, we need to know which lines. Some GIS erase the longest arcs of the smallest polygons. But this procedure is not necessarily correct. Sometimes the longest line represents the firmest boundary. For example, it might be a river meander on which might lie a short soil boundary defining a very small delineation. The most modern GIS allow us to choose the data layer containing the boundaries that must not be touched, for example those corresponding to the topographic base map.

Error propagation

Let us consider as example the following problem (Legros and Bornand, 1989): suppose it is desired to combine a soil map, a slope map and a vegetation map to answer the question 'Where does one find deep soils used for pasture on level land?'. Let us take the case of a small portion of the area, a secton, which answers the question. In our drawing (Fig. 6.7), the section is defined by arbitrary boundaries in conformity with all three maps and chosen so that they will not be the map boundaries crossing it. This portion of the study area, however, contains impurities that the mappers have not seen or have chosen to neglect.

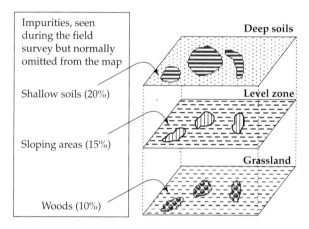

Fig. 6.7 Portion of the study area showing impurities. They are traced on the drawing but do not appear on a small-scale map.

The relevant break up is given in Table 6.2.

Table 6.2 Break up of areas in a portion of the study area

Section	Principal component	Impurities
On the soil map	Deep soils (80%)	Shallow soils (20%)
On the topographic map	Level area (85%)	Sloping area (15%)
On the vegetation map	Grassland (90%)	Woods (10%)

The computer system (or the mapper working at a light table) will select this area as relevant. Actually, the impurities do not appear on the maps; this secton is thus fully composed of deep soils situated in the level zone and under grassland. To what extent is this accurate? There are several possible answers.

- If the impurities match on the three maps, that is if there are shallow wooded soils on sloping lands, it can be estimated that 20%, 15% and 10% are the approximate measurements of the same reality as perceived by different natural-environment scientists. In this case, the area in answer to the question is $100 - (20 + 15 + 10)/3$ or 85% of the secton. In the same perspective, it can be estimated that impurities represent at the most 20% of the total and that the relevant area is therefore 80%.
- If the impurities do not match, that is if the woods are never established on shallow soils and on slopes (because of difficulties in mechanized agriculture, for these woods are for Christmas-tree nurseries), the area in reply to the question tends to become $100 - 20 - 15 - 10$ or 55%.
- If the impurities have no link amongst themselves, the probability of finding them superposed is equal to the product of the probabilities of their occurring independently. In this case, the area in answer to the question is $80 \cdot 85 \cdot 90 / 10,000$ or 61%.

To sum up, in the secton the area that answers the question thus depends on the kind of relationship amongst the variables considered. This leads to two important comments. First, it is preferable to know the relationships amongst the manipulated variables for working properly. This is a general thought for all scientific study; it is valid not only for the example chosen. Second, the superposition of several maps brings down the quality of the result rapidly. With four maps or three maps with accuracy a little lower than that in the example presented, it would be easy to fall below the 50% mark, that is to select sectons the composition of which would be predominantly different from what is expected!

Calculations in the vector mode. The digitized maps give rise to three kinds of calculations. To begin with, it is essential to do the simple manipulations some of which are possible when the map is a single sheet of paper. Then, it must be possible to do certain statistical calculations that directly enhance the usefulness of the stored information. Lastly, spatial calculations directly linked to the GIS concept can be tackled.

Data handling

The procedures cited are those of the ArcInfo software:
- Removal of distortion,
- Projection changes.
- Scale change,
- Rotation and translation of coordinates,
- Suppression of polygons that are too small (ELIMINATE),
- Line dropping between two adjacent polygons with similar composition in the framework of a given application (DISSOLVE),
- Edge-matching to join adjacent maps (MAPJOIN),
- Zooming,
- Smoothing.

Data characterization

Area of elementary soil areas. In the vector mode, the system uses the formula below, where x_i and y_i are the pair of points defining the ends of the vectors concerned:

$$S = \frac{1}{2} \sum_i |(x_i y_{i+1} - x_{i+1} y_i)|$$

Number and size of the elementary soil areas. It is sometimes useful to examine if a Map Unit is cut up into many separate pieces. This is the concept of *spatial integrity*. It can be estimated by calculation of the average size of the elementary soil areas of the unit considered or by display of the histogram linking the size of soil areas and the corresponding frequency.

Shape. There are various shape parameters. Table 6.3 (Fridland, 1976) gives the best known.

Table 6.3 Shape parameters for elementary soil areas

No.	Formula	Name	Author
1	A/L^2	*Shape quotient*	Horton
2	$4\pi A/P^2$	*Roundness quotient*	Miller
3	$(A/\pi)^{1/2}/L$	*Elongation quotient*	Schumm
4	$\pi L^2/4A$	*Ellipticity quotient*	Stoddart

where A = area of the delineation
 P = perimeter
 L = length corresponding to the greatest elongation.

For interpreting these expressions, it is necessary to refer to the circle, which has the largest area for a given perimeter. Thus, in formula No. 2, a circle will have the value indicated below. On the contrary, an infinitely oblate area will have the value zero.

$$4\pi \frac{\pi R^2}{(2\pi R)^2} = 1$$

Nowadays some scientists try to use *fractals* for characterizing the shape of delineations. Let us postulate, using the same symbols as in Table 6.3,

$$P = kA^{0.5d}$$

or, if preferred, $\log P = 0.5d \log A + \log K$.

This enables us represent the values of P in relation to the values of A in a log/log diagram. This is easy because vector-mode GIS give values of P and A for each map delineation. If the points obtained are nearly aligned, it means that d has a fixed value, the **fractal dimension**. In such a case, the complexity of shape of the delineations is preserved, whether they are large or small (Arnold, 1990; Burrough, 1993).

Spatial analysis

Buffering. Buffering involves tracing the line that connects points located a given distance from a given point or line or even an area. In the first case we get a circle, in the second a corridor and in the last, a fringe. In sum, this is a *buffer*. There are numerous applications as shown in Fig. 6.8.

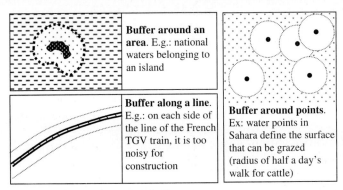

Fig. 6.8 The three different types of buffer.

In mapping, it is possible to represent the boundaries of Map Units with a buffer of size that increases with inaccuracy (Lagacherie et al., 1995).

Contiguity analysis. Such analysis involves expressing the fact that units appearing on a map are often or even always next to each other. This poses no problem for calculation if the map has been entered in a computer system that can construct an appropriate topology. We have seen this above. This results in a *contiguity matrix* for the units (Fig. 6.9), which enables us mathematically define the catena concept.

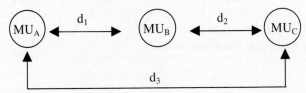

Fig. 6.9 Expression of contiguity in a catena. MU_A, MU_B and MU_C represent the Map Units A, B and C in the order they occur in the field; d_1, d_2 and d_3 are the numerical estimates of contiguity, for example the length of arcs in common between two units divided by the perimeters of all the polygons belonging to these units.

The value of d_1 and of d_2 must be high and of d_3 very small if not zero. Thus the proportion in which the catena described—and a little bit idealized—by the mapper actually occurs in the field is measured. Another application of this concept of contiguity was proposed by Grzebyk and Dubrucq (1994). This was to verify that the soil units at the scale of the map presented were roughly ordered west to east, in relation to the morphopedological history of the region studied (basin of the upper Rio Negro in Amazonia).

It was Fridland who first saw the use of this analysis of the structure of the soil mantle. But at that time (1976), the method of calculation did not allow him to complete the procedure by presenting numerical applications.

6.2.4 Outline of the raster mode

Principle

The raster mode, we have seen, consists of cutting up the map into small equal elementary areas, square and at times hexagonal, termed pixels, then recording the corresponding information in matrix form. Remote-sensing data are presented this way. The system looks simple. It is easy, for example, to count the squares that in the matrix correspond to such or such Map Unit to know its total area. In reality, this procedure has one major disadvantage: much space is required to store the information. For example, a large homogeneous polygon will contain many identical pixels, all of them described, whereas in the vector mode, it will suffice to characterize the perimeter of the polygon by a few points.

Solutions have been found to overcome this disadvantage. Suppose the lines in the matrix of data are presented in the form (A, B and C being the Map Units):

A, A, A, A, A, A, A, A, B, B, B, A, A, A, B, B, C, C, C, C, C, C, C, C, C,

they can be replaced by expressions of the type

8A, 3B, 3A, 2B, 9C

called *run-length codes*. But there are many more sophisticated methods, in particular the *quadtree* and *septtree* methods. Explanation of these solutions will fall beyond the framework we have set for ourselves. Let us only give the principle of *quadtrees*. The map is divided into large pixels, then in the sectors where the pixels are too large to fall in a homogeneous area they are divided into quarters, which operation can be repeated many times. Fig. 6.10 shows how one can draw by this method (dots) the shape of an *elementary soil area* (thick line).

A system of numbering these square delineations has been devised

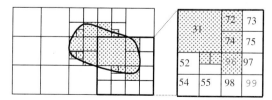

Fig. 6.10 Use of the quadtree and numbering of pixels in a raster system.

(Fig. 6.10, right). These are the *Morton numbers*. They can actually be organized in different ways. They are conceived so that, by using a special addition table (the use of the word 'addition' is restricted here, one must say 'relational'), addition of the value '1' to a box number gives the number of the next square to the right. Thus we find what we would have got by referring to a conventional matrix with the rows and columns numbered in order. This serves to make a translation of the map, to move it on the display screen, for example. In the same way 'multiplication' tables also have been devised to enable us go from one square to another on a different row without recourse to trigonometric functions to rotate the map. There is genius indeed in the concept of quadtrees! Furthermore, new ideas regularly appear (Chang et al., 1997).

Spatial analysis

Moving window. It is often useful to define on a map a small rectangular or, better still, square area representing say 9 (3 × 3) pixels or 25 (5 × 5) pixels (Fig. 6.11). This is the *moving window* that is moved in such a way that its

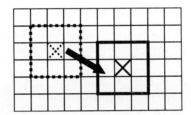

Fig. 6.11 Moving window.

central pixel passes through all the pixels of the map.

This operation is also termed *convolution*. We will see again and again the ways such a device can be made use of.

Contiguity analysis. In raster maps, the notion of contiguity is used for different purposes pertaining to remote sensing. It is outside our concerns to examine them in detail. We should thus be content with mention of digital filtering. The principle is as follows (Caloz, 1992). A moving window of at least 9 pixels is moved across the original digital image (= matrix of pixels, each having a coded value). Each pixel of the mask contains a weighting. Then the new value that will be taken in the original matrix in the centre of the window is defined, taking into consideration the weightings and the value of the pixels around it. In practice, each weighting on the mask is multiplied by the value located under it in the matrix and the values are summed; this total is then divided by the sum of the weightings and the result obtained is introduced into the pixel located at the centre of

0	0	0
0	1	0
0	0	0

X

1	1	1
1	8	1
1	1	1

Y

1	1	1
1	-8	1
1	1	1

Z

Fig. 6.12 Types of filters for the processing of pixel matrices: X is a neutral filter (no modification), Y a low-pass filter and Z a high-pass filter.

the mask. For example, in Fig. 6.12, the mask X does not change the original matrix, whereas the mask Y will give as much weight to the environment of the central point as to the point itself.

Low-pass filters (Y) result in a smoothing and *high-pass filters* or *edge filters* (Z), on the contrary, serve to bring out boundaries (Collet, 1986). In this last case, far from the soil boundary, that is when all the pixels of the matrix below have almost the same value, the weighting proposed leads to a result close to zero. But if the window is found to span a boundary with differing values on either side, a high value is obtained, positive or negative, which is injected in the matrix below. It is sufficient then to neglect the sign and introduce a threshold for finding out a matrix containing only 0's (map delineations) and 1's (boundaries).

It is clear that all sorts of windows-weightings can be used to give various results. For more information, the reader may refer to a specialized book (Schowengerdt, 1983).

Connectivity analysis consists of finding all the pixels confirming a defined neighbourhood property for a reference pixel, for example: pixels located an hour's drive by road from a given starting point. Actually, two applications appear more particularly important in the domain of our interest:

- Searching for all pixels situated upstream of a given pixel and likely to feed the pixel with diffuse or concentrated runoff (this is the concept of catchment); the principle of calculation may be found in the book by Collet (1992).
- Searching for all pixels that limit visibility in all directions around a given pixel. This determines the horizon, an idea used in calculations of solar energy when it is desired to take masking effects into account. Another application deserves mention. This is the representation of each pixel by a colour depending on the number of neighbours that can be seen from (or around) it. This enabled distinction of open country (plains) from closed landforms (hills or mountains), the latter being picturesque landscapes (Falque, 1994).

A Digital Elevation Model (DEM) is required (Chap. 8, § 8.6.1) for processing the connectivity problems cited. In this we leave soil mapping only to take interest in information very useful for planning land use.

6.3 PROCESSING NON-GRAPHIC DATA PERTAINING TO THE MAP

We shall now examine the different steps in development of the data bank for storing and managing non-graphic data pertaining to the soil map. These data are sometimes termed semantic data. Semantics being the study of the meaning of words, the term is not ideal. Tabular data, which closely matches the reality of things, would be the preferred term. We shall leave aside purely computer-related problems but will emphasize problems pertaining to entry of data in the bank or use of the bank.

6.3.1 Definition of technical objectives

It was shown in Chapter 4, § 4.3.2 that the first thing to do is to define the contents of the bank to be established and to indicate how we can go from the object-soil (reality of the soil) to the image-soil (collection of variables and states of variables that describe that reality). We saw that options and variables were very many: synthetic or analytical approach, advanced or primitive language, quantitative or ordered qualitative data, etc. Combination of all the possibilities leads to the impression that there certainly are millions of (slightly) different ways of describing a soil. Undoubtedly there is no ideal solution.

The system finally accepted must be a compromise proving its practical efficiency in different areas:
- *Capability to give a satisfactory account of reality*. The mapper must have the feeling that he can describe in a simple way what he sees without repetition and without unintentional omissions. The problem of repetition is quite difficult to solve. For example, if the following variables are used:

Vegetative cover	**Crops**
Tall woody species	Perennial
Short woody species	Annual
Herbaceous plants	Bare soil
Bare soil	

We risk facing descriptions defining 'bare soil' twice. Yet it is necessary, for each variable, to make provision for instances where the characteristic tested is absent! This example, rather simplistic, is enough to make one aware of one essential fact: the variables constituting the descriptor system are indeed independent in the mathematical sense, but complementary. They form a coherent whole. To wish to add something or subtract something from such a system presumes a thorough knowledge of it to avoid the risk of general disorganization.
- *Possibility of easily processing the information*. The pre-imposed choice (see the two variables above) can appear as a severe constraint, but it

is necessary. Without the pre-imposed choice there will be no standardization and uniformity possible in soil descriptions. The information entered freely on paper, later even stored in a computer, cannot be processed easily and the data bank would lose a large part of its usefulness. But though pre-imposed choices and structured descriptions are necessary, it is advantageous to check that processing is possible. In this matter, experience is irreplaceable and the first data banks established in soil mapping allowed putting up with initial problems.
- *Other criteria*. Other criteria have also to be considered. For example, the descriptor system finalized must avoid fatigue in the user. It must be designed in a manner appropriate to preparation of efficient soil data sheets and ergonomic input screens.

To sum up, design of the descriptor system is an essential step in establishment of a database. It must be carefully done. There is often a tendency to underestimate the time this step will take.

6.3.2 Structuring the data

A computer system that stores and manages relational data is called *Relational Database Management System (RDBMS)*. In this RDBMS the data have a particular organization and are related amongst themselves in such a way that one can directly obtain data using these relationships. For example, which are the soil profiles described in a particular region having a clay content greater than 30%?

There is no point in developing software because they are available and have been perfected by very large groups of high-level specialists. Thus what remains is to select a software and teach it to recognize and manage as desired the data entrusted to it.

To understand the organization of a database, we have to return to the structure of the first data banks. In them, the data were stored in a single file. File design allowed reserving space for each kind of data within it, which could be considered the characteristic record of the bank. Let us take the example of mapping studies done by special-purpose companies; the content of the CARTE file was the following:

MAP
— Map No.
— Map references
— Name of company
— Company references
— Other data

But this procedure leads to duplication of information. Thus the references for the company that has done the work (address, fax and telephone numbers, etc.) have to be repeated in each case although the company in question was entrusted with the preparation of three-fourths of the maps in a batch of 400. There is, therefore, loss of time during data entry. Still more serious, if the company were to change address, it would be necessary to correct 300 entries! This is why it was decided to structure the information and to create several files amongst which correspondence would be established. In this specific case, the following pattern is first taken up:

COMPANY	MAP
— **Company No.**	— Map No.
— Name of company	— Map references
— Company references	— Other data
	— **Company No.**

Linking of the two files is done, for example through the company number indicated on each soil map. But this presentation is not yet complete. Actually, it must now be asked if the names of companies that have not yet done any soil survey but are liable to do so will be introduced in the data bank. This is not absurd, it results in entry from the beginning and in one step the list of companies listed in a particular professional registry. If we follow this procedure, a given company would have done between 0 and n soil surveys. The numbers 0 and n represent the *minimum cardinality* and *maximum cardinality*. Put in another way, if we wish to express the fact that a soil survey is done by at least one company and by at the most two companies, if the companies agree to collaborate in the work, the cardinalities are indicated as follows:

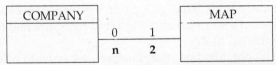

But the pattern has still not taken its absolutely final form. Actually, some data do not strictly characterize the company or the soil map, but are relational properties, that is when taken two a time. An example is the name of the company employee who did the survey. Thus a third file should be introduced, placed between the two mentioned above, when all is said and done, we end up in the *entity-relation model*. In the example given, it takes the form in Fig. 6.13.

Fig. 6.13 Entity-relation model in an RDBMS.

In practice, we are not limited to three entities and two relations. It will be seen later that the model, also termed *conceptual data model* or *relational data model*, is often much more complex. Also, there are many ways of breaking down a particular system into a model of this type. Then the best possible breakdown must be found: the most efficient for computer processing and the most natural for the user. Lastly, breakdown is one thing, reconstruction is yet another. For example, to draw up a table listing surveys with corresponding authors and companies, we must be able to simultaneously use the three files presented above. This has led to developing operations such as linking, projection, restriction, etc., on the tables. To learn more, the reader should refer to a specialized book (Tardieu et al., 1987).

6.3.3 Example of the French 'DONESOL' system

To present an RDBMS concerning soil maps, we shall take as example the DONESOL system developed at INRA-France (Gaultier, 1990; Gaultier et al., 1992b) and later revised to DONESOL.II. This system was conceived to attain three objectives:
- recovery and conversion to relational form of the data earlier collected in the STIPA-1979 system;
- routine processing of the non-graphic data pertaining to soil maps;
- general structuring of map data in a manner to ensure relation of graphic and non-graphic data.

Conceptual data model
The DONESOL system is designed by application of the method explained above (entity-relation model). Its structure is organized in three parts (Fig. 6.14). The tabular data describing the soil map were rearranged at the centre of the relational data model. Profile data were placed on the right and data on the environment were placed on the left.

Organization of graphic data in the model
Graphic data are described by a sequence of four levels:

Survey. The survey defines the project in question. In some countries, USA for example, surveys are classified by *State* and now by *Major Land Resource Region*.

Map Units (MUs). Some people refer to these units as *Soil Map Units*. They are generally correct even though the accuracy seems totally useless in the context. But they are wrong at times. We have seen that the object delimited and represented on the map can be, at least in some cases, not a *Soil Unit* but a *Landscape Unit*, in the definition of which the soil occurs only partly or not at all.

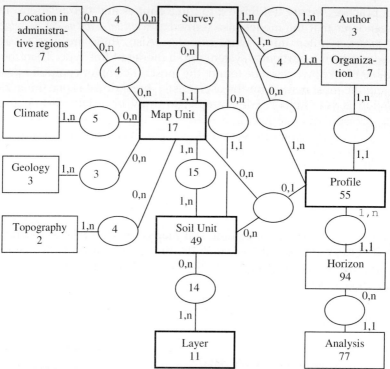

Fig. 6.14 Conceptual data model of DONESOL (Gaultier, 1990) updated in DONESOL.II. The numerals appearing in the different entities correspond to the number of variables in question. The names of files represented by ellipses are not defined because they do not give any useful data here.

Soil Units (SUs). Map Units might be complex and contain several Soil Units in association. In some countries (USA, Canada), this category is termed *Components*; it also specifies the soils in question and their proportion within the Map Unit. But the Map unit may also consist of just a single SU (cf. consociation, Chap. 1, § 1.4.3). In this single case the SU shows all its boundaries: they are those of the MU! However, in the database, MU and SU continue to be distinguished. Some characteristics are described at the MU level and others at the SU level. This is of no importance.

Strata (layers). The concept of strata is close to that of the horizon and pertains to the superposed layers appearing in the profile. But strata are distinct from horizons in two ways:
  • Strata correspond to data at the scale of the entire unit (mean or modal values with associated ranges. In other words, the stratum allows characterization of spatial variability.
  • Strata help to simplify description of the profile. A stratum may correspond to a horizon or even a group of several horizons appearing

in superposition. For example, the various B horizons of the profile are grouped together under a single stratum.

This procedure results in some flexibility and enables use of DONESOL for very detailed studies (one stratum per horizon) or for synthesis studies (a stratum corresponds to a group of horizons), or for an entire profile defined as a whole by some major characteristics. *Surveys*, MUs, SUs and *layers* represent a kind of hierarchy in pedology. But the corresponding information is not processed hierarchically within DONESOL. The entities are only distinguished by their content and by the relationships that bind them to such or such neighbour. The direct relation between Surveys and SUs is strictly conceptual. An SU is conventionally identified by a specific number associated with the Survey number. Thus, to state that an SU belongs to a particular MU is, by the same token, to state that the SU belongs to the Survey whose number is contained in the identifier of the MU mentioned.

Organization of profile data in the model

The structure adopted in DONESOL can be surprising. Mappers have the habit of thinking that their profiles comprise two blocks of data of equal importance: the horizon descriptions and the corresponding analytical data. In fact, one could see the relevance of the relational pattern: an analysis or, more correctly, a specific group of analyses correspond to one horizon and to only one. On the other hand, a single horizon can be analysed many times, every year for example, to follow the change in levels of nutrients.

The problem of connecting profile data with map data must also be considered. Examination of the entity-relation pattern shows that it is obtained at different levels:

- relation with the MU to indicate in which geographic unit the profile in question occurs;
- relation with SU to indicate exactly what type of soil is considered, when the MU is complex and includes several SUs;
- relation with the Survey so that one could enter it in DONESOL and thus protect a collection of profiles already made, without being obliged to describe at the same time the geographical distribution of soils (perhaps there is no map, the map has been lost or the report does not include all the data necessary for a proper description of the MUs and SUs). But generally, it is often very difficult to establish a data bank with old data. It is much better to enter the data when they are collected in the field. This is more efficient and less tiresome. A profile can correspond to several Surveys. This is because a profile done in a detailed survey is often reused in synthesis mapping pertaining to the same region.

Organization of environmental data

Once the choice is made as to the types of environmental data to be introduced in the base map, it must be decided at which level in the Survey/MU/SU hierarchy they should be placed. For example, we can define climate at the level of the Survey or at that of the MU. By being placed at the Survey level, it is accepted that the climate does not vary in the entire map area and one is obliged to never introduce subtle differences at that level. On the contrary, by placing it at the scale of the MU, we retain the possibility of encountering complex cases (concomitant variation according to altitude of soils and climates in mountains). In a traditional data bank, the second case will result in increasing the volume of data to be stored. Actually, it will be necessary to repeat the same climatic data for all the neighbouring MUs influenced by the same climate. In a relational system, however, duplication is almost totally avoided. It is sufficient to enter the characteristic parameters of the climate of each meteorological station in the study area and then to identify, for each MU, the reference station through a simple code. After all is said and done, the environmental data in DONESOL are linked to the MUs. This enables the system to adapt to processing of survey projects whether conducted at large scale or done at small scale. It should be noted that it was not logical to go further and link the environment to soil units (SUs). Let us remember that SUs are actually not delimited spatially within the MUs except when the MU and SU are identical. Thus we cannot classify the geographical environment at this level. The data pertaining to location according to administrative units form a particular case. Actually, the link is to the Survey and to the MU at the same time. This means that we are given the option of saying which administrative units the study area (large scales) or each map unit (small scales) falls in.

It is not pertinent here to present the content of each variable in DONESOL. Let us remember that they number as many as 447 (Fig. 6.14). The volume of data is especially large in some cases, particularly for climate, because many characteristics have to be provided for it. Descriptions of MUs, of SUs and of horizons also require use of many terms pertaining to numerous variables. Lastly, the catalogue of analyses liable to be done is very large.

Computer organization

DONESOL was developed on ORACLE, workstation version. A PC version runs on ACCESS.

Development of DONESOL required considerable work in the INRA laboratories. For example, there are 25 input screens. Other screens are needed for corrections and querying. The total number of screens is nearly 80!

Querying of the data bank can be done in three ways:
- using the input screens (the system is asked to find all the records having a particular value for a specified variable entered in a given input field);
- using pre-established queries (the system can process about forty type queries);
- constructing a specific query by using SQL macro language.

Linkage with graphic data
Non-graphic data (on ORACLE or ACCESS) should be linked up with graphic data (on ArcInfo) so that they can be used together. Let us examine some types of problem that we might have to resolve.

From non-graphic data to graphic data. Suppose, for example, we wish to draw the map of areas with pH lower than 7.0. For this, a query is set up in the non-graphic data so that the list of relevant MUs is returned. A file is created and transmitted in appropriate form to ArcInfo (Gaultier et al., 1992a) and then the corresponding polygons are located. The arcs separating adjacent polygons belonging to different soils but in the same pH category are suppressed. The procedure can be automated if the software are interfaced, or they can be executed step by step by an operator. The latter procedure takes a long time. But in mapping, since many months and at times several years are necessary to do a survey, we are rarely in a hurry!

Furthermore, the problem is complicated by MUs that comprise several SUs. We are then asked to give complex answers, for example to produce a map of the polygons in which more than $x\%$ of the area has pH lower than 7.

From graphic data to non-graphic data. The first useful application is to identify the nature of all the characteristics of a map delineation identified on a screen. When the delineation has been marked, the graphics software gives its coordinates. These are memorized by the user or furnished to an interfaced program. The information is transmitted to the semantic database, which then returns all the required details on the delineation in question. In fact, the most useful applications are the following:
- *Profile coordinates*. Soil mappers should, as a rule, determine the coordinates of their profiles and record them. This is easy with a GPS. But it is a bit tiresome and is often neglected. In fact, it can now be largely automated. The representative profile locations are marked on a topographic base, then digitized on a digitizing table. The graphics program will then calculate the exact coordinates with reference to x- and y-axes. These coordinates are organized in a file, then transmitted with the profile numbers to the RDBMS for entry in the proper place.
- *Locating the profiles in the MU*. When profiles are described in the field, it is not always known which final numbers will be taken in the

database for the Map Units in which the profiles had been observed. These numbers are not required from the field workers even though they are advised to set up Table 5.2 (Chap. 5, § 5.2.1) as soon as possible. In STIPA for example, the profiles are identified in the field by means of two codes showing just the pit number and Survey number. The number of the MU is not foreseen at this stage but must be introduced later. This can be done automatically by superposing in the graphics database the information layer corresponding to the points representing the profiles and the layer representing the MUs. The file obtained is transmitted to the RDBMS.

- *Administrative location of the soils.* We saw earlier that it was possible to record in the RDBMS the administrative units corresponding to the MUs. In fact, the administrative boundary lines often intersect the soil map boundaries in very complex fashion. Thus it is very difficult to establish the correspondence by visual means, especially in quantitative terms (percentage of a given MU in a particular administrative unit). Therefore, the two types of graphic data must be merged using a suitable GIS. Certain countries make available the administrative-units digital file. The result of the graphics merging is then transcribed as a file and sent to the RDBMS. This enables value-addition in soils data by presenting them in a form useful to responsible persons and decision-makers.

6.4 PROCESSING OF PROFILE DATA IN FRANCE

6.4.1 Recovering old data

It is often essential to recover old data that had been collated at a time when computer processing was not possible. The RECUP software program was created at INRA-Montpellier for this purpose.

It starts from profile data in clear on paper. The principle consists of communicating, in clear, data of the type 'fine prismatic structure' to the computer. Each word is recognized, coded again and entered in its place in the RDBMS. The advantage is that the operator is not compelled to keep to a sequence in entry of terms, and can thus follow the order found in the documents used. Naturally, the list of authorized terms must be known. To facilitate training, the most traditional terminology was respected. For example, 'blocky', 'sand' and 'gravel' are acceptable terms. Also, in any difficulty the system displays the authorized terms. The early stages are laborious, but the user gets the hang of things after a single day of manipulation.

6.4.2 The STIPA-2000 system

The STIPA system was developed starting from 1979 (Bertrand et al., 1984; Bonneric et al., 1985). A second version was created about 20 years later (Falipou and Legros, 2002). STIPA is the standard system in France and its domain. It has been in use for two decades in France (Bornand et al., 1994; Robbez-Masson et al., 2000) and also in Togo, in Burkina-Faso (Ouattara, 1980), in Algeria (Djili, 2000) and elsewhere in Africa (Delecour et al., 1978; Legros, 1990). It will therefore be the example for us.

But other attempts deserve to be noted, particularly those of R. van Driessche and A.M. Aubry at ORSTOM (the French organization that has now become IRD). Actually, these authors had proposed a system of direct input of soil descriptions in clear. It was a failure at the time. But with progress in speech recognition the idea could well turn out to be fruitful in the future.

Data acquisition

The general structure of STIPA-2000 has already been presented and also the field data sheets (Chap. 4, § 4.4.2). The data are captured by using input screens, one of which is presented in Fig. 6.15.

The system applies various kinds of checks when data are entered. For example, a code not present in the range of authorized values is rejected. Also, the CHECKUP program (Legros et al., 1992) presented earlier

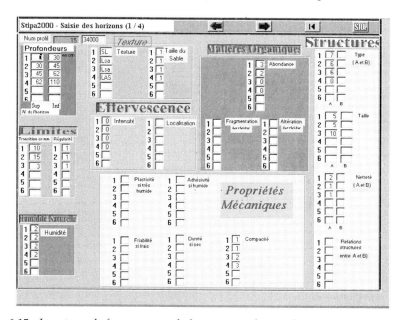

Fig. 6.15 Input mask for postponed data entry of a profile, starting from paper data forms (portion relating to the beginning of horizon descriptions).

(Chap. 4, § 4.5.2) can be used for scrutinizing the data already stored and for detecting inconsistencies. The screens can be recalled any time to correct the information entered.

Data management

STIPA-2000 manages its data through ACCESS-1997 or ACCESS-2000 (Microsoft) by using three 'tables' one each for the environment, the horizons and the analytical data.

Data utilization

The descriptions and analytical data can be printed out. As an example, Fig. 6.16 shows the beginning of a printout in French of a profile stored in the system.

Analytical data are automatically restored in different forms, one of which is given here (Fig. 6.17).

A few specialized programs have been developed to complement the management programs. Thus, one program is used to sketch the profiles

---

Profil n° : **1** *Etude :* 34225 (**Puisserguier**)
Auteur : Falipou P - Coulouma G   INRA MONTPELLIER
Date:02/04/01

Description de l'environnement :
Géomorphologie : profil au bas de la parcelle
Résumé :
Classification francaise (CPCS 1967) 5 121 , Séquence horizon :A/B
texture limoneuse   , . *Discontinuité :* compact (le couteau pénètre incomplètement même avec un effort important). *Effervescence :* forte. *Couleur :*brun-jaune (teinte équivalente au brun-jaune de la planche 10 YR du code Munsell). *Structure :* peu structuré ; profil différencié par la texture

Commentaires : Sol homogène sur toute l'épaisseur, assez compact, peu poreux et très peu structuré. Les racines exploitent cependant toute l'épaisseur. Quelques graviers sur tout le profil (cx rapportés par l'homme au cours du temps)

Description des horizons :
0 - 15cm   -   *Transition de* 5 *cm* ; frais; lsa à sable fin (50 à 100 micromètres) ; *Effervescence* forte ; *Structure* continue ou massive *et structure* polyédrique subanguleuse de 5 peu nette mm ; *Couleur de l'horizon :* 25Y 54 ; *Matière organique* faible (<1%) ; *Propriétés mécaniques :*meuble friable; *Eléments grossiers :* 5%   dont 3% Grès-Calcaires, graviers (0.2 à 2 cm), arrondis, peu transformé et 2% de quartz , graviers (0.2 à 2 cm), arrondis, , non transformés; *Racines* peu nombreuses (8 à 16 / dm2), très fines (diamètre < 0,5 mm), dans la masse, horizontales ; *Porosité :* moyennement poreux; *Pores* pas de pores ; *Identification de l'horizon :* AP1

15 - 35cm   -   *Transition de* 1 *cm* ; frais; lsa à sable fin (50 à 100 micromètres) ; *Effervescence* forte ; *Structure* continue ou massive *et structure* polyédrique subanguleuse de 5 peu nette mm ; *Couleur de l'horizon :* 25Y 54 ; *Matière organique*   faible (<1%); *Activité biologique :* racines décomposées, peu nombreuses ; *Propriétés mécaniques :*compact friable; *Eléments grossiers :* 5% *dont* 3% Grès-Calcaires, graviers (0.2 à 2 cm), arrondis, peu transformé et 2% de quartz , graviers (0.2 à 2 cm), arrondis, , non transformés; *Racines* très peu nombreuses (< 8 / dm2), très fines (diamètre < 0,5 mm), dans la masse, horizontales ; *Porosité :* peu poreux; *Pores* peu nombreux (1 à 50/dm2), extrêmement fins (< 1 mm) ; *Identification de l'horizon :* AP2

Fig. 6.16   Printout of a soil description with STIPA-2000.

Etude (survey) : **595**						Profil : **122315**		
Prof. cm		*% eau*	*Granulométrie en %*					*Ca %*
			A	LF	LG	SF	SG	total | actif
0	20	3,7	15,1	15,2	5,8	14,3	49,6	
20	65	9,1	8,2	22,7	6,2	14,4	48,5	
65	100	2,8	6,1	19,2	6,4	14,5	53,8	
100	120	1,8	5,5	20,1	8,2	16	50,2	

pH	C %	M. Org.	N tot %	N org.	N. Nitr.	N. NH4	Fe lib	Fe tot
5,2	6,63		61,9				1,2	
5	4,47		37,8				1,4	
5,1			8,1				7,2	
5,3							0,8	

Ca Ech (meq)	Mg Ech (meq)	K Ech (meq)	Na Ech (meq)	CEC (meq)	Da	Humidités en %		
						champ	PF2.5	PF4.2
2,2	0,39	0,33	0,05	22,7			30,2	21,9
0,3	0,1	0,05	0,04	14			27,7	15,6
0,1	0,05	0,04	0,02	9			20,7	7,2
0,6	0,05	0,03	0,02	8,2			19,4	5,8

	Commentaires
0 - 20	
20 - 65	
65 - 100	
100 - 120	

Fig. 6.17 Printout of analytical data from the STIPA-2000 system.

using the corresponding recorded information. There is also a program enabling representation of soil textures in practically all triangular diagrams (Falipou et al., 1981). Yet another serves to calculate the similarity between two soils. We will have occasion to return to details of the uses offered by this possibility. Lastly, a complete library of programs for specialized statistical treatments (LOGOS library) has been developed (King and Duval, 1988).

6.4.3 Assessment of processing of profiles with STIPA

Use of STIPA for more than 20 years enables us today to draw up an assessment of computer processing of descriptions and analytical data of soils.

Scientific advantages
Quality of information is considerably augmented. Introduction of computer processing enforces very rigorous work. The terminology is well defined (glossaries). Organization of data capture is more rational. Computer processing enables us detect errors and correct them at source. For example, surveyors are provided the facility of systematic comparison of their field textures with particle-size distribution determined later in the laboratory. This helps them improve their diagnostic techniques.

Quantity of information collected greatly increases compared to what it was when there was no computer processing in support. This is essentially the result of using field data sheets as guide and check-list. Omissions are fewer. To demonstrate this we compared two groups of profiles belonging to two soil surveys (Table 6.4). In the Lodève study (Bonfils, 1992), the profiles were described on STIPA-1979 data sheets. The profiles of the Costière du Gard survey were described twenty years earlier on sheets that did not allow direct computer entry. A later computerization using the STIPA-RECUP program, however, enabled storage of these profiles in the STIPA-1979 system, whereby the corresponding quantity of information could be calculated.

Table 6.4 Quantity of information captured with and without computer-compatible data sheets

Survey	Number of profiles	Number of horizons	Proportion of data filling	
			Soil envir.	Description
Gard (without computer compatible data sheets)	154	505 (3.3 per profile)	20.8/110	16.9/135
Lodève (with STIPA data sheets)	192	634 (3.3 per profile)	66.1/110	29.4/135

The two surveys are almost equivalent (comparable Mediterranean milieus, same number of horizons per profile). But use of data sheets designed for computer processing (Lodève) results in increase in the quantity of information collected. As for the environment, data per profile rose from 20.8 to 66.1. This is three times as much! In theory, STIPA-1979 would have allowed storing 110 parameters. But in each case, many columns could not be filled because they were of no use, for example the variables used to describe the characteristics of a forest... when one is in a cultivated field! The difference is no less clear in horizon descriptions because the degree of filling goes from 12.5% to 21.7%.

Important information to be preserved. Taking into account the cost of redoing a profile (digging, description, analyses), this represents more than Euro 4 million just for the STIPA system (12,000 profiles at Euro 350 each). In

France, these STIPA profiles are stored at INRA, in CIRAD and in some organizations that contribute to soil mapping of the country at scale 1/250,000. In other countries, many thousands of profiles are stored in the STIPA format in the Service National des Sols of Togo, at the INA in Algeria and of course, elsewhere too. As most countries have established similar data banks, the information stored overall represents colossal totals.

Possibilities of data processing are considerably heightened. Such is the usefulness of computerization of soil data. In Belgium for example, statistical analysis could be done of soil characteristics over a large part of the country by means of the data bank that had been constituted (van Orshoven et al., 1988).

Scientific drawbacks
But along with these advantages, computer processing of data has at least one major scientific disadvantage. There is no point in repeating that the equipment installed is unwieldy. It is therefore impossible to update frequently the glossaries, data sheets and the underlying concepts. Also, updating will not be desirable. For the data bank to be utilizable, it must contain homogeneous data. So the type and structure of the data stored cannot be modified too often. A certain rigidity results, not to say some impediment to progress. We had thought that updating should take place every five years. But experience has shown that few establishments have manpower sufficient to proceed with the revisions in question when the time comes. Twenty years would have been necessary to go from STIPA-1979 to STIPA-2000!

Technical assessment
Development of computer science has made easy the tasks that seemed difficult earlier. The pioneers of computer technology thus were recompensed for their efforts. Their old and at times angry detractors no longer remember the time when one of them wrote in 1976 '*a typist could do the job at a lower price than the computer*'. Thus there was progressive gain in calculating capacity, speed of execution, quality of presentation, etc. Above all, the data could be made available to those who needed it. This does not go without counterarguments. In particular, computer systems are becoming complex. Specialists are necessary. But in many establishments, even large ones, computer scientists are few. Furthermore, they are specialized for obvious reasons of efficiency. In short, it is not rare that a system runs because of one single competent person. If that person is ill or is called away for other work, everything could collapse! One solution to contend with this risk is to work with the same tools in several

organizations or laboratories. Thus the laboratories of the 'AGROPOLIS' high-tech zone (Montpellier, France) agreed to manage their data with ArcInfo software when GIS entered the scene. This agreement enabled from the very beginning mastery over this software by many. Training sessions were organized in common.

Also, developments in computer science lead to creation of more and more efficient, but also increasingly costly systems. In the past, setting up a private survey department did not pose any problem of investment in equipment. Today it is different. A complex technical establishment must be organized and financed.

Collaboration and distribution
A data bank is established so that there would be users. Therefore, it must represent a system open to the outside world. This involves various things.

First, the bank must publicize itself, announce that it exists and distribute summaries presenting its contents. This is the concept of ***metadata*** (data on data). In fact these acts of promotion can take time but they do not present any difficulty because many organizations try to list data banks and the volumes of data available.

Next, the data bank must be easy to use, hence the need for developing properly documented systems, that is, systems accompanied by clear and complete instructions for use.

In the third place, the setting up of an RDBMS and loading it with data is a heavy task for an institution whether it is a research laboratory or a private mapping firm. In France, this was primed by development of STIPA because at least three laboratories participated in building it. But this willingness to collaborate found its zenith when the DONESOL project (§ 6.3) was initiated. In this new action plan, the INRA laboratories in Montpellier and Orléans worked in a very complementary way under the direction of a coordinator located at Versailles.

Lastly, it is often necessary to organize specialized sessions to train the users. In ten years more than 100 mappers from 24 different countries were initiated in the use of STIPA. The courses were set up in two ways. First, persons were trained bit by bit in our laboratory for periods ranging from 3 days to 6 months. Secondly, we organized training classes for groups of 10 to 15 persons. This was specifically done at Lomé in Togo in 1986, at Tunis in 1988 and at Bujumbara in Burundi in 1989. Thus STIPA served as technical support for better mutual understanding of French-speaking teams with the help of l'Agence de Coopération Culturelle et Technique (ACCT).

6.5 SOME INTERNATIONAL EFFORTS

We cannot exhaustively cover here all the international efforts undertaken in managing soils data. We shall only illustrate by a few examples the principles examined above. The work presented below has often been found on the Internet. But we shall not give the addresses of the relevant pages as we said in the general introduction. Actually, these addresses change too frequently! They can be found almost instantly by searching for them by key words adding the name of the country in question and the systems presented.

6.5.1 Organization of soils data in USA

Until 2001, information on soils in USA was organized in eight principal types of databases, not too well linked amongst themselves (Fig. 6.18).

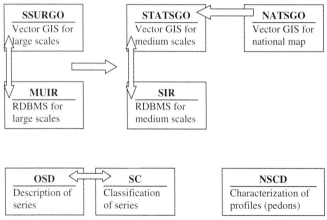

Fig. 6.18 Organization of soils data management in USA within the Natural Resource Conservation Service (NRCS). The arrows show the logical or physical links between the databases.

SSURGO (*Soil SUrvey GeOgraphic database*) pertains to large-scale maps digitized in vector format. The scale varies from 1/12,000 to 1/63,360 with intermediate scales of 1/15,840, 1/20,000 and 1/24,000. The maps are available in DLG-3 format. They belong to different Surveys but are cut up or merged to form sheets of 7.5-minute topographic quadrangle units. A new version SSURGO II has been developed and data will be transferred to it by the end of 2004.

MUIR (*Map Unit Interpretation Record database*). MUIR is the RDBMS corresponding to SSURGO. Its structure resembles that of DONESOL described above. It is organized in five levels:

- (1) County,
- (2) Soil survey,
- (3) Map Unit,
- (4) Map Unit Component (equivalent to Soil Unit in France),
- (5) Soil profile layers.

Compared to DONESOL, the objectives are more general because forest applications (woodland management, existing woodland species, potential woodland species for planting) and environment applications (wildlife habitat, native woodland plants) are examined. On the other hand, the number of variables and properties taken into consideration is limited (88) and therefore reasonable. Almost 80% of the territory of USA is covered. The database is updated every year. It is the responsibility of the *National Soil Survey Center* of the U.S. Department of Agriculture.

STATSGO (STATe Soil GeOgraphic database) corresponds to maps that have been digitized in vector format and are on scale 1/250,000. They were prepared under the supervision of the *Soil Survey Division* of the NRCS by going back to the more detailed surveys stored in SSURGO and simplifying them. They are organized for each state in sheets of *1 by 2 degree topographic quadrangle units.* They are available in DLG-3, ArcInfo 7.0 and GRASS 4.13 formats. Such maps contain 100 to 400 polygons of size no smaller than 1 cm^2 on the map representing 1,544 acres or 625 ha on the ground.

SIR is the non-graphic database corresponding to STATSGO.

NATSGO (NATional Soil GeOgraphic database). The scale is 1/5,000,000. At this level soil boundaries can no longer be shown. Only *Major Land Resource Areas* (MLRA) appear. This map covers all of USA. It also serves to describe the STATSGO data. The database is being developed.

OSD (Official soil Series Descriptions). All the known series of USA are described here: name, location, classification, detailed soil profile description, location of the typical soil profile, range in characteristics, competing series, geographic setting, geographically associated soils, drainage and permeability, use and vegetation, distribution, extent.... They number about 18,000.

SC (Soil series Classification database) corresponds to the classification of each series and its revision when needed.

NSCD (National Soil Characterization Database) contains the analytical data of more than 20,000 profiles corresponding to as many pedons. The link to spatial data is not automatic, but can be done through the name of the series in question or the coordinates of the point (E. Benham, personal communication). All the states are covered. Three-fourths of the data are less than 20 years old, which proves the dynamism of soil survey in USA.

Access to databases
Americans are seeking to organize these databases in a complete, coherent system, the *Warehouse Database*. Particular attention is given to setting up the database on the one hand and to utilization of the data on the other. The system is complex; it relies heavily on SSURGO. But new tools have appeared, especially NASIS and WSDV.

NASIS (**NA**tional **S**oil **I**nformation **S**ystem) is the system used to create, manage and maintain soil surveys. It pertains to non-graphic data, here called *tabular data*. The mapper's motto is 'to do the job better in less time'. The system is under development. It will function on the X Window platform. The chief originality in this due to the fact that the *Map Unit* entity is supplemented by another entity, the *Legend*. Considering the properties of RDBMS, this allows several legends for a single Map Unit and hence, a single map. There are two advantages in this:
- foreseeing in advance the definition of the soil for maps on different scales, thereby enabling later recovery of the data when doing more detailed or less detailed mapping;
- preserving the track of successive legends when the maps are updated and revised: this is very useful when the American soil classification is used because in this classification, modification of details of concepts at times results in great changes in soil names.

WEB SOIL DATA VIEWER is a specialized software presently being developed for enabling, via *Internet Explorer,* access to the data stored in SSURGO and NASIS as well as processing the data downloaded at the user's end. It is a large project led by the *National Resource Conservation Service* (NRCS) with the assistance of the companies ESRI, Microsoft and Compaq Solutions Provider. There is already a prototype through which users can utilize free of charge the data pertaining to some survey areas. It is possible to examine the map by moving across it and simultaneously to display the corresponding orthophotographs. For each Map Unit, the various parameters pertaining to the Soil Unit with the largest area can be consulted, for example the most limiting characteristic for the application in mind. The weighted (by area) average of values for the various Soil Units found within the Map Unit can also be calculated. When the project is completed, distribution of map information will be done almost exclusively through this channel [search the Internet for the site USDA NRCS WEB SOIL DATA VIEWER].

Development is in progress. The project is big and its expected cost is well in excess of one million U.S. dollars. The indications given here are liable to change. Perfection of NASIS and WSDV will certainly lead to rearrangement of the databases at the heart of the system (SSURGO, MUIR…).

6.5.2 Organization of soils data in Canada

In Canada, the work started in the framework of CanSIS (*Canada Soil Information System*) with, in particular, the participation of J. Dumanski, B. Kloosterman and L.M. Lavkulich. At present this country has a *National Soil Database*. The data are distributed free of charge in ArcInfo format. Only the time necessary for sending them is charged for. The logical structure is the same as in USA and France: *Map Unit > Components (up to three) > Soil Name > Soil Layers*. Scales vary between 1/10,000 and 1/250,000. There are 1300 map sheets. Abstraction to 1/1,000,000 was proposed as early as 1991. This abstraction is not purely pedological. It pertains to *Soil Landscapes of Canada* (SLC).

The program NATUREL was developed in Canada (Laflamme and Goyette, 1994) for recovering data earlier stored in natural language using applications such as Word. Once the data is structured, the user queries the data bank by means of the index or keywords that are found in the texts in question. It is thus an original effort and probably very efficient for recovering useful data at low cost.

6.5.3 Organization of soils data in the Netherlands

From 1986 to 1992 our Dutch colleagues developed the SOTER system (*SOil and TERrain database*) suitable for very small scales such as 1/2,500,000. It is a project executed in cooperation with the FAO, UNEP and ISSS (FAO, 1995). The system was locally tested in various regions of the world, a scientifically correct and at the same time promotional approach: LASOTER for South America, NASOTER for North America and SOVEUR for countries of Central and Eastern Europe.

Soils cannot be distinguished on the map on continental scale. The procedure is closer to *physiographic mapping*. This is why the Map Units, called *Terrain Units*, are subdivided into *Components* that are also landscape units. The soils only come in only the third tier. The structure of the attribute database is given in Fig. 6.19 (Batjes, 2000).

The legend of the FAO-UNESCO Soil Map of the World (1988) is used as basis for characterizing the *soil components*. For enabling RDBMS specialists and at the same time persons knowing the soils best to consult the database, *the SOTER summary file* repeats the essential information on 27 variables. This information is duplicated but becomes more easily accessible.

The same country also developed ISIS (ISRIC **S**oil **I**nformation **S**ystem). This is a software that enables storing, editing, printing out and selecting profile data. It uses dBase III and runs on a microcomputer with hard disk and MS-DOS operating system. Various documentations have been published (van Waveren and Bos, 1988). In this system, the soils are described in a simplified way, which may seem to be a disadvantage. But, on the other

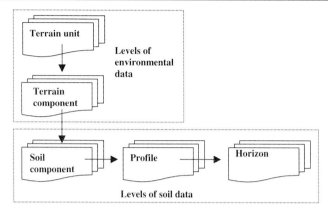

Fig. 6.19 Structure of SOTER (Batjes, 2000).

hand, this enables setting up a data bank with simple means and at reduced cost. The user retains all independence and can work on a microcomputer away from large computer centres.

Conclusions and perspectives

Our review is far from being exhaustive. Many important efforts have not been mentioned. For example, those of England started very early by P.H.T. Beckett with the help of R. Webster, P. Burrough, J.M. Hodgson and C.C. Rudeforth. Also those of Belgium with the work of M. Kindermans and F. Delecour. And there are others in many countries of the world!

However, the work that has been reviewed clearly shows the current state of the problem: the most advanced countries endeavour to save their soils data by structuring and synthesizing them in the RDBMSs made available to the public. The logical structure is the same everywhere: it is accepted that the *Map Unit* may include several *Components*, that is several soils.

But, on the other hand, we have to progress in many directions.

First, computerization will be of greatest benefit to soil mappers. It would be necessary to implement veritable desk-top publishing to help prepare reports accompanying the maps. Actually, in France and elsewhere, compilation of these documents delays publication of many projects whose purely mapping part was completed long ago.

Then, it is seen that computer science is mainly introduced post-mortem in mapping. Mapping is done conventionally, then the data are captured before being processed in the computer. To gain from the current possibilities in computer science, this procedure must be used early on, in the field. Of course, efforts have been taken in regard to description of profiles and auger samples on the microcomputer, but automation has to penetrate further. One should think of automatic recording of the coordinates of observation points and of access from the field through the Internet to specialized libraries offering the surveyor a corpus of knowledge on the geology, botany and climate of the study area.

In some countries, too much time has been spent in data storage. It must be taken care of with American pragmatism, avoiding entry of minute details that overload the database, demand great effort in data entry, while remaining of limited relevance. We must know how to simplify!

Some scientists, noting the parallel growth of storage and computer processing in all disciplines touching upon the natural environment, have the ambition to develop a *Systemic Land Representation in Geographical Information Systems*. The objective is to ensure that all the graphic and non-graphic databases created could exchange data and simultaneously be utilized for solving complex problems relating to agriculture, forestry and environment. Attempts have been made, for example by the GERMINAL team at Lausanne in Switzerland (de Sede et al., 1991; Prélaz-Droux, 1995). Also in line with this thinking is the creation of the multidisciplinary and multithematic database ECORDRE at Montpellier in France (Soto and Bouche, 1993). Canadian efforts (*soil landscape* concept) are also in the same direction.

Lastly, the fundamental problem of all these databases must be solved. It is not possible to describe everything and to store all data, as was said earlier. Under these conditions, the potential user who comes asking for data pertaining to geographical distribution of a particular soil insect, an extremely rare metal or degree of polishing of pebbles 5 mm in diameter is very likely to hear the reply 'we do not have that'. That person might well go back convinced that the data bank is émpty and of no use! Nevertheless, even though a ready answer is not given, of good knowledge of the geography of the natural environment often allows suggesting sites that it would be wise to study because they are different, represent large areas and are probably relevant in the context of the question asked. One is then in the position of the polling organization that cannot predict the result of an election without prior inquiry, but would know how to organize the sampling necessary because it understands the social structure and geographical structure of the population studied. In short, it will be necessary in the future to further master what we could call 'going back to the land', the way was opened up in a test of the natural radioactivity of soils of the White Mountains in New Hampshire, USA (Morton and Evans, 1996). Although the authors concluded that systematic measurement of radioactivity was useful, they proved that one could go beyond that since their interesting study was induced after considerable soil mapping. But they could insist that the initial state of pollution by radionuclides should be established before a nuclear disaster and not after! This is another problem....

CHAPTER 7

MODELLING AND AUTOMATION

As stated in the preceding chapters, mapping procedures, enriched by the practical experience of several years, are now well codified and consolidated. The fact remains that various efforts are being made to improve them further. They are mainly attempts in modelling, computerization and automation. We are going to present methodological studies conducted in the author's laboratory and in associated French laboratories. This is not to neglect efforts made elsewhere in France and the whole world. These efforts are cited in places and discussed in Chapters 2, 6, 9 and 10.

Some key points in mapping are presented in Fig. 7.1. They are dissected in this chapter in an attempt to improve them.

All these elements are not necessarily considered in a given mapping exercise and the order in which they are taken into account may vary.

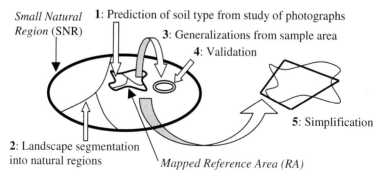

Fig. 7.1 Key points in mapping.

7.1 SOIL PREDICTION FROM AERIAL PHOTOGRAPHS

In many countries, several aerial photographic coverages are done one after the other, a few years apart. The photographs are mostly black-and-white (panchromatic). These aerial photographs are often taken at the same time of the day with the sun high in the sky. Where there is no human habitation, they resemble each other. But in a zone where habitation is dense, the landscape changes slightly from one photograph to another taken later. Ploughed parcels in particular, which allow observation of the soil in place, have changing sites. One may wonder if the simultaneous use of photographs of the same zone corresponding to different years could help soil mapping.

We made a trial in Switzerland (Legros et al., 1997). This country was covered since 1960 by several flight missions (a total of 9). The test pertained to drawing the boundary of the l'Orbe marsh. It is an old reclaimed and cultivated marsh, with black organic soils. Soil observations and the *Best Possible Map* (BPM, see Chap. 5, § 5.3.4) were done by several successive year groups of students of l'École Polytechnique Fédérale de Lausanne (EPFL). Calculations and measurements were done by means of a *Geographical Information System* (GIS). The area of the marsh is 424.5 ha and the perimeter 15,630 km.

7.1.1 Identification of the marsh by its boundary

Principle
Measurements showed that 34.6% of the boundary of the old marsh is seen on average in each coverage (crossing from a light-coloured soil to a black soil). The boundary was discontinuous in the observation windows formed in the plant cover by recently ploughed parcels or those with crops whose slight growth did not hide the soil underneath (Fig. 7.2). We can write $visi$ = 0.346 (this parameter actually ranges from 0.23 to 0.43 in different coverages that are also photographs, to put it simply).

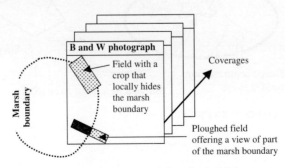

Fig. 7.2 Identification of a soil boundary in a ploughed field.

Modelling
But using a second photograph after the first does not double the amount of information because there are cross-checks: certain ploughed fields on the photograph used second are already on the photograph used first, although it is presumed that the position of these fields is random from one photograph to the other. It is only in the portion of the boundary not observed at a given stage (photo n) that it is useful to see something in the next stage (photo $n + 1$). Consequently, we should think of an iterative calculation according to the formulae:

$$SEEN_1 = visi$$
$$SEEN_2 = SEEN_1 + visi \cdot (1 - SEEN_1)$$
$$SEEN_n = SEEN_{n-1} + visi \cdot (1 - SEEN_{n-1}),$$

where $1 - SEEN_n$ is the proportion of the boundary seen with n photographs.

But this formulation leads to excessively optimistic results because the position of the ploughed fields cannot be considered as fully random: they have a slight tendency to appear in the same place from one photograph to the other. Finally the following is checked:

$$SEEN_n = SEEN_{n-1} + (1 - link) \cdot visi \cdot (1 - SEEN_{n-1}) \qquad (1)$$

where *link* represents the degree of dependence between images. The value is 0 for independent photographs and 1 when the photographs are identical. In the latter case, the first photograph will provide all the information available.

Results
In the example studied, the value of *link* was about 0.3, as obtained by trial and error. In this situation, *link* seems to be a calibration factor. Table 7.1 compares the results from the observations (using a GIS for the calculations) with the values predicted by modelling.

Table 7.1 Proportion of boundary observable by using in succession several aerial photographs with a degree of dependence $link = 0.3$

Number of photographs used	1	2	3	4	5	6
Proportion of boundary observed (photos)	34.8	46.6	64.5	70.3	76.2	81.2
Proportion of calculated boundary (model)	34.6	50.4	62.4	71.5	78.4	83.7

It was seen earlier (Chap. 5, Fig. 5.2) that these values are quite sufficient for tracing with high accuracy a complete soil boundary even if only two or three photographs are available. With 5 or 6 photographs used simultaneously, one may even be tempted to avoid setting foot on the land! But this will be a bad idea, actually:
— as pointed out above, it is necessary to identify right from the start what is seen on the photographs; a priori, a dark colour on a black-

and-white photograph is not synonymous with peat; it could be a black clay, wetness or something else!
— simultaneous minute examination of 5 or 6 aerial photographs is a slow and costly exercise; generally, it is more efficient to devote oneself to real mapping and not just to a laboratory exercise! The above condition must be insisted upon: for years scientists have been seeing development of *remote-sensing* methods, the elaboration and implementation of which take more time than a field traverse and always lead to a more approximate final result!
— lastly, the exercise was done in an easy case: the boundary to be drawn was particularly sharp, marked by contrast in colour of the soils on either side and also by an obvious relationship with topography (the marsh is in a depression).

As the exercise showed that *visi* varied from 23% to 42% on photographs, it is useful to examine it closely before ordering for photographic coverage: one could be more useful than the others when there are several of them.

7.1.2 Identification of the marsh by its content

Principle

Mappers not only draw their boundaries by following them exactly, they also look for and distinguish what is on the left and what is on the right. For example, a ploughed field with light-coloured soil in totality is outside the marsh, whereas if it is ploughed and totally black, it is inside. Unfortunately there is often vagueness in black-and-white photographs. Actually, a light-coloured field in temperate climate may correspond to, say, a mature stand of wheat in the marsh instead of being a bare soil outside the marsh. At the same time, a dark-coloured field could correspond to a young crop, for example maize, growing outside the marsh instead of a bare soil in the marsh (Fig. 7.3).

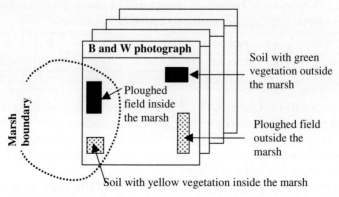

Fig. 7.3 Confusion between colour of soil and colour of vegetation in black-and-white photographs.

Lastly there are grey fields, neither black nor white. But the human brain knows to choose and assign these intermediate fields to one or the other of the two categories black or white, a bit arbitrarily but intelligently by considering the neighbourhood.

The problem often gets simplified. Actually, the growth stage of the crops when the photograph was taken must be taken into account. If the crops were mature and deprived of chlorophyll, many light-coloured fields will be seen, inside and outside the marsh, but black colour retains its significance and mostly pertains to bare marsh. Conversely, if the crops were still chlorophyll-rich when the photograph was taken, there will be many dark-coloured fields inside and outside the marsh. In this case it is light colour that has significance and very probably belongs to bare soils outside the marsh.

Modelling

Let us take the case of a photograph taken early in the season. This is the most classic case. The crops will be green, therefore dark on the black-and-white photographs. Therefore black colour is not always synonymous with marsh. We shall calculate the risk corresponding to, that is, taken in considering black as equivalent to presence of marsh:

P(M/B) is the probability of a field being in the marsh if it is black. This is the value sought;

P(B/M) is the probability of having a black fiels in the marsh; this probability is 1 because the field is black in the marsh whether it be bare or covered;

P(M) is the probability of being in the marsh; this is the relative area of the marsh compared to the study area, on condition that frequencies are equivalent to probabilities, which poses no problem if the study area is large and the fields many; this parameter is known because the map has already been drawn;

P(B/noM) is the probability of having a black field outside the marsh; this is thus the proportion of black fields outside the marsh.

Let *P(B/noM) = (Black A outside the marsh)/Total A outside the marsh)*, where *A* is area.

Then we can calculate the probability sought by applying the formula for total probabilities:

$$P(M/B) = \frac{P(B/M) \cdot P(M)}{P(B/M) \cdot P(M) + P(B/noM) \cdot P(noM)} \quad (1)$$

which, in the present case, immediately gets simplified to

$$P(M/B) = \frac{P(M)}{P(M) + P(B/noM) \cdot P(noM)} \quad (2)$$

The formula can be generalized for several photographs. It is enough if we replace *P(B/noM)*, the only variable term, by the proportion of area outside the marsh that always remains black, therefore covered by vegetation when several photographs are considered. Let us call it P(always B/noM). Then it becomes:

$$P(M/B) = \frac{P(M)}{P(M) + P(always\,B/noM) \cdot P(noM)} \quad (3)$$

We can calculate P(always B/noM) by iteration:

P(alwaysB/noM) = P(B/noM), then

P(alwaysB/noM) = P(always B/noM) − (1 − link) · [1 − P(B/noM)] · P(always B/noM)

Formula (3) can be verified by some logical tests given in the original article (Legros et al., 1997). We shall only mention one here: if the number of photographs is very large, P(alwaysB/noM) tends to 0, then P(M/B) tends to 1, no error is possible: black reveals the marsh in full (woods, geographically fixed and dark-coloured but easily identifiable, are excluded from the calculation)!

Results

The calculations were done with a computer using a FORTRAN program of about ten lines. The results obtained are given in Table 7.2.

Table 7.2 Probability that black corresponds to the marsh [P(M/B)] taking into account the normal proportion of black fields outside the marsh [P(B/noM)]

Number of photographs	Values of P(M/B) for 4 values of P(B/noM)			
	0.20	0.55	0.75	0.90
1	0.78	0.56	0.49	0.45
2	0.89	0.65	0.54	0.46
3	0.94	0.73	0.59	0.48
4	0.98	0.80	0.63	0.50
5	0.99	0.85	0.68	0.52
6	0.99	0.89	0.72	0.54

In the table, the value of P(B/noM) = 0.55 corresponds to what we actually observed on the processed photographs. The risk of error in considering black as equivalent to marsh is 1 − 0.56 = 44% when just one photograph is used (row 1). So, theoretically grey density on black-and-white photographs is of very little help! In fact, this is not totally true because the mapper does not only take this into account. The cluster effect of the black fields that locally crowd together is also considered. Also, we have seen that the mapper perceives 34.6% of the marsh boundary directly (Table 7.1). With a few direct soil observations, the mapper will succeed in tracing the soil delineation.

On the other hand, if three photographs taken at different times are available, one has about a 3 in 4 chance that black pertains to the marsh. At the same time, 62% of the boundary can be seen directly. All this should enable one achieve high certainty in delimiting soils in poorly accessible areas.

Let us now suppose that colour photographs are used. Crops, whether yellow or green, will not be confused with the black colour of the marsh. The information conveyed by a black field is a certainty. If it is agreed that the proportion of the marsh boundary seen in ploughed fields is an approximate measure of the relative area of these fields (Table 7.1), this means that 1/3 of the area of the marsh will be directly seen with just one photograph and 2/3 with three photographs. A test of colour photographs will suffice to show that their use greatly improves the quality of work and ease of mapping.

7.2 TRIAL OF SEGMENTATION OF SPACE

In the section above, the soil boundary was examined in detail. This pertains to large-scale survey. But often the issue is less about finding a boundary than about allocating portions of space to such or such *Map Unit*. This is often the case in small-scale mapping of *Landscape Units*. Modelling and automating the procedure has been attempted in a project for a thesis (Robbez-Masson, 1994).

7.2.1 Principle and method

The area in which one wants to work is divided into pixels and the corresponding environmental information is recorded in matrix form: altitude, slope, nature of rocks, etc. Here, to simplify the problem and present it more easily, we limit ourselves to consideration of just slopes. Let us presume now an observer is present in one of the pixels in question. The observer is found, say, at the centre of a dissected plateau area presenting shelves as well as very steep slopes. At very large scale it would be possible to separate these shelves and slopes by mapping and to distinguish the corresponding soils. But at small scale, this is impossible. At this level, it is the association of the two units observed that is characteristic of a certain type of landscape to be identified in its entirety and separated from other types. The method proposed by Robbez-Masson extends earlier work (Girard et al., 1991, 1992). The principle is as follows. First we fix a window (moving window), say, of 5 × 5 pixels at the centre of which the pixel in question is positioned. The slope classes occurring in this window are listed and the data presented in the form of a distribution histogram. Thus the landscape in which the central pixel at issue occurs is characterized. Let us now assume that some characteristic histograms of the local landscapes had also been prepared. It is then sufficient to compare the histogram now calculated with the reference histograms to find the most similar ones and to state, say, that the environment of the area where we are enables placement of the pixel in question in a dissected plateau landscape (steep slopes and shelves predominant) and not in a hill landscape (gentle slopes predominant) (Fig. 7.4).

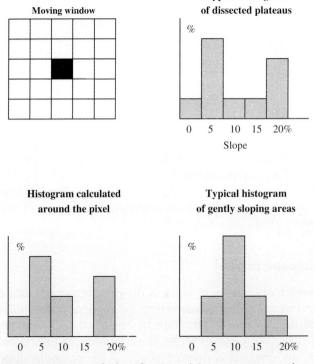

Fig. 7.4 Characterization and identification of the environment of a pixel (Robbez-Masson, 1994).

For implementing such a procedure, it is necessary to have available a statistical method capable of comparing two histograms. The author, after having gone through all the possibilities, chose the Manhattan-distance method (the same as the *city-block metric* method). Also, even though the landscape observed is attributed to the closest of those within the reference group, the distance measured between the histograms is almost never zero. So the attribution is not perfect. At the same time as the pixel is classified one can define the quality of the grouping. Naturally it is planned to process not just one but all the pixels. This is done by moving the window over the entire grid corresponding to the original map (slope map in the example taken here). At the end of the calculation all the pixels are assigned to a specific landscape type. Thus it is possible to visualize the resulting map. The procedure depends less on zonation than on the successive study of each pixel, considered genuine in its environment. Along with this, the map of classification quality is obtained, pixel by pixel.

7.2.2 Discussion

Such a method of classification uses a certain number of parameters one can play with. The size of the window can be changed. When it is reduced, the map obtained is marbled with very many small delineations. If the window is enlarged, the resulting map is simpler: its lines are rounded, delineations are fewer and the smallest of them vanish. Figure 7.5 gives results of some typical attempts taken from among many others not presented here. They pertain to the Caroux mountains (Aude, France). The plateau zones are white, the hill zones black.

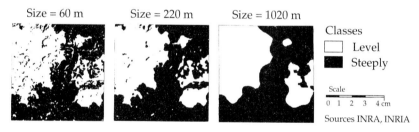

Fig. 7.5 Change in the drawing according to window size (Robbez-Masson, 1994).

It is useful to ask oneself at which window size the best result is obtained. The 220-m window seems preferable and this is why we have reproduced it. Actually the lines on the map are then reasonably simple, but some small delineations remain, which have to be retained to bring out that this mountain environment is heterogeneous. Well, this 220-m window corresponds to a distance ranging from 110 m to 156 m all around the observer (the second figure represents the diagonal of the square). These values are very useful because they are of the order of magnitude of those proposed by Walter (Chap. 5, end of § 5.3.6) when, in geostatistics, one wishes to estimate the distance at which the soils cease to be correlated (concept of range in the variograms). This also corresponds to the distances Lagacherie found useful in the application of his laws of neighbourhood (see § 7.3.2 below). When all is said and done, and even if the available results are still very incomplete, mappers must consider the environment in which they find themselves at the scale of one to three hectometres when the work is done in the contrasted landscapes of Europe. Also, weights may be assigned so that the pixels close to the observer have greater weight than the pixels at a distance in drawing the histogram, that is, in the landscape synthesis that is done.

Robbez-Masson also suggested change in the number of reference histograms, which poses the following questions: how many landscapes are there in my environment and where are the most characteristic landscapes to be found? One can give arbitrary answers based on intuition and experience, or even use a statistical procedure that will provide the best elements on the basis of a systematic examination of the pixels of the original map. It is also possible to introduce histograms for characterizing zones poorly classified when the first trials were done. Construction of the map then becomes iterative.

The same method enables consideration of a complementary problem: knowing the different landscape types found in a *sample area*, what are the boundaries of the region where one or the other of these landscapes is found or, if preferred, what are the boundaries of the regions the sample area is representative of (Lagacherie et al., 2001)?

All this work on simulation is particularly useful. Before they were done, other studies belonging to the same family of thought pertained to remote sensing. But the studies reviewed above open up discussion of the problem of synthesis in soil mapping.

The software, called CLAPAS, was written by J.M. Robbez-Masson. It is available as freeware on the Internet.

7.3 GENERALIZATION OF LOCAL OBSERVATIONS

7.3.1 Extrapolation from pedons

The publications we are going to present (Bonneric, 1978; Legros and Bonneric, 1979) correspond to the very first French attempt at computer-assisted soil mapping.

Principle
The procedure involves copying that of the expert and reasoning out by analogy applying the following postulate relating to pedogenesis: if such soil is regularly found in a particular environment, the same soil will be found in a similar environment. We start from a collection of type profiles relating to a given area. These profiles are defined by their belonging to taxa of a classification and at the same time by the principal characteristics of their environment. We used 38 profiles described during soil mapping of the Pilat Regional Natural Park (PRNP). These profiles, thus constituting

reference profiles, are located in the mountain zone under forest, on crystalline rocks that show arenization. The classification used was the French CPCS (1967) system, but here we give the equivalents in the WRB system. The numbers of profiles were

Podzols	9
Entic Podzols	8
Haplic Umbrisols	9
Dystric Cambisols	12

The Podzols were at higher altitude, whereas the Cambisols occupied the foothills of the Pilat mountains. The others were in intermediate position. The environment of these soils was defined by altitude, slope, exposure and rock. The last is actually characterized by its weathering layer: pH and contents of iron, silt and clay.

The question was 'to what extent is the computer system capable of reconstructing the entire soil map of the Park on the basis of knowledge of characteristics of the environment for the whole zone and knowledge of the soil cover only at the level of 38 profiles?' To answer this question, the Park was first subdivided into 2496 square pixels each of size 25 ha. From this was calculated the 'Gower distance', which will be presented later (§ 7.4). This led to estimating by a number between 0 and 1 the similarity of the environments of the map pixels to the environments of each of the 38 profiles. Then a synthesis was done (calculation of means) grouping the references (profile environments) by soil type (Podzols, Entic Podzols, etc.). This may be sketched by a diagram on which the profiles and corresponding pedons are represented by small circles, the degree of similarity by lines and the abstracted values (means) by black rectangles (Fig. 7.6).

Once the calculations are done, a matrix can be set up to give opposite the pixels, represented by number, the similarity of their environment with that of the soil types considered (Table 7.3).

If a pixel has an environment that most closely resembles the environments in which, say, Podzols are found, the odds are that the particular pixel also has soils of that type. This matrix enables a soil map to be easily inferred.

Table 7.3 Similarity of environments of pixels of the map to those of the reference profiles

Pixel number	Similarities of the pixels and the taxa for their environments			
	Podzols	Entic Podzols	Haplic Umbrisols	Dystric Cambisols
n	0.8	0.7	0.2	0.5

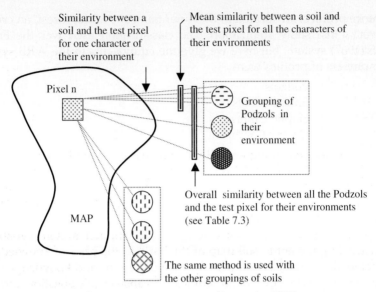

Fig. 7.6 Illustration of comparative reasoning (Bonneric, 1978).

Results and discussion
The above approach has shortcomings that deserve to be mentioned.

The pixels are assigned their maximum similarity on a scale of 0 to 1. In some cases, in the row of the matrix of resemblances, there might be no high value. One may find, for example, a maximum resemblance of 0.15 or 0.10. It must then be accepted that one does not have for reference a situation close enough to what the pixel is in for making a proper diagnosis. In other words, a threshold must be chosen below which the assigned value is not significant. This threshold may be 0.70 or 0.75. This means that the map predicted and drawn generally contains 'blank areas'. The only way to correct this is to find out in these milieus, rather unknown, new references that will be added to the set of profiles. The exercise is then repeated with these additional data. However, one can imagine another application of the presence of these 'blank areas'. This approach may still be of use to find the boundaries of a Small Natural Region (SNR), which may be defined as the aggregate of all the pixels whose degree of similarity to a reference core or to at least one of the sites constituting it is above the threshold value.

The second difficulty pertains to the situation wherein the environment of a pixel resembles equally all the known environments (more or less equal values in the row of the matrix). The highest resemblance, which is very slightly higher than the others, is not necessarily the best. The method then lacks discriminating capability. In this case, quite rare in real life, the automatic calculation has reached its limits.

The third problem is that of weight of the criteria taken into account. The elements used to describe the natural environment do not necessarily have equal importance. Weights must be assigned to the variables that appear significant in terms of pedogenesis. Fortunately, we are not totally helpless in making choices in this matter. The question is, 'How must the variables characterizing the environments of these profiles be combined and with what weights for getting the profiles grouped in clusters corresponding exactly to the recognized taxa?' To answer this, it is necessary to implement a kind of discriminant analysis. In short, one utilizes the information provided by the profiles before using them for starting the process of measuring similarity.

Having drawn this map by automated means, it was easy to compare it pixel by pixel with the results of soil survey. Table 7.4 gives the results of this comparison in the form of a confusion matrix for the 817 pixels under forest, that is, in a practically undisturbed environment.

Table 7.4 Confusion matrix pertaining to prediction of soils in the PRNP (Legros and Bonneric, 1979)

Confusion matrix		Test of computerized generalization			
		PZ	EP	HU	DC
Conventional field mapping	PZ	43	1	0	0
	EP	25	252	45	0
	HU	10	52	90	4
	DC	3	4	13	375

PZ = Podzol, EP = Entic Podzol, HU = Haplic Umbrisol, DC = Dystric Cambisol

The results are not at all bad (overall accuracy of 83%). The comparative procedure then has some advantages in spite of the shortcomings reviewed above. First, it utilizes very completely the information available. When all calculations had been done, the simulated map was constructed on the basis of nearly 570,000 elementary comparisons. The proposed method thus uses a big 'bundle of presumptions'. Because of this, it simulated rather well the procedure of the expert who constantly makes hypotheses (gambles) by using his own earlier experience. Also, the pixels where there was no agreement between the prediction of soils and the soil map were examined one by one. In the beginning we thought that the map would be an infallible means for verifying our attempt at automated mapping, but it became necessary to ascertain that this was not always true. It was not possible in the case of many pixel groups to state 'The map is correct and the simulation is wrong' without going back to the field. This happened mostly in the scarcely characteristic situations for which any extrapolation procedure (cartographic or algorithmic) is risky. This is the problem with intermediate-altitude zones between the the Podzol zone and the Entic Podzol zone. Thus simulation has the main advantage of pointing out the areas that pose problems and for which it is necessary to be sure that the mapper had indeed taken observations and was not restricted to bold extrapolations.

7.3.2 Extrapolation from a small reference area

Principle and method

The studies that will be summarized here were done for the purpose of a thesis (Lagacherie, 1992). As in the case above, they were aimed at automated drawing of the map by a computer and, above all, at analysing and simulating the mapping procedure. The context of the work was the following: suppose a representative Reference Area (RA) was mapped by conventional means at the core of a Small Natural Region (SNR). An attempt would then be made to extract from the portion of the map already prepared the data necessary for automated mapping of the remainder of the Small Natural Region.

Establishment of soil-landscape laws. One could yet say that one wishes to build a soil-landscape model. The starting point is the same as in the preceding method: the characteristics of the environment of the pixels throughout a region (slope, rock type, etc.) are known and, in this region, the characteristics of the soil also are known for a sector called 'reference sector'. Under these conditions, we can locally link the properties of the soil to those of the environment to establish the soil-landscape laws that are later applied over the entire region. This is done by applying a segmentation algorithm (tree-based model) (Breiman et al., 1984). To understand what this means let us take an example with just two units MU1 (Map Unit 1) and MU2 (Map Unit 2), and a single characteristic of the environment, namely altitude. In this case the algorithm searches by trial and error for the value to be used of the altitude variable to best separate the pixels containing MU1 from those containing MU2. Obviously this is only possible if MU1 and MU2 are not located at the same altitude. The result of the segmentation is presented in a YES/NO form (Table 7.5).

Table 7.5 Example of the results of an elementary segmentation

If altitude < 125 m, one finds			If altitude > 125 m, one finds	
MU1	MU2	and	MU1	MU2
80% of cases	20% of cases		30% of cases	70% of cases

But the algorithm can work with several soil units and with several variables of the environment. The calculation becomes longer and more complicated. It is necessary to specifically compare and hierarchize the various elementary segmentations. If, say, the altitude sorts out the pixels less efficiently than the geological substratum, the algorithm will be applied first to the substratum. Altitude will come in later to improve the sorting in each of subgroups already obtained. The altitude thresholds retained could then be specific in each branch of the dichotomy. Thus, a classification tree will finally be obtained that will enable us deduce rules such as *if the substratum be alluvium, the altitude be greater than 100 m and if the slope be less than 10 degrees, then the Map Unit is MU4 in 71% of the cases, MU2 in 24% of the cases and another in 5% of the cases.* If a pixel having the environmental characteristics listed above is found outside the Reference Sector, we could bet that it belongs to MU4 while knowing that there is only 71% chance of being correct and 29% chance of being wrong.

Use of such an algorithm gives rise to a few problems (Lagacherie and Holmes, 1997). In particular, if the segmentation is allowed to run its course, one could get pixel groups actually composed of a single individual. Theoretically, the classification will be perfect; but in practice it would be unusable. First of all, there will be little chance of later finding another pixel having all the characteristics necessary for being allocated to the group. Secondly, we well know that such rigour in the number and accuracy of the deciding criteria will pertain to no real thing in the field. Thus a 'stop' criterion is necessary. In principle, it is a matter of stopping the division of each of the branches of the tree as soon as further dichotomy no longer reduces the internal heterogeneity of the groups already created. To learn more of the subject, the reader should refer to the papers cited.

Establishment of the laws of neighbourhood. The soil mapper knows that different neighbouring soils follow one another in space in a precise fashion. Calculation of contiguity (Chap. 6, § 6.2.3 and Fig. 6.9) enables us verify this phenomenon. In this context, Lagacherie's procedure is as follows: a soil unit, for example MU1, is selected in the Reference Sector; then a square or approximately circular window is pulled along the pixel field; it is stopped each time MU1 is at the centre of the window and then we examine to see which soils are found in the immediate surroundings; this enables construction of the matrix in Table 7.6 assuming, in the given example, that there are only 9 soil units in the Reference Sector.

Table 7.6 Inventory of the neighbours of each Map Unit

	MU1	MU2	MU3	MU4	MU5	MU6	MU7	MU8	MU9
MU1	987	134	12	0	0	121	78	45	701
MU2…	…	…	…	…	…	…	…	…	…

This table is read as follows: MU8 is found 45 times as neighbour to MU1 when the latter is at the centre of the window. MU1 is found 987 times as its own neighbour (it occupies many contiguous pixels). The numbers can be transformed to %, that is to frequencies and, by extension, to probabilities because the number of pixels is very large.

It is possible to change the size of the window to determine which is the most probable unit taking into account what we find at the centre if a distance of 10 m, 50 m, 100 m, etc., is considered. Moreover, various attempts have shown that exceeding 400 m is of no use because, from this distance on, there is no relation whatsoever between soil units (the examples investigated pertained to large-scale maps). This is to move closer to what we have said above (§ 7.2.2). Lastly, the direction can be taken into account for identifying, say, the most probable neighbour of U7 100-m north of it or, better still, at 100 m toward the footslope. In sum, we obtain a matrix giving the neighbourhood relationships for each distance and in each direction taken into account.

The information thus collected is voluminous and must therefore be managed properly in the computer system. It remains now to utilize it outside the Reference Area. This is done as follows. When an auger hole is examined, say in a soil identified as belonging to MU8, one can declare by utilizing the statistical data collected, 'at 50-m from the observation made, along the steepest slope, the greatest probability is of finding the unit MU4 (73% of cases). I therefore bet on it, knowing also that I have 27% chance of being wrong.'

Implementation of the simulation process
Summarized in Fig. 7.7 are the five steps in calculation of the soil-landscape rules. These rules are first established by means of the tree-based model by comparing the soils data extracted from the Reference Area with the corresponding data on the environment. Then, in a second step, these rules are used to estimate the nature of soil at all points in the natural region for which the properties of the geological and topographic environment are known. This results in a predicted soil map. At the same time, the risk of error in this prediction is estimated, since the method used permits it. Thus a second map that defines the uncertainty in each pixel of the first map is obtained. Lastly, field work is undertaken to check out these two maps. It is then not necessary to do systematic mapping. It is sufficient if some validation sectors are chosen and surveyed.

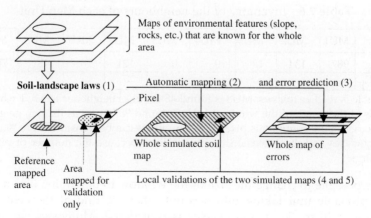

Fig. 7.7 Establishment and use of soil-landscape laws in mapping by simulation; adapted from Lagacherie (1992).

It is seen that methods of this type result in a kind of *disaggregation* of the soil map (Bui and Moran, 2001). In other words, the *tree-based model* enables us rediscover what criteria the mapper had used for getting the soil delineations. In some cases, the mapper takes care to himself reveal the secrets and to provide a manually constructed dichotomic key to indicate on what criteria the units were differentiated (Bruckert, 1987). But then, the probabilities of success are not quantified.

Figure 7.8 shows how the laws of neighbourhood could be implemented. The laws are first established in the Reference Area. But to apply them, auger holes must be available around which extrapolation can be done. It will certainly be impossible to go to the field just to make an auger hole, then to return to the office to do a portion of simulation and then to start all over again. Actually, the soil map prepared in the conventional manner is already available for the entire zone for which the map is desired to be prepared automatically. To do a simulated auger hole study goes back to providing the system information on the nature of the soil in the only pixel chosen. On this basis, we can test many strategies for automated mapping. The simplest is to define the grid at the nodes of which the auger holes are given, thereby enabling prediction of the nature of soil over the rest of the map. Another method consists of simulating a few auger holes, constructing the map on this basis, then redoing auger sampling in the areas where the calculated uncertainty is highest.

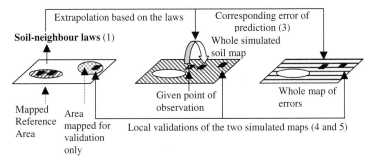

Fig. 7.8 Application of the laws of neighbourhood (adapted from Lagacherie, 1992).

More generally, great attention is now paid to the methods aimed at determining the uncertainty at each point of the prepared maps. A special issue of *Geoderma* (**103**, 2001) was devoted to these questions.

In the case studied, the Reference Area represented 800 ha. The Small Natural Region, middle valley of the Hérault river (France), covered 40,000 ha. The validation sectors were three in number and had areas of 45, 44 and 42 ha.

Results
This study highlighted several conclusions (Lagacherie et al., 1995). First of all, it showed that it is possible to simulate one part of the mapping process

including taking into account experience in the field from studies done previously by the mapper. Then, it is very interesting to ascertain that the neighbourhood laws are more efficient than soil-landscape laws in predicting geography of soils. Of course, this is partly due to the fact that the environment of soil is characterized at information-science level in a very simplistic manner. In the field, the mapper makes the diagnosis on many more finer indices. However, the lesson must be accepted: knowledge of the lateral structure of the soil mantle can facilitate mapping greatly. Last, but not least, this thesis leads to building a tool very suitable for study of sampling. On this topic, the author shows how the risk of error diminishes when the number of auger probes done increases (Fig. 7.9).

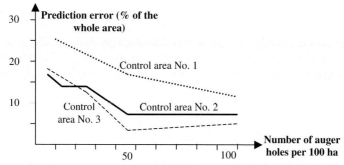

Fig. 7.9 Diminution of errors in prediction of soils with increase in number of auger holes used to implement the neighbourhood laws: control areas 1, 2 and 3 studied in the middle valley of the Hérault river, France (Lagacherie, 1992).

There obviously is a limit (more or less sharp) beyond which further effort is not worthwhile. The residual error still remains very high beyond this limit. This is normal: here observations are not checked (map constructed in the conventional way) but predictions are evaluated (map drawn automatically). The simulation system may also be used to select the location of auger holes: where to do them and in what order? To improve the tool and to enable the simulations to be made easily and rapidly, computer studies were done (Ledreux, 1992). The soil-landscape laws were utilized through an expert system. Following these studies, efficient tools rich in potential are available to dig deeper into study of the mapping procedure.

Supplementary studies
The principle of prediction from the neighbourhood laws and soil-landscape laws was again used to predict for every pixel not only the name of the soil type but also a property such as water content at the wilting point (Lagacherie and Voltz, 2000). The value predicted at each point is the mean

of the values characterizing the different soil types likely to be present, with the corresponding probability as weighting. This method is valid when the transitions from one soil to the other are gradual.

7.3.3 Other studies on computer-assisted mapping

The investigations we have examined above are actually attempts to predict the spatial distribution of soils or soil properties. This is an exercise attempted by many researchers! One clear article summarizes the statistical methods that can be used in this area and indicates how a validation is organized based on calculation of the Root Mean Square Error (RMSE) after the 'jackknife method' has been applied to subdivide the data into two subsets, the first for basing the prediction on, and the second for conducting the validation (Bishop and McBratney, 2001).

It is from such a perspective that Chaplot and Walter (2002) attempted in Brittany to predict the values of a Hydromorphic Index integrating the redoximorphic features of the profile. The prediction was based on multiple regressions relying on topographic parameters. The results were locally good in catchment areas corresponding to the data subset for prediction as well as to the data subset for validation. But they were more difficult to extrapolate beyond the watershed.

By using discriminant functions based on topography and by combining them with some rules concerning the rock and coming close to an expert system, the name of the soil was predicted in the Vosges mountains in France. But success was limited to 55% of the check sites (Thomas et al., 1999).

One approach with philosophy close to that of our first example (§ 7.3.1) was developed in USA (Zhu et al., 1997, 2001). The type of soil and even certain characteristics such as thickness of the A horizon were predicted on the basis of characteristics of the environment. The most original part of the work was that the authors never referred to a reference area or reference pixels. They only referred to series identified in USA. Similarity of the pixel in question to each one of the environments of these series was established on the basis of rules drawn up by experts. It was represented by a value from 0 to 100. The pixel is allocated according to the highest similarity found for it.

The interesting approach of two Australian scientists (McKenzie and Ryan, 1999) should also be mentioned. Relief was given by DEM; climate was reconstructed using mathematical functions tested in the field; geology was extracted from the corresponding maps. This enabled stratified sampling, that is, optimal definition of 165 sites that were studied in the field for an area of 50,000 ha to be mapped. Then the tree-based model of Breiman was used as in the study presented above (§ 7.3.2). Thus the method offers an aid for selecting in the best manner the observation sites (location, number) and then for making delimitations. In addition, it is largely automated and appears reasonable in the methodological effort it involves. But prediction of soil depth is still of poor quality.

Similarly, Gaddas (2001) predicted the nature of the soils in 350,000 ha of the Rhône valley (France) by overlaying the zonation maps corresponding to different pedogenetic factors and by using the expert knowledge gained in the field from just 150 observation points. Then the result obtained was compared with the soil map prepared earlier by other scientists. The degree of success varied between 45% and 80% according to the conditions retained for the calculations (weighting according to the relative areas of Map Units or no weighting, taxonomic level considered for the comparison).

7.3.4 Validity of sample areas

The real representativeness of 'representative profiles' or 'Representative Area' from which extrapolation is done was not discussed in the preceding methods. We must go further.

Calculation of the minimum area to be surveyed
If the study area forms part of a Small Natural Region having definite specificity in characteristics of the natural environment, it may be relevant to ask what area would be advantageous for the mapper to get a representative sample of the whole. In other words, what is the landscape mesh? In yet other words, what is the landscape pattern that, as in a tapestry, is repeated identically to form the entire region?

The method presented here (Pourgaton, 1977; Legros, 1978) was implemented a posteriori on a prepared map. Thus it has no predictive value, except if it is applied to natural environments similar to those it was developed in. The map is presumed to have few complex Map Units containing several soil units. It is then divided into large pixels. It is no problem, however, if several Map Units be represented in each pixel. A moving window (defined in Chap. 6, § 6.2.4) is then dragged across the map. From this the number of soil units found on average in the area of the window can be calculated. Then the exercise is repeated with a window of different size. Finally, curves such as those in Fig. 7.10 can be drawn.

Checks were done in the regions of France where the maps had been generated on 1/10,000. The results obtained (Favrot, 1989) show that, on this scale, if an area of about 1000 ha is judiciously chosen, one can obtain units that represent three-fourths of all those listed in the Small Natural Region in question and cover 90% of the area in that SNR. On this basis the appropriate area to be surveyed for defining in good conditions a Reference Area for Drainage (Chap. 10, § 10.7.6) could be standardized at about 1000 ha.

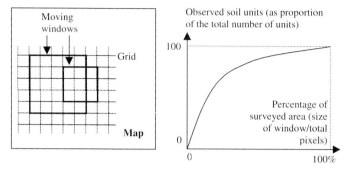

Fig. 7.10 Relationship between the area surveyed, shape of the moving window and the number of units observed.

Representativeness of Reference Areas (RAs)
An extrapolated model is constructed from an RA. This model is based on soil-landscape correlations and may be close to one or the other of those that have been presented above (§§ 7.2, 7.3). Then the identified boundaries of the Small Natural Region (SNR) in which the work was done are compared with the boundaries of the region that the model defines by calculation as similar to the RA. Normally, these boundaries must practically coincide. If the calculated region is smaller or larger than the SNR, the RA is not perfectly representative. It was with this thinking that a study was done in the author's laboratory (Lagacherie et al., 2001).

Representativeness of networks of soil-observation sites (Arrouays et al., 2001)
In France, it was planned to cross-rule the territory by a soil survey network with observation points and measurements established at the intersections of a regular grid. It is certainly a proper approach in certain cases, for example to work out the distribution of elements linked to the Tchernobyl cloud. But to determine correctly the consequences of pollution, the vulnerability of the environments affected must be properly characterized. In other words, systematic sampling must allow establishing the representativeness of the various environments affected. The practical problem is then to determine the ideal size of the grid mesh: if too fine, it will result in development of too costly a network; if too coarse, it will not allow appropriate characterization of the milieu (soil types, land-use type). Three approaches were suggested for such a situation (Arrouays et al., 2001):

- Determine if the grid points will pertain to all the 'Soil × Land Use' combinations that could be found on the national territory. The list of these combinations was prepared by using the soil map on 1/1,000,000 along with the *Corine Land Cover* map. Four hundred and twenty-six combinations were found. Then those combinations not represented by the points of the network were found and the corresponding area calculated (Fig. 7.11).

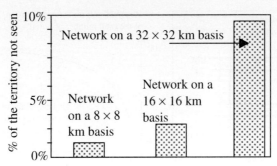

Fig. 7.11 Simulation of the representativeness of an observation-site network (Arrouays et al., 2001).

- A similar approach consists of calculating, for each Small Natural Region (SNR) of the national territory, the percentage of area corresponding to the Soil × Land Use combinations touched by the grid points. Actually, certain SNRs could contain combinations that do not occur at the grid points. Conversely, other SNRs are characterized fully. But it must be kept in mind that this characterization is often indirect: some SNRs do not contain observation sites but simply combinations characterized elsewhere.
- Yet another approach was developed by the authors. It consists of examining if the grid points would be representative of their immediate environment. The question boils down to, 'Will the observation point fall in the soil unit most represented in the mesh of which that point is the centroid?' It is obvious that the coarser the mesh, the lower the probability. But the calculations lead to quantification of this.

In sum, there are conceptual methods for selecting the mesh size corresponding to gridded soil-observation networks.

Representativeness of maps already made (King and Saby, 2001)
France covers 55 million hectares while only 8 million ha have been mapped in the systematic inventory on 1/100,000 (40 sheets surveyed, but all not yet published, of 293 sheets theoretically to be surveyed, cf. Chap. 10, § 10.7.4). It could be asked if this is sufficient for having a correct picture of the soils of the country.

To see how much of the picture is given correctly, King and Saby (2001) used the soil map on 1/1,000,000 available for the entire country, but not very detailed. They then examined if the 40 sheets on 1/100,000 allowed detailed characterization of all the MUs of the 1/1,000,000 map, keeping in mind that all Map Units of the 1/1,000,000 map were only named following the FAO system (Chap. 10, § 10.1.3). To be more exact, the analysis pertained to the following points:

- Comparison of relative areas of various Map Units in all the 40 sheets surveyed and in all of France; in particular, they looked for the Map

Units of France not appearing at all in the sample available at 1/100,000;
- For better characterizing the soil of *xy* type at a given point in the national territory, it will be risky to refer to the same *xy* described in one of the 40 sheets available if this sheet is 1000 km away from the point considered! Under these conditions, it is useful to take into account the geographical distance between a pixel of the 1/1,000,000 map in which *xy* is found and the 1/100,000 map sheet that is nearest and contains the same Map Unit *xy*. A modification of the method consists of studying not only the distance to the nearest map sheet but also the distance to the sheet in which the Map Unit *xy* covers the largest area (Fig. 7.12).

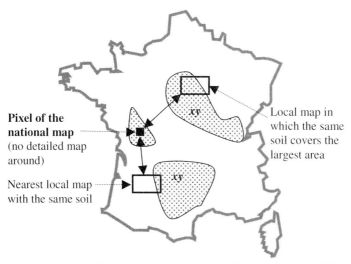

Fig. 7.12 Method of checking representativeness (King and Saby, 2001). External boundaries of the country from Institut Géographique National.

The method has one disadvantage. The map on 1/1,000,000 scale was prepared using specifically the data corresponding to the maps on 1/100,000. At a pinch, a soil that would not be identified at 1/100,000 could not appear on the synthesized map on 1/1,000,000. Therefore, the 'representativeness' of the 40 sheets is artificially exaggerated. However, this map on 1/1,000,000 was constructed using scales other than 1/100,000. So the bias is not too much.

This interesting study led to the following conclusions:
- At first analysis, the existence of these 40 sheets on 1/100,000 gives a very good idea of the soils of France because the most widespread soils appear in the sample;
- But 25% of the area of the country is still poorly characterized in the 1/100,000 maps (Map Units not observed or characterized too far from the area in question);

- Lastly, it can be established that introduction of an additional 1/100,000 sheet greatly modifies the results of the analysis: in other words, the sample is still a little too small, consequently giving unreliable results.

7.4 MEASUREMENT OF THE SIMILARITY BETWEEN SOILS

It was seen, first in Chapter 1, § 1.4.4 and then in § 7.3.1 above, that the concept of similarity must be used in mapping. This pertains particularly to verification operations. They often amount to checking that the soil observed in a particular area is indeed what must be found from the map. On the one hand, the human mind assesses the similarities, approximately of course, but with great ease: a particular object is immediately recognized as of the same species as another similar object. But the computer can only detect family resemblances after very laborious calculations! But we shall here present the latter because their use in our discipline is great. Soil scientists drew inspiration from the work of statisticians who developed the clustering methods, for example Principal Component Analysis.

7.4.1 Dissimilarity between two soil data

Quantitative characteristics
Let X_i be the value of a characteristic of profile i and X_j the corresponding value in profile j. The value $X_i - X_j$ is a measure of the difference or, more correctly, the *distance* between i and j for the characteristic X. Distance in the mathematical sense is thus the opposite of *resemblance*. However, talking of the distance between two objects can lead to confusion with the geographical distance. In many cases it will be better to follow the vocabulary of English speakers and use the terms *dissimilarity* and *similarity*. Moreover, dissimilarities are not taken for verifying the axiom (cf. properties of the sides of a triangle):

$$d(i, j) \leq d(i, k) + d(k, j)$$

In other words, it is possible to define dissimilarities that do not admit of coherent geometric representation in a space of two or more dimensions. This precaution having been taken, the word 'distance' can still be used knowing what it actually covers.

But the quantity $X_i - X_j$ proposed above has several shortcomings: it can give negative values, and it depends on the units used. Thus the dissimilarity calculated for clay content (%) cannot be compared with the dissimilarity calculated for phosphorus (‰). Mathematicians therefore tried hard to find indices devoid of these disadvantages. The problems of sign were avoided by taking the absolute values or by squaring the result. Then, the value obtained is divided by a factor rendering the dissimilarity dimensionless.

The Euclidean distance is the most popular:

$$d(i, j) = \frac{(X_i - X_j)^2}{\sigma_X}$$

where σ_X is the standard deviation expressing the variability of the character X in the population from which i and j are drawn.

The Gower distance is also used (Bonneric, 1978), given by

$$d(i, j) = \frac{|X_i - X_j|}{X_{max} - X_{min}}$$

where X_{max} and X_{min} are the greatest and smallest values respectively obtained for X in the population studied. The Gower distance thus gives a value between 0 and 1; the similarity can then be expressed by subtracting the number obtained from 1, for example if the dissimilarity is 0.8, the similarity is 0.2.

Of the quantitative variables, particle-size distribution is a difficult matter. By applying what was said above, the dissimilarity between two particle-size compositions can be calculated as follows:

$$d(i, j) = \frac{(A_i - A_j)^2}{\sigma_A} + \frac{(L_i - L_j)^2}{\sigma_L} + \frac{(S_i - S_j)^2}{\sigma_S}$$

where A = clay content for i or j
 L = silt content for i or j
 S = sand content for i or j
 σ = standard deviation for A, L and S.

Thus this dissimilarity cannot be confused with the graphic distance measured on the texture triangle, which in the French orthonormal triangle is:

$$l(i, j) = [(A_i - A_j)^2 + (L_i - L_j)^2]^{1/2}$$

The dissimilarity d plays a uniform role with sands, silts and clays. A sand appears as close to clays as to silts, something quite questionable. In the case of the graphic distance (l), silts are particularly far from clays, at least in the French orthonormal triangle. This is also equally hard to accept. Under these conditions, the least flawed formula will perhaps be:

$$d'(i, j) = \frac{1}{2}\left[\frac{|A_i - A_j|}{100} + \frac{|S_i - S_j|}{100}\right]$$

In this case, the dissimilarity between pure sand and pure clay is double that between pure sand and pure silt or between pure silt and pure clay, which seems logical.

Qualitative characteristics that can be ranked
In many cases, qualitative variables take values that can be ordered (ranked variables). For example, redoximorphic mottles can be grouped into (1) very large, (2) large, (3) medium or (4) fine. Then the *rank difference* is introduced, given by:

$$\frac{|R_i - R_j|}{R_{max} - R_{min}}$$

where R_i is the rank in profile i and R_j that in profile j. For example, if profile i has very large mottles and profile j medium mottles, the rank difference will be

$$\frac{|1-3|}{4-1} = \frac{2}{3}$$

On the same principle, one can go back to slightly more complicated formulae (van den Driessche and Garcia-Gomez, 1972). Obviously the rank difference can be used for quantitative variables to the extent that their values are grouped into classes. This allows combined processing of quantitative variables and ranked qualitative variables (King, 1986). But one will not know how to proceed further and arbitrarily rank the values of variables such as 'type of structure', 'colour', 'humus type', etc. Thus it is unusual to use rank difference for dealing with all soil properties.

Similarity between purely qualitative variables
If the properties are purely qualitative, and if no hierarchy can be defined for them, measurement of dissimilarity is impossible. We are satisfied with stating that the two objects are similar or dissimilar. For example, the dissimilarity between calcareous and granitic parent materials will be defined as follows (Table 7.7).

Table 7.7 Choice of dissimilarities between qualitative properties

	Calcareous rock	Granite
Calcareous rock	0	1
Granite	1	0

This table is read as follows. When one profile is on limestone and another on granite, their dissimilarity is maximum and has the value 1 for the property 'parent material'.

7.4.2 Dissimilarity between horizons

When the dissimilarities for all the characteristics taken separately are available, it is possible to calculate the overall dissimilarity between two horizons. But since variables of the three types mentioned are simultaneously involved, the synthesis cannot be done without caution. We can calculate a mean of the dissimilarities after thought and after possible introduction of a system of weighting.

Then a very classic statistical problem arises. What is to be done when one value is missing, for example when the particle-size distribution has not been determined in one of the two horizons compared? One solution is to forget the variable in question and to carry out the comparison on the remaining pairs of variables. Another method is to accept that the dissimilarity is maximum for any pair in which one value is missing. It gives the impression that we have not 'cheated with the data'. But the reality is more complex and all depends on what is being investigated.

Another problem, quite as classic as the first, is that of variables having a zero or absent value for every element of the pairs that are being compared. Absence of gravel is recorded in both cases, calcium carbonate content is zero for each profile, etc. Theoretically, the dissimilarity is zero and the two elements of the pair are similar. But is it correct to consider identical two horizons that have the following properties in common: no salt efflorescence, no ferruginous cuirasse, no banana trees, no periglacial stone pavement, etc.? This kind of difficulty is well known in phytosociology, a discipline in which we often ask ourselves whether absence of an indicator plant in a vegetation survey constitutes useful information or not.

In sum, it is not very reasonable to calculate the dissimilarities with a ready formula without examining all the particular cases, that is, all the variables and all the values they can take! Otherwise, there is great danger of doing an unsupported synthesis.

7.4.3 Dissimilarity between profiles

Since we now know how to measure the dissimilarity in two values of a characteristic and the dissimilarity between two horizons, it is possible to tackle the problem of the dissimilarity between profiles.

Consideration of depth
A method is suggested for comparing profiles at fixed depths. For example, we examine what exists at 30 cm, 60 cm and 90 cm. This brings us back to compare the horizons observed at these depths. But the risk of finding as different two very similar profiles is large (Fig. 7.13).

To avoid this danger, or at least for obtaining a kind of statistical smoothing, the most usual solution consists of multiplying the compared depths. For example, the characteristics of the horizons are examined every 5 cm. Another method is to calculate the similarity between the characteristics of the first profile observed at a given depth to those of the second profile at the same depth, then at a little greater depth and finally a little less (Fig. 7.14). The difference retained as characteristic is the smallest difference found (King, 1986).

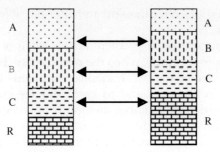

Fig. 7.13 (Flawed) comparison of two profiles at three depths.

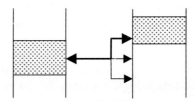

Fig. 7.14 Finding the most similar horizon at a given depth (King, 1986).

Having so obtained the dissimilarities for different depths, it is possible to synthesize, or let us say obtain an average, to summarize by a single number, the overall dissimilarity between the two profiles.

There is a very similar approach that erases the horizon concept. The profiles are first cut up into standardized slices of, say, 20 cm. The characteristics corresponding to each depth are considered to directly represent the profile. For example, a profile with five depth intervals will be characterized by 5 values of pH, 5 values of C%, etc. This permits us to represent the profile by a single row of data and thence constitute a matrix with the data on characteristics in the rows and the different profiles to be grouped in the columns. The grouping is done using the VLADIMIR program (King and Girard, 1988).

Using presence/absence of horizons
Another method is to consider as identical the profiles that have the same sequence of horizons. In this situation, each profile is described as having or not having the different kinds of horizons described. Table 7.8 shows how to classify profiles that can have 7 kinds of horizons numbered 1 to 7.

Table 7.8 Matrix indicating the presence/absence of reference horizons in the profiles (Girard, 1983)

Horizon	H_1	H_2	H_3	H_4	H_5	H_6	H_7
Profile 1	+		+	+			+
Profile 2		+	+	+			+
Profile 3		+	+	+	+		+
Profile n	+	+				+	

From here it is possible to define the similarity between two profiles by using the methods employed by phytosociologists for interpreting their vegetation surveys, for example, the Jaccard index. Then the thickness of horizons can be considered and used for weighting in calculation of similarity. A horizon that is absent is theoretically present with a thickness of 0; a horizon so thick that it occupies the entire profile is weighted with the value 1.

Clustering methods

In such a situation the crux of the problem is to group the horizons observed into a few principal types. For this we can use *clustering* methods. The principle is as follows. A few horizons judged typical are chosen. They will serve as reference horizons. The others will be selected according to their maximum similarity to one or other of these reference horizons for constituting *clusters*. We know that we have the tools for this (§ 7.4.2). Fig. 7.15, left, schematizes the results obtained. The x- and y-axes represent the combination of several variables pertaining to the horizons (say, C%, pH, clay %, etc.). The drawing thus corresponds to what specialists call two-dimensional ranking. But the groupings obtained have all chances of being imperfect the first time. Actually, the references chosen by the mapper to serve as the nucleus for grouping of horizons are not always representative of the average cases. They often represent 'nice cases', that is, extreme cases and not the *central* one. For example, the most clayey horizon in the area is taken. This can be found out by doing a principal-component analysis on the horizon data.

Then it is necessary to recommence the clustering process by choosing references at the centre of the scatter of points previously obtained

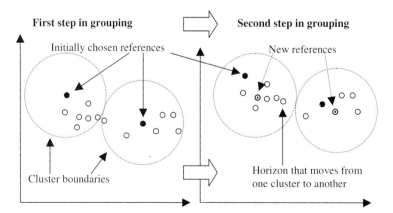

Fig. 7.15 Shifting of nuclei for grouping so that they are representative of the horizons linked to them.

(central points). Some horizons might change their cluster (Fig. 7.15, right). The process should be repeated several times until the groupings are appropriate and stable. Ideally, the iterations are done one by one with human checks at each step. The DIMITRI program was developed for this purpose (Girard and King, 1988).

Generally it is practical to use an entirely automated algorithm. There are several methods for this. We can choose the first references at random because they will almost certainly be replaced by others that will be revealed to be better [cf. the *nuées dynamiques* (dynamic clouds) method earlier developed by Diday]. We can also choose the references based on various statistical methods such as Principal Component Analysis.

7.4.4 Introduction to fuzzy logic

Fuzzy k-mean
All the methods presented above are already old. But they have been perfected in the context of *fuzzy set theory*. Let us return to the problem of grouping of horizons.

Let d_{ic} be the dissimilarity between the *i*th horizon and the cluster c composed of horizons and represented by its centroid; let k be the number of clusters. We can now think of calculating the quantity given by

$$m_{ic} = \frac{d_{ic}}{\sum_{c=1}^{k} d_{ic}} \tag{1}$$

Now m_{ic} will express the dissimilarity between i and c in relative terms as a proportion of the dissimilarity between i and all the clusters. If the dissimilarity between i and each cluster (c = 1, 2, 3, ... k) is calculated the same way, we get the membership relation of i to each cluster. We find here the concept of *fuzzy vector* examined earlier (Chap. 2, § 2.1.4). In other words, the classification of i becomes *fuzzy*, hence less simplistic, the approximations being explicit (example: where i is simultaneously close to two clusters). Actually, one prefers to calculate the quantity

$$m_{ic} = \frac{[(d_{ic})^2]^{-1/(q-1)}}{\sum_{c=1}^{k} [(d_{ic})^2]^{-1/(q-1)}} \tag{2}$$

This expression being a little strange, let us see how the values of m_{ic} vary with values of q. Table 7.9 gives some examples. It is assumed that there only are two clusters, the one at 0.1 and the other at 0.9 dissimilarity (total = 1).

Table 7.9 Calculation of m_{ic}

q	$-2/(q-1)$	d_{ic} (cluster at 0.1)	d_{ic} (cluster at 0.9)	Σd_{ic}	m_{ic} (first cluster)	m_{ic} (second cluster)
1.5	−4	10,000	1.524	10,001.5	0.999	0.001
2	−2	100	1.234	101.2	0.988	0.002
11	−0.2	1.588	1.021	2.609	0.608	0.400
101	−0.02	1.047	1.002	2.049	0.510	0.490

When q tends to infinity, we are led to the fuzzy vector with value [0.5,0.5]. The grouping becomes totally vague: the point i belongs to two of these clusters equally. When q tends to 1, the fuzzy vector tends to [1,0]; i is then completely attributed to the nearest cluster. Generally, values between 1 and 2 are given to q. This popular method is widely employed in soil science. It is known by the name *Fuzzy k-mean*. It must be taken for what it is: a very ingenious and very arbitrary mathematical procedure because the degree of fuzziness (expression 2) is chosen instead of even measuring it approximately (expression 1). This is justified when contrasts are exaggerated for drawing soil boundaries (Odeh et al., 1992; Lagacherie et al., 1997).

Graphical representation
It is possible to measure the membership of a horizon to each of the clusters grouping the horizons studied. Let us assume that these clusters are named Togo, Nundi and Wewak. We can then use the following representation of the profiles. It was suggested very colourfully (Fig. 7.16) by Triantafilis and McBratney (1993).

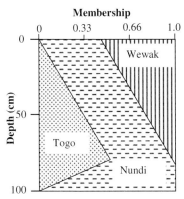

Fig. 7.16 Representation of a profile as function of the membership of its horizons to reference clusters (Triantafilis and McBratney, 1993).

7.4.5 Validation of the calculation of similarity

All the calculations presented above are quite complicated. At each stage of the operation the problem is arbitrarily simplified. We must then ask ourselves if, at the end of the calculation, the algorithms for comparison of profiles lead to results that can compete with the diagnosis easily provided by a soil mapper seeing the soils or examining their description sheets. This is why validation tests were done by our team.

Principle of validation

In what follows, we shall distinguish the mapper (author of the map) from the experimenter (person returning to the field for devoting himself to a certain number of checks and observations).

(1) An experimenter goes to the field to dig auger holes in the known units corresponding to a soil map already prepared.
(2) The auger holes are recorded on STIPA sheets (Chap. 4, § 4.4.2) and computerized, that is, stored in a digital file.
(3) In addition the auger holes considered perfectly representative of the units defined by the mapper are computerized. These *reference auger holes* often actually exist. But they also may correspond to average cases reconstructed on the basis of data provided for each unit of the legend and report of the map (*type auger holes*).
(4) Lastly the computer is asked to which reference auger hole, that is, to which unit of the legend each auger hole newly dug in the field by the experimenter is most similar. One knows the result that must be found when the soil unit that has been examined is identified. Then one has the means for verifying if the automated determination of similarity is correct or not. The algorithm developed for this has been named SOSIE. It gives values of similarity between 0 and 1.

Extension of the method

Although the principle of the experimental procedure is simple, putting it to practice comes up against one major difficulty. Actually, the mapping units that serve for testing the method cannot be considered pure (they are at best *consociations* and sometimes *associations*). The new auger hole often cannot be recognized by the computer system because it is quite different from what characterizes the unit (type auger hole or reference auger hole). This is why we were led in the second experimental phase to compare not only two but four methods of diagnosis giving the soil type found at a specific point in the control sector:

- *'Experiments' diagnosis*: the observer in charge of trying out SOSIE was taken along. Before digging the test auger holes, this observer visited and examined each Map Unit (MU) of the already-drawn map and set

out to understand how to recognize each one of them. He is then capable of expressing an opinion on the attribution to such or such MU of the auger holes dug and described.
- *'Map' diagnosis*: the proposed unit is that marked on the map.
- *'SOSIE' diagnosis*: the unit identified is that given by the algorithm. This algorithm determines the similarity with each MU of the legend, then selects the highest similarity and thence the corresponding unit.
- *'Identification key' diagnosis*: During mapping, the soil boundaries are established on the basis of a certain number of precise rules that can help establish an identification key. For example, *'if the soil is very red, very gravelly and located on a flat summit, the Map Unit is MU4'*. But, though the rules for defining units are well established by the mapper, he is the first to have to disobey them locally for drawing the boundaries with necessary simplifications! Actually, repeating the mapper's reasoning accurately for each auger hole may lead to results different from the map consulted.

To sum up, one is led to construct and interpret tables similar to that presented below (Table 7.10).

Table 7.10 Consideration of four methods for identification of Map Units

Map Unit name ⇒	Observer's opinion	Given by the map	Given by the SOSIE algorithm	Given by the identification key
Auger hole 1	No. 4	No. 4	No. 3	No. 4
Auger hole 2	No. 5	No. 2	No. 5	No. 5
Auger hole n	No. x	No. y	No. x	No. z

The experimenter must give his own diagnosis at first to avoid being influenced by knowledge of the results of other determinative methods. Besides, the ideal is to work in pairs: one person serves as guide and consults the soil map, the other is positioned as expert working 'blind'.

Application and synthesis
This validation protocol was tested in and near the Reference Area of the Jura mountains of Bresse, France. The map, drawn at 1/10,000, enabled distinction in this sector of 16 Map Units differentiated on Plio-Quaternary deposits (Lagacherie, 1984). The auger holes studied for trying out SOSIE numbered about 200 (Pavat, 1986; Simonneaux, 1987).

Before any interpretation is done, the system must be calibrated by defining the value of dissimilarity above which the two soils compared will be considered to be completely different. In fact, if one does not proceed in this manner, SOSIE assigns the newly dug auger holes to the closest reference unit, even if the similarity is actually very limited and we are obviously facing a new soil. This calibration decides the sensitivity of the method. It always remains somewhat arbitrary.

The results finally obtained are more or less correct but full of lessons to be learned:
- Diagnosis with the key and diagnosis by SOSIE both have their specific advantages. Use of the key appears natural as a supplement to use of the map. But this key is totally unsuitable if a soil unit not listed is found. The identification may or may not end up as near rubbish. On the other hand SOSIE always finds the known unit the least different from the new unit and measures the dissimilarity between them; this is a datum.
- When one takes the precaution of digging many auger holes in a theoretically homogeneous map delineation, they are not always attributed to the same Map Unit or Soil Unit. Thus the diagnosis by SOSIE is not perfect. The algorithm shows itself to be incapable of choosing between two units that are similar to each other. So SOSIE must be considered as a means of quickly finding from the data bank the 3 or 4 units that can be suitable. From here on, the final identification belongs to the expert who alone is capable of subtly reasoning out by considering such or such detail important in one place and negligible in another.
- These attempts have pointed out the flaws in the manual for soil description. For example, a 'blue horizon with 50% reddish brown mottles' is actually the same as a 'reddish brown horizon with 50% blue mottles'. But this is not obvious to the computer system that records that neither the matrix nor the mottles have the same colour! This led to modification of description of mottles in STIPA-2000.
- Last but not least, we should ask ourselves if comparing two objects by means of a single number expressing their overall dissimilarity is not to simplify reality very grotesquely. It is no doubt practical for statistical calculations but is terribly simplistic! We must be conscious of this. But, on the other hand, calculations of similarity have one advantage: they are essential for defining the membership of an auger hole to a taxon or category not just in terms of 'yes' or 'no' (1 or 0) but by an intermediate value, for example 0.7. From this point of view it is more realistic.

7.5 SIMPLIFICATION OF BOUNDARIES

7.5.1 Statement of the problem

There are two principal circumstances leading to simplification of natural boundaries and their distortion until they coincide with some pre-existing artificial boundaries:

- when we want to transform a *vector map* to *raster map*,
- when we want soil boundaries or boundaries linked to an environmental characteristic to coincide with administrative boundaries to show the information on this basis; for example, soil acidity shown by counties. A large number of maps in atlases are presented on the basis of administrative divisions. We are used to this, but it is often linked to great inaccuracy.

Metral (1997) and the author of this book wanted to study and quantify the errors corresponding to these kinds of simplification. The results are not yet all published. Also, detailed presentation of the method will require very long elaborations. We shall limit our purpose here to a few principal essential points.

7.5.2 Basis of modelling

Figure 7.17 presents the principle of the method. The *Large Spatial Units* (LSU), for example soil delineations, are replaced by groups of *Small Spatial Units* (SSU), for example pixels or small administrative units such as 'communes' in France.

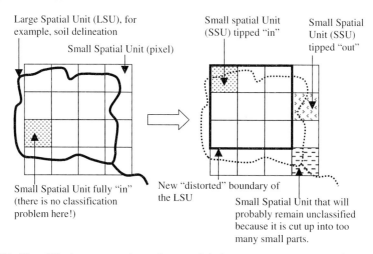

Fig. 7.17 Simplified presentation of a model for reorganizing map boundaries on the basis of existing *Small Spatial Units* (pixels, small administrative divisions).

It is easily seen that there are three principal parameters in the model:
- the average size of the Large Spatial Units (LSUs) or their number if a given area is considered;
- the average size of the Small Spatial Units (SSUs) or their number if a given area is considered;

- the 'tipping' threshold. Intuitively, this is 50% (the pixel is given the value to which most of it corresponds). But in fact this may go from 100% (the pixel falls in the given group only if it is completely of that type) to 0% (the pixel is grouped according to its major component, even if the latter only corresponds to 20% or 30% of the area).

7.5.3 Principal results

The model obtained is a bit complex insofar as its implementation involves some simple mathematical formulae. The following conclusions were reached (Metral, 1997):
- The error is lower the larger the LSUs and the smaller the SSUs. Obviously! In detail, the *accuracy* is greatly altered if the LSUs are not at least four times larger than the SSUs. Above all, there is the possibility of predicting exactly the numerical value of the error by using the model. This had not been obtained by specialists in *rasterization of vector maps*.
- If the tipping threshold is chosen as zero, all the SSUs are classified, the *indeterminacy* is zero but the distortion of the original map is maximum and the *inaccuracy* is large compared to the reality. In the reverse case, if the tipping threshold is 100, the inaccuracy becomes zero but the indeterminacy tends to be high. In practical terms, many SSUs will remain unclassified, that is, blank on the map. We can show that the curves have the appearance shown in Fig. 7.18. The *inaccuracy* does not increase very rapidly when the *tipping threshold* decreases. Thus there is no use in fixing the latter at 50% as one could intuitively think, but 33% or even zero might be suitable. This is linked to the fact that the SSUs spanning 3 or more than 3 LSUs are very few (there are rarely more than 3 lines on the maps meeting at a point).

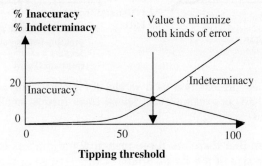

Fig. 7.18 Change in inaccuracy and indeterminacy in relation to the tipping threshold when natural boundaries are forced into artificial ones (pixels, small administrative regions).

- If a map with large blank areas is not desirable, it is difficult to drag the inaccuracy below 10%.
- *Rasterization* of vector maps is an operation with distortion of boundaries that often costs heavily in accuracy!

In sum, maps in Atlases have acceptable inaccuracy where their objectives are essentially didactic. Figure 7.19 shows an example of a map of Small Natural Regions included in Côte d'Or, France (Chrétien, 2000) when the natural boundaries are artificially adjusted to commune boundaries with a tipping threshold of zero.

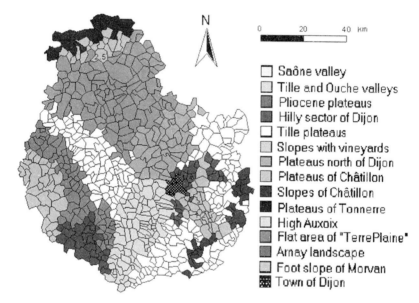

Fig. 7.19 French Department of Côte d'Or in which boundaries of the Small Natural Regions (corresponding to the LSUs of the model) are modified to match the administrative boundaries of communes (corresponding to SSUs). From data of Chrétien (2000).

Conclusion and perspectives

The examples discussed in this chapter show that the models developed and the simulations done have helped in synthesis of the numerous data and introduction of prediction of geography of soils. All this has led to very useful thinking and allowed a better analysis of the mapping process. But the future is full of questions. Computer science has overcome the obstacle represented by the mass of calculations required. By the same token, it has freed the imagination of modellers. The latter have set out on suitable laboratory exercises that first of all involve acquisition of a colossal volume of data for each pixel of a study area, for example obtaining and recording various characteristics of the environment of soils. Without neglecting the fact that different tools can help in acquisition and manipulation of these data, the fact remains that it would be good to stop some day. Otherwise the so-called automated calculation of soil map will be much longer, costlier and more approximate than its preparation by a mapper working in the field. The mapper wants to say that it is essential, following the tasks discussed above, to prime a deeper thinking in order to show in what way they can improve or modify the procedures of those who make maps without access to sophisticated technology. Are practical rules going to be highlighted? Will very simple-to-use software be developed? It seems important to proceed practically. Certain papers that we have examined are aimed at organizing and limiting the field observations, and appear to be very promising from this point of view.

CHAPTER 8

PRINCIPLES OF THEMATIC MAPPING

The soil map can be made use of from two very different perspectives (Leparoux, 1988; Legros and Bornand, 1992) as shown in Fig. 8.1. Firstly, it is the base for constructing derived maps useful for solving problems of agronomy or the environment; this is *thematic mapping*. It is said that *land characteristics* enable prediction of *land properties* or *land qualities* through a process of *land evaluation*. The procedure is deductive. Secondly, the soil map can be compared with an independently obtained thematic map to examine if the soils are related to the phenomenon studied. For example, what is the role of the soil and its hydric regime in the geographical distribution of bacterial wilt of peach trees (Vigouroux et al., 1987)? This second point of view represents a causal study and corresponds to an inductive procedure. Its role is totally fundamental. Actually, it is from an approach of this kind that one can understand the importance of soil in a specific problem. But this kind of investigation is not done directly from mapping; hence it will not be discussed here.

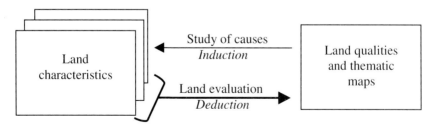

Fig. 8.1 Ways of using soil maps.

In the area that concerns us (land evaluation), development of GIS has resulted in storage of heterogeneous spatial data that can be combined. For example, data pertaining to climate, soil, slope and land use can be combined to construct a map of suitability for production of dessert grapes. But how do we proceed in detail? What weight is given to the various factors? How can one utilize the map data that are mostly qualitative and not quantitative? How can one separate the essentials from the non-essentials and, in a given case, retain just the relevant data from all those available?

The present chapter endeavours to answer these questions and to suggest procedures as rigorous as possible to avoid pitfalls. It relies in particular on advances made in *Multicriteria Analysis*. A synthesis is attempted at the end of the chapter.

8.1 PROCEDURES OF THEMATIC MAPPING

8.1.1 Objectives and methods

The *objectives* of thematic mapping as it concerns us are very many. They pertain to productivity, security of this productivity, protection of the environment, viability of the planning and its social acceptability (Smyth and Dumanski, 1993, 1994). For example, a map of susceptibility to erosion by water can be drawn; this is a *vulnerability map*.

There are three main approaches possible (Fig. 8.2):
- *First*, how to make a choice for each area, between many possible uses? For example, in which area of the zone will we allow construction of buildings, maintain green areas or encourage agriculture? Drawing a *capability map* is done in this situation. The problem mentioned can be treated within the framework of the alpha problem of specialists in multicriteria analysis.
- *Second*, how to group the delineations in the map into categories of equal value in relation to a given objective? For example, how can we

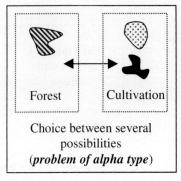
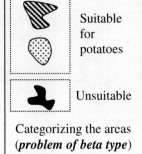

Fig. 8.2 Various kinds of problems to be solved.

find the areas very suitable, moderately suitable or not suitable at all for potato cultivation? This is a matter of categorization of areas. We have here what specialists in multicriteria analysis term the beta problem (Schärlig, 1985). This is, in particular, a matter of conceptualizing a *suitability map*.
- The *third method* is of less general use. Its aim is to rank all the areas in relation to a given objective. For example, to classify all the areas, from the best to the worst, for their resistance to erosion hazards. This is the gamma problem of multicriteria analysis. Through later introduction of partitioning in the hierarchy by applying one or several thresholds we obtain aptitude classes. This path thus enables resolution of the problem of degree of aptitude. The advantage is that we can change the thresholds and thus find, for example, 20%, 25% or 30% of the areas the most suitable for such or such use of the soil. But the procedure presumes establishment of a very complete hierarchy.

In reality, things are still slightly more complicated, particularly in the case of the alpha problem. Certain objectives might be conflicting. For example, forest OR crop production. Others cannot be excluded, at least locally, for example, forest suitable at the same time for recreation AND wildlife.

8.1.2 Treatments and models

For handling the problems mentioned above, the data are utilized according to two procedures. In the simplest case, it is a matter of extraction of data. The necessary data will be extracted from the soil map. If this operation pertains just to one factor (for example, soil depth), it results in preparation of a *monofactorial map*. But very often it is necessary to utilize jointly several kinds of data that are then studied at the same time. They are combined by *map overlay* (Chap. 6, § 6.2.2). In all or nearly all cases, thematic mapping is the art of selecting, from a huge mass of available data, *only* the pertinent data for appropriately combining them and thus obtaining the solution to the problem posed.

Different kinds of models can be used to solve a given problem by involving cartographic data. To classify them is dangerous because there is necessarily no agreement amongst authors regarding terminology. We will here retain three categories of models that represent, from the first to the third, a less and less empirical and more and more accurate approach: diffuse (Delecolle, 1989), statistical and mechanistic models.

Diffuse models
These models are used when the factors involved in a given problem are known and in addition it is known if these factors have a positive or negative effect in the given situation but with no idea of the exact modalities of

their action. It could be, for example, construction of a map of suitability for maize cultivation without knowing quantitatively the relation between yield of the crop and the pedoclimatic factors of the environment producing it. However, it is known that a certain number of factors promote the yield while others have a negative effect on it. One will then presume favourable for maize cultivation a site where the favourable factors are combined together and unsatisfactory another where the favourable factors are absent. Under these conditions, a deep, fertile, non-gravelly, well-structured soil will be considered appropriate. The diffuse model thus works on a *bundle of assumptions*. This procedure comes close to common sense and so has possibilities of giving useful results.

The different types of diffuse models widely used in our discipline will be studied in detail and their limitations will be defined. They will be presented in the framework of *multicriteria analysis*. For full information on the subject, one should refer to the papers of Roy (1985), Schärlig (1985), Simos (1990), Maystre et al. (1994) and Maystre and Bollinger (1999).

Statistical models and black boxes
This kind of model is used when the environmental characteristics are involved in the question studied in a manner determined by a statistical law. For example, the annual plant biomass production of maize under optimum growth conditions is linearly correlated with incident solar energy. We can then calculate what biomass will result from interception of a given amount of energy, with no knowledge of photosynthesis. We are thus facing a **black box** that directly links the inputs (here, energy) to the output (plant biomass). Let us consider an example. To be able to say 'my car consumes 8.2 litres for 100 kilometres', we are led to construct in a way everyone knows a model of the black box type, knowing nothing of the energy value of the fuels. This, however, is enough to accurately estimate the fuel that we will consume in travelling from Paris to Zurich.

Statistical models are much used in soil science. They are particularly useful in predicting the value of certain unestimated characteristics on the basis of others whose values are known. They also enable us to go from certain easily measured properties (such as particle-size distribution) to certain properties measured with greater difficulty (permeability, for example). In our discipline we have picked up the habit of terming such formulae for calculation, whose main object is to compensate for a shortage of data, as *pedotransfer functions*. Actually, in many instances it is better to use the term *pedotransfer rules*. In fact, the models used can come out of the field of pure statistics to include, for example, expert-system elements. A new thinking on the question was suggested by McBratney et al. (2002). This progress from a soil characteristic to a property including position in a classification has also been described as *target-property model* (Hewitt, 1993).

These essentially statistical models must not be confused with the so-called *stochastic* models into which a component of *randomness* is introduced, actually a perturbation. For example, in the model for conversion of energy into plant biomass, the incident energy could be modulated randomly and within determined limits to account for absence of sun during bad weather. Reality is approached but, by distorting more or less obviously the entered data, interpretation of the results often gets complicated. All the models studied here are built to give one result (and only one) from manipulation of the data. Reproducibility of their results is a quality that we judge to be essential and must be obtained whichever expert manipulates them. They are thus *strictly deterministic*, the statistical models included.

Mechanistic models (process models)
Here the problem is tackled with care to model the phenomena studied. Let us again take up the example of suitability for maize cultivation. We will attempt to examine scientifically if the essential requirements of this crop can be satisfied in each milieu where we are thinking of introducing it. For water requirements, the answer will be obtained by using a water-balance model. If irrigation is not available, all the sites with calculated water deficit of zero will be considered suitable. Models of this type can be called *mechanistic models* or even *process models* insofar as they attempt to reproduce the phenomena at the very root of the function of the system on which a diagnosis is based. The approach is as scientific as possible. It has the advantage of solving *ipso facto* the problem of weighting of the factors to be taken into account because it is known how and why they act. We will present later a short review of some mechanistic models particularly useful in our field. There are many more of them, a few of which are mentioned in a European document (EUR-17729-en, 1998).

The mechanistic approach thus appears more accurate and more scientific than the approach through a diffuse or statistical model. But it is based on hypotheses that are often very simplistic. The processes are simplified and the parameters assume empirical values. It is necessary to ask ourselves, for example, if it is really acceptable to reduce the growth of a maize crop to just satisfying its water needs. Many factors are forgotten such as danger of attack by pests and diseases and non-optimal agricultural practices. There is some pride in thinking that one is going to properly simulate nature in a computer! This is why some scientists very correctly prefer the more modest and less universally accepted term *functional models* (Lark, 2000).

Summary
To sum up, we have a choice of expert diagnoses (diffuse models), methods at times accurate but empirical (statistical models) and apparently scientific methods but very simplistic in nature (mechanistic models). This is why no

one kind of modelling can claim to have decisive advantages so that the others can definitely be rejected.

8.1.3 Fundamental scientific limitations

Organization of the map
We have indicated several times that Map Units (drawn on the map) may contain several Soil Units (not distinguished on the map because of scale). This situation, rare in detailed studies, is very common in synthesized maps drawn on medium or small scale. Obviously this complicates preparation of thematic maps. For example, it is difficult to draw a pH map if many complex Map Units are found each covering Soil Units that are not distinguished and have different pH. However, it is advisable to keep the following in mind:
- In each of the soil delineations corresponding to a given Map Unit (MU), the proportions of the different constituent Soil Units do not vary much; for example, one will have [MU3 = 30% SU2 + 70% SU4]; actually, if the proportions were very different from the reference values over large areas, it would become necessary to introduce another unit corresponding to, say, [MU7 = 70% SU2 + 30% SU4].
- If the various SUs of an MU are contrasting, it is obviously impossible to draw a map of average values having practical significance, but one can always draw frequency maps; for example, it would be possible to bring out a map of MUs in which pH is higher than 6.5 over more than 60% of the area; in fact, for each delineation, one knows exactly the pH of the various SUs and the approximate corresponding areas.
- In such a context, the diagnoses do not have local value (one is not sure the pH is higher than 6.5 at a given point) but are statistically valid: a particular MU surely will have pH higher than 6.5 over more than half its area.

Dangers of a strictly pedological approach
The soil map is rarely a tool sufficient for sorting out management problems. Other data are very often necessary and knowledge of soils is just a fulcrum for embarking on a multidisciplinary decision process. In most cases, we thus pass through a map overlay procedure. We will return to this question later (Chap. 9).

Consideration of all elements of the problem
Even in the pedological aspects of the problem, one may forget an essential factor. Let us take a real-life example. In the limestone terrain of the southern Alps, the growth was measured of young cedar trees on the hills where gravelly summits with shallow soils alternate with marly saddles with deep

soils (Algret, 1983). The idea was to thus check and quantify the diffuse model consistent with asserting that growth is good in the depressions when several favourable factors are combined: deep, fine-textured, not very calcareous soils in a landscape position favourable for water to collect. Yet it was observed that the trees grew better on shallow soils than on deep soils! In very gravelly shallow soils the young roots were forced to explore the entire volume available and to penetrate the fractures in the rock. This saves the plants when summer droughts occur. On the contrary, the cedar trees growing on fine-textured soils are more sensitive to shortage of water. This example is significant: no model can be constructed without a very good knowledge of the problem under consideration. In particular, nothing will exempt the workers from field knowledge and direct observation.

Relevance of pedological segmentation
It should also be asked if the problem studied is a matter requiring an approach leading to the distinction of pedologically homogeneous or ecologically homogeneous zones. For example, an underground iron pipe corrodes when the pH changes rapidly, that is, in places where we pass from one Soil Unit to another. It is not certain that the soil map available for a given sector was done in the spirit appropriate to tackling this sort of phenomenon.

Indirect or direct procedure
Does use of the soil map constitute an obligatory step for approaching the question in mind? This question is posed in all instances where this map is used, but it is more particularly apparent when the characteristic, the value of which is desired to be defined at all points in the study area, is not one of the criteria used to delimit the Map Units. Let us take an example. Suppose the issue is to estimate the cation exchange capacity at all points in the study area. Actually, this property is only determined on a few profiles in routine mapping. Under these circumstances, its estimation at all points in the study area by using the soil map calls upon an indirect approach in three stages (Fig. 8.3):
- establishing the *pedotransfer function*, often a simple statistical correlation for linking, in a set of available data (profiles), the cation exchange capacity to the mapped properties, in particular clay content and organic content;
- extracting from the soil map all the information relating to clay and organic matter; possibly drawing corresponding monothematic maps;
- constructing the derived map of cation exchange capacity by applying the pedotransfer function established for this.

But errors accumulate in such an indirect procedure. Firstly, the available data have a certain margin of inaccuracy. Secondly, the pedotransfer function that uses them has very poor predictive capability. The resulting map

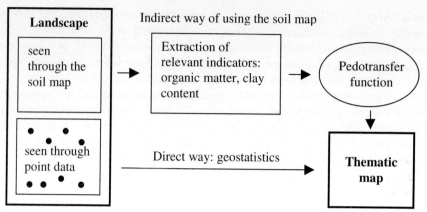

Fig. 8.3 Construction of a monofactorial thematic map with or without use of a soil map.

risks being of insufficient quality. Thus we ask if it would not have been better to use a procedure that avoids the roundabout method through the soil map and consists of directly estimating the property in question at each point in the study area. The latter can be achieved by using geostatistics. We first go to the field. Then we do the observations or measurements at random or following a regular grid. Then we interpolate between the points by means of a suitable mathematical procedure.

Both procedures were locally compared by Voltz (1986) and Voltz and Webster (1990). The direct method has statistical rigour but it is not without flaws. Interpolation is done blindly, disregarding indications provided by the environment (spatial aspects). Because of cost, the study grid is often too open for properly understanding detailed variation over a short distance. Lastly, it is not routinely applicable because it will be unthinkable to implement as many field plans as there are problems to be solved! On the other hand, the soil map, rapidly constructed at limited expense appears to be a suitable means for predicting the value of certain parameters, even hydric. This is particularly obvious in natural settings where the soils differ greatly over short distances (say, sandy soils adjoining clayey soils). This is then expressed by considerable verified variation in the parameter for which the values are sought. But one should be very careful not to jump to hasty generalizations. On the same soil map some boundaries will be revealed to be pertinent to the problem under consideration and others not at all!

Some scientists suggest combining both approaches. For example, at a point in the area, the value of the parameter in mind is the mean of the values predicted by the soil map and by kriging (Rogowski and Wolf, 1994). Another mixed approach consists of asking an expert to identify the soil at the selected sites. The value of the parameter normally characteristic for that soil is assigned to the sites (without direct determination), and then the predicted values are interpolated using geostatistical methods (Voltz et al., 1997). A third approach consists of using the map for identifying the soil boundaries in order to organize kriging soil type by soil type (Utset et al., 2000). Many others could certainly be invented....

8.2 DIFFUSE MODELS AND COMPLETE AGGREGATION

8.2.1 Organization of data

Evaluation matrices
Suppose one wishes to treat *Map Units*, *pixels* and *polygons* within the framework of a given application. Recourse to diffuse models (§ 8.1.2) means that we use several criteria that can be combined into an *evaluation matrix* or *decision matrix* (Table 8.1). Actually, GIS specialists will easily find in this matrix what is called the *Polygon Attribute Table* (PAT) in ArcInfo.

Table 8.1 Example of evaluation matrix

		DIAGNOSTIC CRITERIA			
		Quantitative *data*, e.g.	Ranked *data*, e.g.	Qualitative *data*, e.g.	Expert's diagnosis, scores [0, 20]
Individuals to classify : MUs, pixels, polygons...	1	120 cm	Low	Cubic	15/20
	2	80 cm	Moderate	Granular	17/20
	3	160 cm	High	Prismatic	5/20

This matrix has as many columns as there are criteria, and as many rows as there are individuals to be considered. The values of the criteria for the individual in question are found at the intersection of rows and columns. The values are measurements (for example, soil depth), scores (for example, 11/20) or at times even qualitative estimates (for example, good or bad). In short, the evaluation matrix summarizes the main points of the data useful for making proper choices or desired groupings.

The criteria for diagnosis have to satisfy various conditions: be *non-redundant* and hence *unrelated*, covering the question *exhaustively*; be *operational*, that is practically useful. Lastly, they must be *decomposable*, to adopt the language of specialists. Rather let us say that they must each represent a scale of values that is at least with clearly graduated rungs if not continuous. Thus clay content is an acceptable criterion. This is not so for soil colour for how are we to state that one colour is better or greater than another? In fact, these criteria must enable us make choices, thereby to express preferences!

Definitions of utility scores or 'utility functions'
Suppose we wish to introduce in the evaluation matrix a factor that takes soil depth into account, this criterion being significant for the problem under consideration. We can introduce this depth directly for each soil individual concerned (concept of *raw score*), but this procedure has two disadvantages:
- the values are not normed, that is, they are not graduated from 0 to 1 or from 0 to 100; it will thus be difficult to use depth at the same time as other criteria whose numerical values are not of the same order of magnitude;
- the properties that we want to take into account do not necessarily enter proportionately and linearly in the problem considered; pH, for example, is not more favourable the higher it is; it goes through an optimum value; it is necessary to introduce a function that enables replacing the data by their value in the context studied.

It is relatively easy to surmount the first obstacle. For example, we can norm the values of y in profiles i as follows (*standardized score*):

$$y'_i = (y_i - y_{min})/(y_{max} - y_{min})$$

where y_i is the original value, y'_i the value sought, y_{max} the maximum value in all the profiles i and y_{min} the minimum value in the same set.

The second difficulty can lead to calculation of the *utility function* as suggested by American researchers based on *multiple attribute utility theory*. Consider, for example, the role of stoniness in the context of a problem in agricultural potential. We ask the expert (that is, one who understands the effect of coarse fragments) as follows:

'Mr Expert, we suggest a game to you. You toss for it. If it is tails, you earn one hectare with 100% coarse fragments (sic!). If it is heads you get one hectare without any stones. But you also have the option of refusing the draw, in which case you directly obtain one hectare with 50% stones. Will you toss?'

The expert estimates that one hectare with 0% stones has the maximum value, that is, '1'. On the other hand, one hectare with 50% stones is hardly worth more than one hectare with 100% stones. The expert gives both the value '0'. So he finds it more advantageous to toss. But as it is a game, one is not allowed to do so. The question is again put changing the proposition intended to be substituted for the toss: 'if we propose one hectare with 30% stones in place of the draw, will you take it?'

Suppose the expert refuses the toss for 15% stones and prefers the assured gain. Then a fundamental point has been marked on the utility curve, with ordinate of $(1 + 0)/2 = 0.5$ and abscissa of 15% (Fig. 8.4).

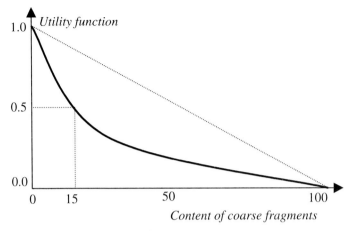

Fig. 8.4 Utility function for coarse fragments.

The game is continued to refine the line of the curve above and below utility of value 0.5. For example, for low values we give the expert the toss between one hectare with 15% and one hectare with 100% with the option of directly preferring one hectare with x% stones. If the expert's opinion turns to a value a, we find the point with the following coordinates:

$$x = a;\ y = (0.5 + 0)/2 = 0.25,\text{ and so on.}$$

This kind of exercise is enlightening. It shows that many utility functions are not rectilinear at all.

Evaluation of constraints

In the area of our interest, diffuse models are often used empirically, that is, without specific reference to multicriteria analysis. However, one will very quickly recognize that the idea of utility score is close to that of constraint, which is more familiar to soil scientists. A limiting factor for the application in mind is called constraint, for example, stoniness or excess water. Thus, constraints are unfavourable factors while utilities are more generally favourable factors. But it comes to the same thing. For example, if soil depth can be used as a utility score, lack of depth evaluated with reference to a maximum or threshold can quantitatively express a constraint.

For doing a rational study without forgetting any element of the problem, we generally construct a conversion table that states to which level of the constraint each value liable to be assumed by each property of the soil corresponds. This table is not always provided to the user; we are sometimes happy to only present to the user the results that can be deduced from it. This table, whether seen or not, summarizes interpretation efforts. Sometimes it is just 'opinion of experts'. In other cases it corresponds to delicate agronomic experiments or to actual investigations (Arrouays, 1987). The example presented in Table 8.2 pertains to establishment of constraints for movement of forest tractors in Quebec, Canada (Robitaille and Grondin, 1992).

Table 8.2 System for conversion of characteristics to constraints for forest tractors

Type of constraint	Proportion of area affected		
Slopes > 30%	<15%	15-25%	>25%
Outcrops	<25%	25-50%	>50%
Organic materials	<25%	25-50%	>50%
Clay on moderate slopes	<50%	>50%	—
Resulting constraint level	Slight	Moderate	High

E.g.: if 10% area has slopes > 30%, it constitutes a slight constraint

8.2.2 Data processing

Implementation of a complete aggregation

Suppose we have, by application of the method presented above, drawn up an evaluation matrix comprising only normed numerical values acting more or less linearly in the problem under consideration (for example, constraints estimated on a scale of 0 to 4). It is now necessary to synthesize all the criteria used and express one overall judgement on each individual to be classified. The method of considering the criteria as comparable and combining them in a mathematical formula which allows their substitution by a single value is called *complete aggregation* in multicriteria analysis. The single value, if preferred, is the *weighted linear combination*. The simplest, obviously, is the arithmetic mean. Methods of complete aggregation are of various types. One can add up the criteria, multiply them if necessary or even use mixed systems (Vink, 1975, p. 306). A *choice function* is used, according to the vocabulary of the IDRISI software. Then the score calculated for the individual to be classified and is compared to a reference threshold. If it is above the threshold, the soil is evaluated as suitable. If it is below, it is unsuitable, at least where only two classes (suitable and unsuitable) have been defined.

We should keep in mind that students are evaluated in this manner in their examinations. For English-speaking agronomists, this is the *rating system* or *parametric system*. It was first developed in USA in the late 1960s by the team of Ian McHarg under the name *Land Suitability Analysis*. These studies, done before Geographic Information Systems made their appearance, may seem outdated. This is not true, however, because in their details, they show all the questions that must be solved for treating these problems of suitability or potential.

Calculation of weights

Weights might be introduced; there are means of calculating them, for example the method of Saaty introduced in 1977 (Eastman, 1993). The expert compares all the criteria in pairs and first evaluates their relative importance on a *continuous rating scale* going from 1 to 9. Table 8.3 helps in this.

Table 8.3 Evaluation of the relative weights of criteria taken in pairs

How much less important is the first criterion than the second?			
Extremely?	Very much?	Strongly?	Moderately?
Corresponding scores			
1/9	1/7	1/5	1/3

Or, how much more important is the first criterion than the second?			
Extremely?	Very much?	Strongly?	Moderately?
Corresponding scores			
3	5	7	9

Or, is the first criterion as important as the second? (score = 1)

When there are 4 criteria C1 to C4, this results in a *pairwise comparison matrix* (Table 8.4), which indicates, for example, that C1 is 'strongly' less important than C2 (score 1/5). But, contrary to what could be understood by using the said document, the matrix is not mathematically symmetrical! Thus it is also necessary to indicate that C2 is 'very much' more important than *C1* (score 5). An intermediate matrix is worked out by replacing each value by the quotient obtained by dividing it by the total of each column and expressed as a percentage. Lastly, the final weights are determined; they correspond to the mean of each row (Table 8.4).

Table 8.4 Calculation of weights by complete aggregation

	C1	C2	C3	C4		C1	C2	C3	C4		
C1	1	1/5	3	1/7		8	4	25	8		11
C2	5	1	3	1/3		37	22	25	20		26
C3	1/3	1/3	1	1/5		2	7	8	12		7
C4	7	3	5	1		53	66	42	60		55
	Pairwise comparison matrix					Intermediate matrix					Weights

One can check and even quantify coherence of the results. When all calculations are done, the ratio of the weights must be reasonable. For example, for C1 and C2 we must have, at least very approximately:

11/26 (result obtained*) *is not really equal to 1/5 (original hypothesis)!

A careless expert is in danger of suggesting aberrant weights and saying at the same time, say, that C1 > C2, C2 > C3, and C1 = C3. There are methods for detecting this.

For some applications, when there are many experts, we are led to taking the mean of the weightings they propose. To sum up, Saaty's method seems more scientific than a simple ranking of criteria from the most important to the least!

In contrast to this approach, there is what consists of giving the experts small cards on which the names of the criteria are written. They are then asked to arrange the cards in order of decreasing importance (Fig. 8.5). But they are also given blank cards for insertion in the hierarchy to indicate that the criterion is much more important than the one following it (Maystre and Bollinger, 1999).

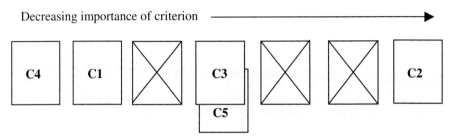

Fig. 8.5 Empirical hierarchization of five criteria by an expert using the pack-of-cards technique.

The weights can be calculated easily from this, say in the following manner (Bojorquez-Tapia et al., 2001):

$$W_i = n - r + 1$$

where W_i is the weight for criterion i, n the total number of criteria (8 in Fig. 8.5, including the blank) and r the ranking of the criterion (for example 4 for C3 in Fig. 8.5).

It must be noted that multicriteria analysis is often used as a negotiation technique between experts who may be representative of social, administrative or political forces and defend the interests of the appointing group. This is the case, for example, when deciding location of a site for treatment of urban wastes. Therefore the heat must be taken out of discussions and simple methods proposed to favour dialogue. But all systems depend on the honesty of the experts. For finally eliminating any possibility, it is enough that an expert, having fully understood the principles of multicriteria analysis, introduce here and there a deliberately too high or too low a weighting... .

Management of conflicts in land capability
When a land-capability map is constructed, it is necessary to manage the conflicts and make choices when a given area is suitable for many uses at the same time. The 'Mola' procedure of IDRISI suggests modelling in the following manner (Fig. 8.6) the choice between two possible uses of the land at a given site.

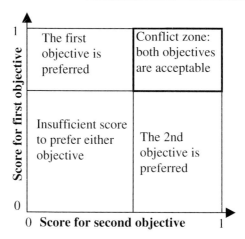

Fig. 8.6 Choice between two possible objectives when preparing a land-capability map.

But the objectives are not necessarily mutually exclusive. For example, a profitable forest may also form an area for recreational activity (hunting, fishing, etc.). McHarg suggested drawing up the table crossed with all the objectives for examining them in pairs and evaluating, in each case, their possible compatibility.

Limitations of the complete-aggregation approach
A soil having many shortcomings, normally unacceptable, may be salvaged and declared capable if it otherwise has many good qualities. There is compensation, which is almost absurd. For example, a soil located on a very steep slope is not suitable for mechanized agriculture even if it is otherwise very fertile. Actually, for this system to function usefully, we must introduce the concept of elimination score. Every soil not reaching the minimum required for any criterion is eliminated. The total score is calculated for soils that have passed the hurdle of such elimination. This total then helps in two things: firstly, to again eliminate the individuals whose score is insufficient because they have weaknesses in many areas; secondly, to create a hierarchy of those that have not been eliminated. British work (Thomasson and Jones, 1989) correctly uses this two-stage procedure.

But whatever be the efforts undertaken to improve the method, using complete aggregation is seldom justifiable because data of different kinds are often mixed. For example, it is stupid to total up, from a decision matrix, the age of the farmer, price of lavender flowers and the minimum winter air temperature, even if these criteria are all to be used to characterize the feasibility of converting wine-producing farms to crops used as base for perfumes. To take another example, it is abnormal to consider average and above all as equal two students one of whom has obtained at the examinations

20/20 in mathematics and 0/20 in English, and the other 0/20 in mathematics and 20/20 in English. In all the cases we face *incompatibility of criteria*. Using a *weighted linear combination* is, as they say in France, 'to mix up dish cloths and towels'. How are we to do the desired synthesis under these conditions? We will see in §§ 8.3 and 8.4 valid recommendations in regard to diffuse models.

8.3 DIFFUSE MODELS AND SYSTEMS OF DOWNGRADING

8.3.1 Theory

Non-compensatory models (Jacquet-Lagrèze and Siskos, 1983) can be used. This term means that the satisfactory or excellent value of one criterion does not enable compensation or making up for the unsatisfactory value of another criterion. This system is often used in agronomy, by the FAO in particular. Every pedoclimatic environment that falls below a specific level of a given constraint is automatically downgraded and assigned to a lower category of potential. This procedure is quite rigorous. In this context, crop production is not supposed to be the resultant of a group of interacting factors but the yield allowed by the factor that appears least favourable (law of the minimum). Models of this type can also be called, quite appropriately, *models with dissatisfaction levels*.

Lastly, we should mention the existence of *lexicographic models*. The criteria are ranked according to their importance. The first criterion is used to classify the individuals, the second and the ones following serve to decide between equally ranked individuals. One can thus obtain an almost complete hierarchy of individuals even if each criterion only enables two or three classes to be established. Everything takes place as in a dictionary, in which the words are ordered by the first letter, then the second and so one.

8.3.2 Application of downgrading in thematic mapping

Downgrading is implemented using a table that enables us go from already calculated constraints (see, for example, Table 8.2) to capability or even to the hazard to be calculated. Table 8.5 provides a simplified example. It has been extracted from the papers by Arlidge and Wong (1975) cited in *Practical Pedology* (McRae, 1988), The data pertain to sugar cane in the tropical zone.

Table 8.5 System for converting constraints to suitability for sugar cane (McRae, 1988; simplified)

LEVELS OF THE CRITERIA			RESULTING SUITABILITY
Water availability	Nutrient availability	Susceptibility to erosion	
High	High	None to slight	S1 (very good)
High	High	Moderate	CS1 (medium)

We have seen earlier that two broad types of objective can be pursued: construction of suitability maps or construction of capability maps. Tables are also available for conversion of constraints to capability; they will be found in the papers cited. With the help of these tables, it is possible to answer the question, that is, to draw the desired thematic map.

This procedure has variants, The constraints that form an intermediate step between the data and the final interpretations can be shown. Then a map for each constraint (monofactorial map) or a map for all the constraints taken together can be drawn (Duclos, 1980). The procedure can also be automated whereby errors are avoided. For this the ALES system of the FAO enables us build a *decision tree* as desired.

8.3.3 Difficulties in downgrading

Let us consider an example. Suppose we have identified five constraints each showing three states (absent, present at a disturbing but not unacceptable level and present at a level that prohibits the use planned for the soil). Then we must consider $3^5 = 243$ combinations representing as many cases it is necessary to look into for establishing a diagnosis! In theory, each one must be separately studied because no two are equal. For example, a soil delineation with the sole defect of presence of water in excess is not comparable to a delineation with just one constraint of excess of stoniness. Naturally, if certain constraints are linked between themselves, for example if the sloping soils are at the same time gravelly, the actual number of combinations represented will be reduced further but will still be large. The solution of taking as suitable only those areas not having any defect is seldom appropriate. In fact, it represents a single favourable combination out of the 243 considered. Without pretending that all the cases are equally probable and equally represented in area, this shows the danger in thus selecting not the suitable soils but the ideally suitable soils, which probably are very few. To sum up, using a 'constraint-to-suitability' (Table 8.5) or 'constraint-to-capability' conversion table is thus a solution to the combinatorial problem, but could mask oversimplification and quite often not taking into account all the possible cases. In this context, displaying all the constraints on a single map is very useful because it enables us realize the complexity of the problem.

Another difficulty related to thematic mapping on the basis of constraints is due to the sharpness of the thresholds defined for passing from one level to another. For example, we will consider as suitable a soil with 14% stones and eliminate another with 15% stones, which is perhaps better that the first in all other aspects. When all is said and done, thematic mapping on the basis of constraints and downgrading is rarely a correct method. We will see later how improvements can be made.

8.4 DIFFUSE MODELS AND PARTIAL AGGREGATION

When the criteria are too heterogeneous to be used together and directly (sum or product), and when the downgrading becomes too complex, we implement a *partial aggregation*, also known as *outranking method*. For showing how one proceeds with this, it is necessary to review some concepts. The central idea is to compare in pairs and criterion by criterion all the individuals to be classified. Software has been created for this, in particular the ELECTRE program (Maystre et al., 1994).

8.4.1 Fundamental concepts

Reference individual
Most diagnoses are done on the basis of accumulated experience supported by observations and actual cases. For example, a given soil delineation with known characteristics is revealed to be good for potato cultivation. This result summarizes the opinion of farmers or even the opinion of experts who have conducted agronomic trials. The delineation then constitutes a *reference individual*.

The reference individual contrasts with ordinary individuals, which we do not know are suitable or unsuitable for the application in mind. But the reference individuals and ordinary individuals are defined on the basis of the same criteria that have a specific value in each case.

Outclassing (dominated/undominated individuals)
If a test individual has, for every criterion, a value higher (in the sense of 'better') than that of the reference individual, the former is said to outclass the latter (the test individual is *undominated*). On the contrary, if the test individual is weaker in every criterion, it is said to be *outclassed*. It is *dominated*. This suffices to divide the ensemble of individuals into three groups: those that outclass or equal the reference, those that are outclassed by it and those that are unclassifiable because they are superior in some criteria and

worse in others. Two or three judiciously selected references are thus enough for ranking the individuals into three or four major categories and for solving one problem of thematic mapping (cf. the method known by the name *trichotomy of Morcarola* (Schärlig, 1985). But for this to happen, it is still necessary that unclassifiable individuals be few, which is not always the case. Thus we have to examine how the latter are to be treated.

Slight and strong outclassing
The concept of outclassing does not intrinsically involve manipulation of numerical values. One can, for example, compare states: good, medium, poor. But numerical values enable refinement of the outclassing by taking their magnitude into account. For example, a 70-cm deep soil outclasses a 60-cm deep soil but less so than does a 120-cm deep soil. In other words, we measure the level of outclassing for a given criterion by taking the difference in values corresponding to test individual I and the reference individual R. This leads to three concepts.

- The *no-difference zone*. If the criterion has very slightly different values for the reference R and the individual I, the outclassing or downgrading has no real significance. For example, is a soil said to be 65-cm deep different from a soil of depth 60 cm? For postulating that the difference is significant, does it have any implication whatsoever in the problem considered? The no-difference zone is symmetrical with respect to zero since the individual can dominate the reference or be dominated by it.
- *Zones of strong preference*. If the difference between individual I and the reference R is large, it is judged significant. There is strong preference in both cases: when the reference is strongly outclassed by the individual or the individual is strongly outclassed by the reference.
- *Zones of slight preference*. These are intermediate between the above-mentioned zones and are limited by the no-difference and strong outclassing thresholds. All this can be represented in a diagram (Fig. 8.7).

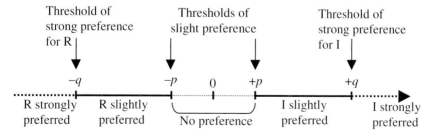

On the x axis: values of score = *[value of criterion for I – value of criterion for R]*

Fig. 8.7 Diagram of outclassing levels and preference zones.

Theoretically more than two levels of outclassing could be defined, but the manipulation of data and calculations would have become more complex. In such case, it is better to use measures of dissimilarity (Chap. 7, § 7.4). Actually, multicriteria analysis is not designed to get great refinement. It does not want to drift toward an excessively numerical approach associated with very many calculations. The original data are seldom known with very great accuracy. The models used, we have seen, are often empirical. Under these conditions, we must be realistic: nothing will be served to do calculations of very high intrinsic accuracy.

Consistency index

Here we are concerned with characterizing individuals that outclass the reference in just some criteria. For this, we propose to count the number of criteria considered. More correctly, we look for the number of times there is consistency with the hypothesis 'the individual examined outclasses the reference'. This helps calculation of the *consistency index* given by

$$CI = \frac{\text{Number of Consistencies}}{\text{Number of Criteria}}$$

It is noted in passing that outclassing is a concept considered, as the case may be, at the level of one criterion or even at the level of all the criteria, that is, at the level of an individual.

The criteria need not be of the same significance. Weights can be assigned to them. In this case, the formula for the consistency index is slightly modified. Each consistency in the numerator is multiplied by its weight and summed. In the denominator, the number of criteria is replaced by the sum of the weights of the criteria. In all cases, values obtained for CI are between 0 and 1. For the individual to outclass the reference, the consistency index must obviously exceed 0.5. But we can consider that this is not enough and a higher threshold can be fixed. We can also introduce, as will be seen later, a *veto* threshold.

Credibility

Now it is necessary that the concept of strong or slight outclassing be practically usable in the calculation of consistency index. For this, the *credibility* of the hypothesis 'the individual outclasses the reference' is examined. Strong outclassing corresponds to a credibility of 1. Slight outclassing corresponds to an intermediate credibility, say 0.5. But it is also possible to make the value of credibility vary linearly between the no-difference threshold and the threshold of strong outclassing. Below the no-difference threshold ($+p$), the credibility logically has a value of zero. It is zero over the entire zone of negative values corresponding to the fact that the individual is classified lower than the reference (Fig. 8.8).

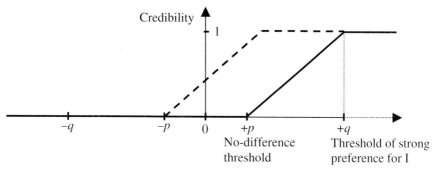

On the x axis: values of score = [value of criterion for I − value of criterion for R]

Fig. 8.8 Definition of credibility.

There is another way of looking at this. If the individual is very slightly outclassed by the reference (the zone between -p and 0), we can consider that the hypothesis 'the individual outclasses the reference' is not unquestionably invalidated. On the graph, this leads to the credibility curve taking off from the x-axis not at the upper no-difference threshold (+p) but at the lower no-difference threshold (-p) as indicated by the dashed line. This can be justified in certain applications.

Once the credibility has been determined, we can redefine the consistency index as follows:

$$CI = \frac{\sum \text{Credibilities}}{\text{Number of Criteria}}$$

Credibility is actually not independent of outclassing level and we can question the practical use of this concept.

The veto threshold
Outclassing on the basis of the CI takes into account most of the criteria but not all. We can say that an individual outclasses the reference while it is inferior to the latter in one or several criteria. Thus the consistency index is not enough. It is also necessary to examine the data in greater depth. If required, a *veto threshold* must be introduced corresponding to the point at which it is not possible to declare as superior an individual very inferior in one criterion. This is the concept of qualifying grade in examinations taken by students. By this principle they are prevented from passing although they are above the threshold required (average) for all the tests taken together.

Graph of outclassing
Once the values of outclassing have been determined, it is possible to represent them by a graph (Fig. 8.9) on which the individuals ($h, i, j, k, \ldots$) and their hierarchical position appear.

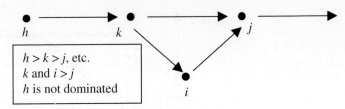

Fig. 8.9 Graph of outclassing according to one expert.

This kind of graph thus embodies the results of the study, expert by expert or, if preferred, weighting system by weighting system. In Fig. 8.9, the individual h is in top position. It represents the best pixel according to the expert.

But the graph is not always easy to construct or interpret, above all if the individuals to be ranked are very many. It is possible to obtain different graphs by trying to classify the individuals or solutions from the best to the worst or from the worst to the best. Suppose R = I for one criterion (no-difference zone). If the question 'is it credible that R > I?' is asked, the reply will be 'yes'. Thus R is positioned before I on the graph. But if the question 'is it credible that I > R?' is put, the reply is always 'yes' and the graph is modified. Specialists therefore distinguish between the *Top-Down* approach and the *Bottom-Up* approach. Some scientists prefer to say that they do an ascending distillation or descending distillation. This reminds me of my childhood when a vendor at a fair shouted *'Come, Ladies and Gentlemen, to my stall to see an extraordinary animal measuring 2.5 m from head to tail and 2.3 m from tail to head!'*

The final results of the study can be presented in graphs that Maystre and Bollinger (1999) call *'Surmesure'* graphs (Fig. 8.10).

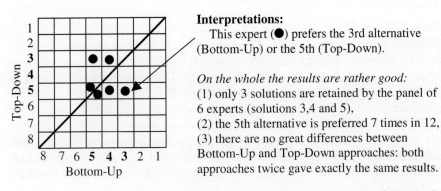

Fig. 8.10 Example of *Surmesure* graph for making the final synthesis of the results (12 hierarchies) corresponding to 6 experts taken to examine 8 possibilities in a problem of land management, for example, locating a motorway (Maystre and Bollinger, 1999).

Stability of the results
What was presented above shows that there can be all kinds of variants of the method: veto threshold or no, greater or smaller weightings, considering degrees in the outclassing or not, Top-Down or Bottom-Up distillation, etc. A good way of working is to test several approaches and see if this greatly changes the final result obtained by the *'Surmesure'* graph.

8.4.2 Limitations of partial aggregation

The concepts reviewed allow us to solve various problems of thematic mapping involving diffuse models. Adjustments may be necessary in every case. But the apparent simplicity of these procedures harbours a few traps. Three of them deserve mention here.

Problems of slight or very slight preference. Consider the case of soil depth. A 60-cm soil is equal to a 59-cm soil (cf. the concept of no-difference threshold presented earlier). Similarly a 59-cm soil is equal to a 58-cm soil and so on. We can go on till, say, 10 cm. But it is obvious that a 60-cm soil is not equivalent to a 10-cm soil. We are up against a kind of intransitivity: A equals B and B equals C does not imply that A is equal to C. This can lead to difficulties when we wish to hierarchize all the objects to be classified.

Condorcet's paradox. This is another case of intransitivity highlighted by Condorcet, the French philosopher, mathematician and politician of the 18th century. For presenting the case, let us take the example formulated by André Warusfel, repeated in the book by Schärlig and revised slightly. Suppose the members (40) of the Académie Française are relocated by enforcing decrees issued earlier by the government of Edith Cresson. They must now have their meetings in the provinces and no more in Paris. They are allowed to choose from among Lyons (L), Montpellier (M) and Quarré-les-Tombes (Q), a very small village in Morvan. Each member votes by classifying the sites in order of preference. For example, L, M, Q or L, Q, M, etc. Since it is a matter of permuting three objects in all possible ways, the Academicians have 3 × 2 × 1 = 6 types of responses available. Through voting, they distribute themselves into six groups as given in Table 8.6.

Table 8.6 Vote of the Académie Française according to Warusfel; Condorcet's paradox

Type of vote	Order of preference	Number of votes of this type
1	L Q M	11
2	Q M L	9
3	M L Q	6
4	M Q L	7
5	L M Q	4
6	Q L M	3

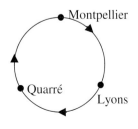

It is easy to verify that Montpellier is preferred to Lyons (22 against 18), that Quarré-le-Tombes is preferred to Montpellier (23 against 17) and that Lyons is preferred to Quarré-les-Tombes (21 against 19). This can be represented by a cycle (vicious), as drawn to the right of the matrix of data. In other words, outclassing of A by B with outclassing of B by C does not always imply outclassing of A by C. This amounts to saying that the diagnosis reached could be partly illogical. It is necessary to be mindful of this difficulty.

Volume of calculations. In the methods using outclassing, the individuals or solutions studied are all compared in pairs. This is a logical step when it is a matter of choosing among some possibilities given a priori, for example, deciding from six routes for a motorway. But this method lends itself poorly to systematic comparison of all the pixels of a map for preparing, say, a suitability map. With 10,000 pixels there will have to be 50 million comparisons. The synthesis is not necessarily valuable. Some authors emphasize this difficulty (Joerin et al., 2001). What is left is the possibility of choosing reference individuals and comparing all the others only with them. For example, with 5 references and 10,000 pixels there are only 50,000 comparisons to be made. This is a traditional solution in statistics. We have examined it earlier (Chap. 7, § 7.4, Fig. 7.15). In multicriteria analysis, it can be adapted by calculating, criterion by criterion, the dissimilarity between each individual and the two references corresponding respectively to the *ideal solution* and the *non-ideal* solution (i.e. the worst solution). The method, developed by Hwang and Yoon in 1981, is named TOPSIS for *Technique for Order Preference by Similarity to Ideal Solution* (Malczewski, 1996). The two extremes are at times named, rather poetically, zenith and nadir (Jankowski et al., 2001).

8.4.3 Exercise

In the Appendix is given an exercise (A3) to familiarize the user with the concepts presented in §§ 8.2, 8.3 and 8.4.

8.5 DIFFUSE MODELS AND COMPARATIVE REASONING

The method consists of applying the following principle: 'If studies or observations show locally, with certainty, that the suitability for such or such crop is high, it will also be high in similar milieus having the same soils and the same environments.' This is *transfer by analogy*. In practical terms, this amounts to extrapolating agronomic results obtained in some experimental or demonstration parcels. But we will not develop this approach. In fact, the reasoning used is, in principle, exactly that studied in Chap. 7 (§ 7.3.1). Remember that this is reasoning by analogy for reconstructing the whole soil map from knowledge of just a few *pedons* well characterized pedologically and environmentally. In general, simulation of this kind is useful as it is close to how the expert intuitively uses his earlier experience. This same approach is used under the name *compromise programming* (Pereira and Duckstein, 1993).

8.6 SEQUENCE OF SPATIAL MODELS

It is useful to quickly show, using just a few examples, how different types of calculations leading to spatialization of data important in thematic mapping can be put together. These calculations implement models, many of which are mechanistic.

8.6.1 Calculation of topographic parameters

Preparation of a slope map is an almost indispensable prerequisite in many applications. For example, slope is a constraint for mechanization. It also enables us estimate erosion hazards and hence to say if the land can be put to the use in question without deteriorating the soil (Bonfils, 1989).

Slopes were earlier calculated manually. A square-mesh grid was laid on the topographic map and the altitude contours were counted in each secton. Nowadays a *Digital Elevation Model* (DEM) is used. This is actually a magnetic record giving the altitude at the intersections of a regular grid. The slope can be calculated in different ways from regularly distributed points with determined altitudes. An excellent paper by Skidmore (1989) has reviewed the methods and discussed their accuracy.

Other parameters can be obtained from the same grid of altitude values. Table 8.7, based on another more complete one (McSweeney et al., 1994), gives the most important.

8.6.2 Solar-energy map

Solar energy can be measured with suitable accuracy using specialized equipment. But it is obviously impossible to do these measurements everywhere. Hence the problem arises of spatial extrapolation of point data. Models for this have been designed by meteorology specialists and have later been adapted for agronomic needs (Durand and Legros, 1981). These

Table 8.7 Topographic parameters derived from a DEM (McSweeney et al., 1994, simplified)

CALCULATED PARAMETER	DEFINITION	USEFULNESS
Slope	Altitude gradient	Land capability classes, runoff rate, solar energy
Aspect	Slope azimuth	Modelling of solar radiation, evapotranspiration
Profile curvature	Slope profile curvature	Runoff conditions
Catchment area	Area draining to catchment outlet	Runoff volume
Catchment length	Sum of distances between pixels and their final outlet	Flow acceleration
Relief irregularities (RI) on a transect	$RI = X_i - (X_{i-1} + X_i + X_{i+1})/3$ where X is the altitude at the points $i-1, i, i+1$	Understanding water and solute movement on slopes

models (Fig. 8.11) were recently revised for expanding their possibilities and improving accuracy. The sequence of calculations is as follows:

(1) Calculation of the position of the sun in the sky at a given moment.
(2) Determination of the energy available at the boundary of the earth's atmosphere as a function of the position of the earth, that is, the earth-sun distance.
(3) Estimation of the quantity of energy absorbed by the atmosphere as a function of the sun's elevation above the horizon and sky conditions (clouds). Sky conditions obviously are very variable. It is on this that the greatest inaccuracy in calculation rests. So it is necessary to know

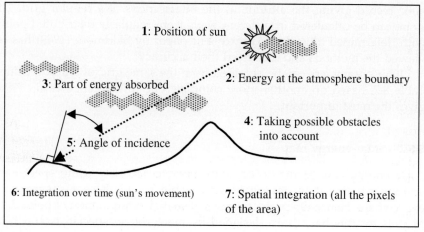

Fig. 8.11 Principle of calculation of solar energy.

the climate of the place and to have statistics on hours of sunshine. The results obtained only represent the most probable values.
(4) Accounting for possible interception of light rays by high relief (morning and evening in high mountains).
(5) Determination of the angle of incidence of light rays with respect to the normal = perpendicular to the soil surface, and calculation of the energy received (slope maps are useful for this).
(6) Repeated calculation every quarter hour or every half hour for summing the total energy received during the day.
(7) Repeating the entire procedure for each pixel constituting the map.

On the above basis we try to predict the quantity of energy to be received on the south flank of a mountain, parallel to the slope, while only knowing the energy received parallel to the slope on the north flank (Legros, 1986). A check done on Mount Ventoux in the foothills of the Alps showed that the error generally was less than 5%. Thus this type of model shows itself to be efficient for generalizing, as a function of the slope, a point determination of the energy received at a meteorological station. But one must be careful! The meteorological station and the extrapolation zone must belong to the same climatic unit, that is, must be affected alike by cloud systems.

8.6.3 Spatialization with crop models

Objectives
Models have been designed for simulating the functioning of crop plants. These are *crop models*; they have all kinds of applications (Legros et al., 1996):
- Prediction of consumption of water by plants and thence calculation of *soil-water balance*; this is useful for directing irrigation, for predicting the number of days suitable for working the soil, for examining change in water-table levels, for defining *runoff* from the soil surface, etc.
- Prediction of crop *yields*.
- Calculation of economic *profitability* of the crop.
- Change in yield should there be climatic change, etc.

As a consequence, these models are very many. To mention the best known: AEZ (FAO), CENTURY, CERES-MAIZE (Jones and Kiniry, 1986), CERES-WHEAT (Ritchie and Otter, 1984), CORNGRO, EPIC (Sharpley and Williams, 1990), STICS (Brisson et al., 1998), etc. There are many others such as those developed by a group of European laboratories in the framework of the EU: EuroAccess II and IMPEL (Loveland et al., 1994; Legros et al., 1996; Rounsevell, 1999). Internationally, there certainly are more than a hundred of them.

Principle of functioning of crop models
There are various types of crop model. But schematically all consider the plant as a machine for transforming solar energy to carbohydrates to the

extent that the plant is not otherwise subject to stresses relating in particular to availability of water, temperature and salinity. With stress, growth reduces and the potential yield diminishes. According to the objectives of the modellers the model is focused on one particular aspect of the biological functioning of the plant.

The models that we have used and even developed (MIRABEAU, *Modèle Informatique Relatif Au Bilan en Eau*) concern more particularly the fate of soil water in the dry Mediterranean environment. In this situation, the major input is rainfall, possibly supplemented by irrigation. But it is at times necessary to take into account lateral surface runoff that reduces or augments the amounts of rainfall actually percolated. Water losses are many more: possible deep drainage but mostly evaporation (E) from the soil surface and transpiration (T) by the plant. Evaporation and transpiration are often grouped and calculated together (*potential evapotranspiration (PET)* and *actual evapotranspiration (AET)*). But in other models they are separated. Evaporation and transpiration depend on solar energy received (cf. the PENMAN formula). This partly explains the greater interest we have in solar energy, followed by slope. For proper estimation of E and T we have to understand the growth of the plant:

- *growth of leaf area* for estimating the corresponding transpiration and determining the proportion of bare soil left and liable to evaporate water;
- *root growth* for determining the extent to which the reserves of soil water can be exploited.

Figure 8.12 shows in a simplified manner how such a model can function (MIRABEAU, based on EPIC and EuroAccess II) But there are all sorts of other options for calculating stresses and their effect on growth.

Validation of the calculations

Validation is done by comparing the prediction made by the model to the actual thing, which for example may involve an agronomic experiment. The results are more or less correct and risk of error large. But contrary to what one could intuitively conclude, error propagation through the model is not always amplification. To take an example, bulk density (BD) is a data input of these models. It serves to define the size of the water reservoir in the soil. This parameter is associated with about 10% uncertainty. But if the soil is very deep the reservoir in it is very large and so will never be filled. Thus we can calculate exactly the volume of water that enters or leaves the reservoir without the size (unknown) of the latter coming into the picture. Inaccuracy in the BD will thus have no effect. On the contrary, if the reservoir is overfilled during the rainy season, the losses will be a function of the size of the reservoir and, therefore, of the BD. This simple example shows three things:

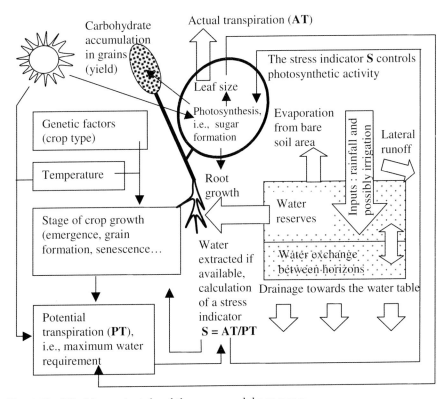

Fig. 8.12 Working principle of the crop model MIRABEAU.

Firstly, filling up of the pores of the soil with water is generally occasional, the consequence is that the error of 10% in the BD will be reduced in the output from the model and will represent about 5% of the water balance; this clearly shows that these crop models are quite 'robust'; they give relatively correct results; other examples could be given to demonstrate this.

Secondly, it is seen that the consequences of the error in one parameter, here bulk density, can be more significant or less according to the values taken by another parameter, say, soil type or the value taken by an input datum, rainfall; otherwise, before using the model for an application, it will be necessary to test it several times with all sorts of juggling of the parameters and data.

Lastly, in such a situation, a particular model will be perfectly successful in dry climate and less so in humid climate; none is universal: there are only better or worse models in given conditions.

Leenhardt (1991) has studied error propagation related to pedological data inputs in a water-balance model.

8.6.4 Machinery work-days

Soil cannot be worked when it is too wet. It is at times even impossible to have agricultural machines moving on it. There is the danger of compacting the soil so strongly that it will later be difficult to restore a normal structure. Under these conditions *trafficability* (possibility of movement of machines) and *workability* (possibility of working the soil) are distinguished. They are expressed in *machinery work-days*, that is, days available per year for exploitation of the land.

Understanding of the water balance, by means of a program such as EuroAccess II, thus enables us to predict the periods during which the soil will be saturated with water. During these periods it is impossible to sow or harvest the crop. This can lead to avoiding certain crops that must establish before the end of winter or which must be harvested late when the seasonal excess of water recurs.

Attempts at spatialization with regard to the number of days available have been made (Rounsevell, 1993; Rounsevell and Jones, 1993).

8.6.5 Climatic risks

We determine on the basis of precise agronomic trials for example if such or such temperature is necessary for starting germination and such or such quantity of water to ensure normal growth of the plant to maturity. If the climatic data of the region in question are otherwise known, it becomes possible to examine the number of times in the last 20 or 30 years that the conditions occurred together for satisfying the needs of the plant at the appropriate time. This enables estimation of the probability of success of a sowing or maturation. If climatic and soil maps are available, we can think in terms of area and define the regions that are suitable for germination of wheat, maturation of maize or providing the water requirement (Bonneton, 1986; Leeds-Harrison and Rounsevell, 1993). In short, maps of agricultural risks can be prepared.

Risks of agricultural drought are determined by using rather sophisticated crop models. There is abundant bibliography on this subject.

8.6.6 Risks of erosion

Many attempts have been made for spatializing erosion risks. Two approaches can be distinguished.

The first method concerning erosion is inductive. It aims to find out the principal factors explaining the initiation or intensity of erosion in identified areas. The procedure consists of drawing the map of erosion status and comparing it to maps giving the distribution of environmental factors presumed to be responsible (position on the slope, soil texture, etc.). This analytical and statistical approach is very educative (SESCPF, 1991).

The second method is deductive and predictive. It consists of measuring and mapping, rather accurately, the various factors that traditionally are responsible for initiation of erosion and combining them by means of the famous Wischmeier model (Wischmeier and Smith, 1978) still called, rather immodestly, the *Universal Soil Loss Equation* (USLE):

$$A = R \cdot K \cdot SL \cdot C \cdot P$$

where A = predicted soil loss in tonnes per hectare,
 R = *erosivity* factor depending on rainfall,
 K = *erodibility* factor linked to texture, organic matter, structure and permeability of the soil,
 SL = factor related to the slope angle and slope length,
 C and P = factors accounting for land use (crop, agricultural practices, etc.).

The 'CORINE' maps estimating risks of erosion in the European Union were prepared on this basis (Bonfils, 1989; CORINE, 1992). Still, it falls in the category of *diffuse model* (multiplicative) although the proposed formula has some mechanistic foundation (Chamayou and Legros, 1989; Musy, 1991).

8.6.7 Social and economic factors

All the studies presented here in thematic mapping only concern pedoclimatic potentials and do not take social and economic factors into account although the latter play a major role in land use. When necessary, the mapper works in collaboration with agronomists and specialists in human geography.

Let us only point out that simple models exist for estimating the profitability of a crop in relative value by comparing its actual production with the maximum production possible in a given region (production potential).

8.6.8 Synthesis

All the models considered only give a fragment of the diagnosis. To fully answer an agronomic problem, it is not enough to use them successively or in parallel. A synthesis is necessary. Let us take the case of preparation of a map of suitability for a particular crop. How do we decide? What is to be thought of a crop suitable for the pedoclimatic environment and profitable as well but risking, in the long term, degradation of the soil by erosion? The choices pertain to very heterogeneous elements. We run into a typical problem of incomparability (cf. end of § 8.2.2). Thus, whatever effort might be taken to conduct a reasoning as precise as possible based on mechanistic models, we reach a stage when disparate data must be combined to reach a decision. This is summarized in Fig. 8.13. Burrough (1994) has briefly but completely reviewed the advantages and shortcomings of using quantitative models in environmental sciences.

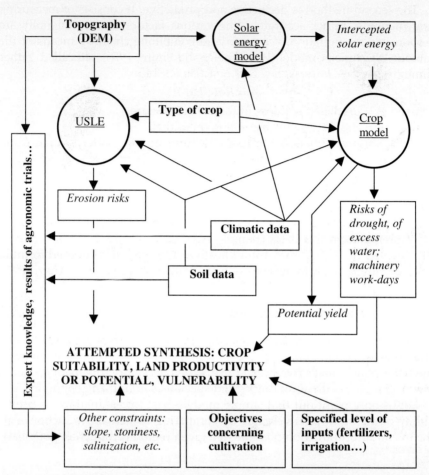

Fig. 8.13 Combined use of various models in thematic mapping. Bold: data; underlined: models; italic: calculated or evaluated indicators; all caps: results.

8.7 CRITICAL ANALYSIS

8.7.1 Drawbacks of thematic mapping

Lack of validation
Often no validation is done at the end of the work that results in a thematic map. In other words, an unverified diagnosis is given. For example, we state that such or such sector is suitable for potato, while nobody has the intention of doing the trial immediately. It also happens that the conclusion

is not verifiable. For example, what validation could a 'map of small natural regions of tourist and aesthetic interest' be subjected to? We know that Karl Popper considers to be without scientific basis a proposition we cannot verify to be true or false.... But empirical diagnoses are at times indispensable, although thematic mapping comes more within expertise than science. Be that as it may, it is useful to systematically seek to measure the quality of work done.

Excessive simplification
Thematic mapping is a kind of modelling that introduces some simplification of reasoning and a reduction in the number of data taken into account. Thus there is risk of disregarding essential data. This has been emphasized at the beginning of the chapter. Supplementary aspects of this question merit attention.

Consideration of the spatial location. Let us work on an example. We met at the city gates of Vénissieux, near Lyons in France, a farmer who told us *'I am giving up arboriculture though my cherry trees are in full production and give excellent fruits on perfectly suitable soils! Do you see these huge buildings into which are packed many families with many children? Do you think one could harvest cherries at the foot of such towers without sleeping in the orchard with a gun and being mad enough to shoot at children? I am forced to stop.'* We know how to take into account such constraints if they are identified. The distance to the houses can express one of them. On the other hand, with the methods presented above, the individuals concerned (pixels, polygons) are treated separately without taking into consideration for one individual the results obtained for the neighbours. This is a big lacuna because there are many examples of undesirable neighbourhood areas. The following map (Fig. 8.14) is the worst the author could imagine! The reader is invited to find out all the errors (solution at the end of the chapter).

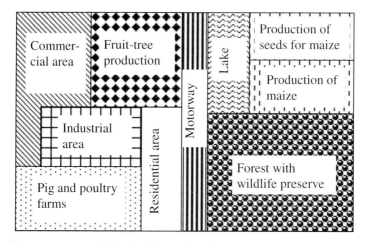

Fig. 8.14 The worst conceivable land-use-planning map.

Without using an iterative approach examining the boundaries and neighbourhoods, one can submit the thematic map produced to experts who will suggest local corrections if necessary. In the Department of Rhône, specialists give great importance to *agricultural landscapes* for planning use of the environment (Boutefoy et al., 1999).

Consideration of temporal changes. In this domain too, we can take recourse to a real-life example. On the southern edge of Lyons (same area as the cherry farmer's), we prepared a capability map for defining the sectors that would be urbanized and those that would be retained for agriculture. The landscape consisted of valleys with good deep soils and narrow high hills covered with coppices and showing hardly any level areas. Under these conditions the scheme for urbanization appeared obvious: leave the valleys for agriculture and build houses on the hills. The new residents would have superb views with huge green areas in front of their eyes. But this implicitly supposed urbanization over a maximum of 10% or 15% of the area.

Now suppose that pressure of urbanization is very high. The valleys, even less residential, will be speckled with buildings and agriculture will be compromised. Some uninteresting vacant lands will be left between the houses. If this constitutes the term 'ineluctable' for utilization of space, it is better to reason otherwise: urbanize the valleys right from the start and conserve the wooded hills that can form superb public parks. Thus correct thematic mapping presumes consideration of the time factor and probable changes.

Variation in inputs
Suitability of the soil for such or such use can be very different according to the efforts one is ready to agree to make for improving the production level. Will one drain the land? fertilize it? go from animal-powered agriculture to mechanized agriculture? It is necessary to correctly understand the starting hypotheses and make sure they do not remain implicit. In some cases we may be led to prepare several thematic maps, each one corresponding to properly identified hypotheses.

8.7.2 Advantages of thematic mapping

Sequencing of data
Use of computer algorithms or at least strictly codified reasoning requires prior sequencing of the data. Well-structured files containing all the necessary information must be available. Those who have used these methods are in agreement in establishing their interest right at the start, that is, even before calculations begin! At this stage we must recollect the lesson from Fontaine's fable *The Labourer and his Children*. Computer-aided thematic mapping is a treasure, not always accessible, but investigation of it generates work and organization, thus progress.

Analysis of the procedure

Thematic mapping is implemented after analysis and approximate reconstruction of the expert's method. This reconstruction often appears simplified, even grotesque. But it has the merit of existing and being recorded on paper. Thus it represents the state of the art at a given moment! It can then be studied critically and improved. This is the way to progress. Also, the procedure, being described in all its details, becomes reproducible. It can be used by several experts who, starting from the same data will, in theory, find the same result. Thus we have enhanced the *reliability* (in the statistical sense of the word) of the diagnoses. In spite of this, there still are experts who are allergic to such approaches and make fun of them, finding them too elementary. To counter them there is a very simple parry, that is to say, *'I admit you are right, here is a pencil. How do you suggest we go about it?'* Discussion usually stops there and the expert escapes! If the expert stays back, you too have gained, because he will help you progress... .

Introduction of simulation

Use of the computer and calculation algorithms enable us do simulation, that is, to produce several thematic maps by increasing or lowering the values of certain parameters. This is obviously very important for examining the consequences of the hypotheses made. Let us take an example. Suppose a map of suitability for sugar beet is desired. We know that a soil is suitable up to 6% slope and that it is unsuitable starting from 10% slope, but we hesitate at the diagnosis regarding slopes from 6% to 10%. Then it is sufficient to prepare a first map with the threshold at 6%, a second map with the threshold at 10% and to compare them using a GIS. We can get an entire series of possibilities between two extreme cases. In the first case, both maps are similar. This leads to the following conclusion: of course, the problem of knowing if one can cultivate sugar beet between 6% and 10% slope is important, but it does not concern me (no pixels show this gradient or the pixels have been eliminated because of some other shortcoming). In the second case, the maps are quite different. The map of the differences between these two initial maps indicates geographical areas where the problem exists. It is there that it is necessary to go for examining if sugar beet is cultivated and to ask the farmers. Simulation also has the advantage of making apparent to the creator of the map the uncertainties attributable to the work. It is characteristic of laboratories with necessary tools available that thematic mappers draw not one but 30 or 40 maps before stopping at what seems to be appropriate and to represent the best possible compromise between all sorts of hypotheses and constraints!

Quantification

The maps, machine-drawn in this manner, enable synthesis of the results in numerical terms. The possibilities of GIS in this area are known (Chap. 6,

§ 6.2). Obviously we can do a balance sheet of the areas appearing throughout the map. But other records are accessible. It is enough to make one think. Still, reading studies done, particularly students' dissertations, leads us to think that the possibilities of quantification are under-exploited. Statistics summarizing the entire question are often not presented.

Conclusion and perspectives

Soil scientists, somewhat like Monsieur Jourdain who wrote prose without knowing it (in the play by Molière), manipulate multicriteria analysis without always understanding it well. However, it has been explored and codified a little all over the world. In the case of English-speaking scientists the major orientation is towards complete aggregation. The IDRISI software and the studies of Eastman are the manifestation of this. In the French-speaking world, the preference is for partial aggregation for avoiding problems of incomparability. In this context, advances have come above all from Swiss researches carried out at l'École Polytechnique Fédérale de Lausanne and French studies conducted at the Université Paris-Dauphine.

We have only lightly touched upon the subject. There are many other methods due to the inventive genius of the investigators. However, advances in methodology remain to be made. Firstly, it is better to include these methods in the framework of GIS when they pertain to partial aggregation. Efforts have been made (Joerin, 1995; Jankowski, 1995; Laaribi, 2000), Secondly, and this follows logically, it would be necessary to implement diagnoses with consideration of the relative spatial position of the delineations or pixels examined. The necessary tools are available and we have presented them earlier (Chap. 6, § 6.2.3). But, to the best of our knowledge, they have not yet been sufficiently integrated with thematic reasoning.

Furthermore, one need not think that automatic processes will one day replace man. The situations to be considered are complex and changing. The expert alone can master them. Also, taking the help of powerful methods of calculation and to codified but simplified methods does not give results necessarily better than using human intuition with its ability to rapidly decide after almost instantaneous selection of the principal elements of the problem posed. Actually, the two paths are different and contrasting, but in some way complementary. This is normal and is favourable to stimulate thinking and to lead to progress.

Solution to the exercise (Fig. 8.14):

- The commercial area is far from the residential area.
- Fruit trees are too close to the residential area.

- The residential area is too close to farms (odours), too close to the industrial zone (risks), too close to the motorway (noise).
- But this residential area is separated from recreation areas (forest, lake) by the motorway. Is there a bridge?
- Production of maize seed is impossible and otherwise prohibited next to a field of ordinary maize (cross-pollination).
- To grow maize at the edge of the forest is to incite stags and roebuck to help themselves to the crop. Is the forest fenced?

It is hoped, without thinking too much about it, that this type of absurd organization of the area does not really exist....

CHAPTER 9

SOIL MAPPING AND MULTIDISCIPLINARY APPROACH

Specialists in soil mapping are often distressed to see that their object of study is not taken into consideration as it should have been in problems of management of the land or protection of natural environments. There are many reasons for this: the complex, even esoteric, aspect of our specialized vocabulary, compartmentalization of disciplines, by which the soil component of a multidisciplinary problem is not always seen, and deficiency in laws that do not always provide for protection of soils and taking them into account in the decision process. Under these conditions, it is only right to more closely examine our interfacing with other disciplines, particularly agronomy, civil engineering, ecology and climatology. For doing this we shall depend on some specific, concrete interdisciplinary approaches that have been successful. These studies could serve as example and provide the material for setting up our thinking. In all the cases the synergies established or even those imaginable in the future are to be examined.

9.1 SOIL MAPPING AND AGRONOMY

The problem is general in developed countries: firstly, mappers prepare their soil maps and, secondly, the agricultural world (farmers, agricultural advisers, agronomists) do a large number of soil analyses for these same regions. One should then understand if an interaction will enable work to be done better and to economize on time and money. We will base our thinking on what has been done from this viewpoint in Belgium and France. Each year in France, very many soil analyses (about 300,000) are done by farmers anxious to check the fertility of their soils and to plan their fertilizer application. The situation is the same in Belgium.

9.1.1 Emergence of the spatial approach to agronomy

In Belgium, 90,308 soil analyses distributed over the entire territory were used (Van den Driessche et al., 1993). This represented, according to the regions of the country, from 0.5 to 18 samples every 100 ha. It was then possible to study the fertility levels spatially and in time.

- *In space*. These analyses were grouped by administrative units and the corresponding average or modal values (cf. statistical mapping, Chap. 2, § 2.2.2) calculated. For example, the P level in soils of Belgium decreased sharply from the north to the south of the country.
- *In time*. The studies are less advanced because it is necessary to wait some decades before changes with time can be established. But there is no doubt that soon we will have a means for recording the behaviour of farmers through their fertilization practices (change towards intensification, towards extension, taking into account a greater respect for the environment, etc.).

At a more detailed scale, in Hainaut for example, van Koninckxloo and Huart (1994) followed the change with time and in geographical space of contents of humus, phosphorus and potassium on the basis of 5000 analytical determinations per year. It was observed that soils were richer close to sugar refineries because of use of residues. Also, good positive correlation was obtained between favourable soil conditions, crop yields and farmers' incomes. Conversely, in certain sectors, degradation of financial capital, diminution of fertilizer application and a tendency for farming to disappear were observed.

In France, soil mappers are counted among the forerunners of spatial studies pertaining to soil analyses (Leleux et al., 1988). Using their competence in information technology, they collated and then processed the results of analyses pertaining to 71,000 soil samples taken all over Brittany. As these analyses were referenced by the name of the concerned communes (administrative subdivisions), it was possible to construct maps showing the mean content of P, organic matter, etc., of these communes (Fig. 9.1).

Similar studies were done in the Midi-Pyrénées region (Mathieu et al., 1993), for which 1200 samples pertaining to 30 communes were used.

Subsequently, the same approach was repeated at the national level. Thanks to the support of many analytical laboratories, the results of analytical determinations (on average 10) done on 300,000 soil samples were recovered (Schvartz et al., 1997; Walter et al., 1997). The data were grouped by canton (administrative subdivision). Then a national picture of the distribution of pH, organic matter, P, etc., in soils could be given.

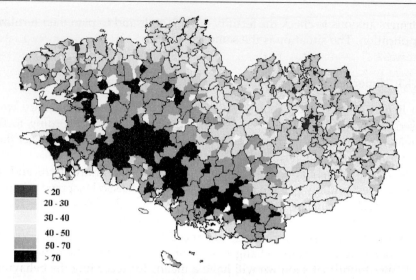

Fig. 9.1 Organic matter content (‰, median values) of soils of Brittany during the period 1980-1985 (Leleux et al., 1988).

These syntheses, taking administrative subdivisions as basis, have the advantage of corresponding closely to the concerns of the farming world and its advisers. But we should not forget that the farmers who get soil analyses done are the dynamic ones who often fertilize the soil better. Thus there is danger of a systematic bias, which did not escape the notice of the authors. All these studies represent a macroscopic view the relevance of which was underlined in our general introduction.

9.1.2 From the soil map to agronomic interpretation

Placing soil analyses in relation to typology of soils
It obviously is useful if sampling of soils for analysis could be placed in the framework of regional typology of soils. Interpretation of the corresponding data will then be facilitated and agronomic advice probably more suitable and easier to be generalized. But this identification of soil type is difficult for at least two reasons. Firstly, even though the entry forms suggested by the analytical laboratories to the farmers make provision for recording the geographical coordinates, the relevant data are rarely entered; so the analyses are poorly referenced and most often attributed only to the administrative unit from which they were taken. Thus we cannot identify the soils to which they correspond on the basis of their location even if we otherwise have a good soil map available. Secondly, the records provided by the farmers or their agricultural advisers are brief; for example, excess of water is not recognized because it rarely pertains to the plough layer. Thus we are not in a position to identify the soils by their properties described even by

using computer tools that serve to find 'what resembles' such or such sample (Chap. 7, § 7.4). In the future, it will be necessary that the analysis forms be complete and that farmers, properly trained, make the effort to fill them up. In the meantime, attempts have been made to speed up the matter (Schvartz and Douay, 1992).

Rationalizing fertilizer application by taking the environment into account
Having taken into account the large number of analyses to be interpreted, some thought of creating software for automatic interpretation of soil analyses. It is in this way that a study was started at the INRA: the 'LIAT' project (LIAT, 1985). It did not lead to anything operational enough to justify commercialization. Actually, the greatest difficulty is that interpretation cannot be done in the absolute without considering the regional climatic, pedological and technical situation. For example, 1.5% organic matter in a soil will be declared insufficient for large-scale farming in northern France but sufficient in the vineyards of southern France, where one must reckon with much more rapid mineralization in the hot, dry climate. Therefore interpretation must be regionalized. Specifically, LIAT planned consideration of the following files:
- file of regional soil units;
- file of regional climatic units;
- file of typical regional cropping sequences;
- file of typical agricultural situations.

It can be well seen that the objective was very ambitious: to build a giant expert system taking into account extremely diverse situations. In Belgium the BEMEX system was developed (Vandendriessche et al., 1993). But then, climatic variability in that country is much less than in France... .

9.1.3 From agronomic data to pedological synthesis

As soil analyses are very numerous, we can ask ourselves to what extent they will be insufficient for preparing the pedological outlines required for planning regional development (particularly on scale 1/250,000). Sporadic attempts have been made starting from the databases in which these analyses are stored and exploiting the possibilities of GIS. The mean or median content of clay or of organic matter for each administrative unit of the region in question is thus determined. Danneels and Schvartz (1993) reviewed with great lucidity the difficulties in such a project. Firstly, the data are supplied without accurate referencing (coordinates), as we have said before. Then, these soil analyses only pertain to the surface layer. Contrarily, a soil map takes into account the middle and deep horizons of the soil, particularly for studying problems of water economy. Lastly, the spatial resolution is not good. If we compare the number of analyses to the corresponding areas, we find about one analysis for every 250 ha in the Nord Pas de Calais Region. Theoretically, this corresponds to an accuracy of 1/250,000 (Chap. 3, § 3.4). But in reality, this is far from the truth. The sampling points were actually not carefully selected as is done in a free survey. Moreover, from one point to the other there is neither observation of the soil surface nor consideration of the environment. Under these conditions, the authors cited consider that the accuracy actually attained is 1/1,000,000. But the soil map of France at this scale already exists! But improvements are possible. It will be necessary for analytical sites to be still more numerous, better referenced and put back at least in a simple geomorphological setting (valleys, interfluves).

To sum up, it is strictly impossible to do away with mapping at 1/250,000 to replace it with automated interpolation between soil analyses! But this is not to say that this interpolation is of no use. Many mappers will love to compare it with their own studies and possibly locate the sector to which they have to return. This will be all the more desirable as, the analyses being done, their machine exploitation from the mapping viewpoint is not very costly.

9.1.4 Towards integrated agronomic and pedological studies

Examination of the extent to which the soil map can be used for agronomic purposes is an essential work that has particularly interested Belgian scientists. Actually, they have available since 1990 an application-oriented (great emphasis laid on texture, redoximorphic features and stoniness), detailed (1/20,000) map of almost the entire country. According to the authors involved in these studies (Neven and Engels, 1993; Engels et al., 1993, 1994), four documents must be used together:
- soil map;
- map of plots so that the results are presented in a manner directly pertaining to the farmers;
- Digital Elevation Model (DEM) so that relief constraints are considered;
- division of the country into ecological zones so that problems are regionalized.

Various land-uses have been tested on 37,000 ha spanning Condroz, Famenne and Ardenne. Actually, these are exploratory trials, numerous but not treated in detail. Of them, we will retain the following because of their originality:
- choosing forest species suited to the environmental conditions;
- crossing of environmental data with socio-economic data to obtain zoning of 'agricultural situations'; for example, there is a bond between relief, soil type and the dispersion of plots in various farms;
- GIS-aided regrouping of lands; the algorithm will consider quality of soils, their area, their position and the crops for suggesting optimized distribution and spatial regrouping.

To summarize, many attempts have been made to date for linking studies of soil scientists to those of agronomists. The former understood the vital need, in soil mapping, of leading to practical applications in the domain of agronomy. The latter finally saw the usefulness of spatialization that enabled them relocate practices and developments in the context of the natural environment and corresponding constraints. Whatever its fundamental explanatory role might be, the farm is not the only spatial level at which the soil could be considered in agronomy! Other levels are revealed to be interesting (Benoit and Maire, 1992; Deffontaines and Lardon, 1993).

9.2 SOIL MAPPING AND CIVIL ENGINEERING

In Switzerland, a mountainous country, good soils are rare and the intention is to save the resource they represent. For this, a collaboration was established between the civil-works engineers and soil mappers. The work to be examined here as a good example, was done with the scientific cooperation of 'Institut d'Aménagement des Terres et des Eaux' (IATE) of EPFL (Lausanne, Switzerland).

9.2.1 Consequences of road building

Swiss legislative procedures
Consideration of the soil in town and country planning projects is governed by a few principal texts:
- legislation by the Confederation, particularly the law of 7 October 1983 on protection of the environment (LEP), the ordinance of 19 October 1988 relating to investigations of environmental impact and the directives issued by the Federal Department of the Environment, Forests and Landscape (OFEFP);
- the recommendations published in duplicated form by the Swiss Association of Transport Engineers; this document constitutes a guide for conducting studies on the impact as far as road-building is concerned; it covers the different points of the law and the technique (OFEFP et al., 1993).

Objectives of intervention of soil scientists
Construction of a road or motorway involves:
- first examining all the immediate and direct consequences of the changes expected in the soils of the neighbourhood: physical, chemical and biological modifications, risks of erosion and possible transformation of the hydric regime;

- predicting, along with specialists, the indirect consequences of the development; these can be negative (for example, drastic reduction in size of a farm that was viable earlier) or even positive (reorganizing the landscape with regrouping of plots, improvement in access roads, etc.);
- drawing up, if necessary, a statement of conditions before the works start so that one could later highlight the negative consequences if any of the development;
- organizing the protection and reuse of the good soils that are salvaged in the area where the projected development will occupy the land.

Works

These principles were established for construction of the motorway that bypasses Geneva (N1A) and the Transjura motorway (N16) in the Porrentruy-Délémont (canton of Jura) and Tavannes-La Heute (canton of Berne) sections. The soil science laboratory of EPFL was approached as the construction progressed. Construction of the motorway was monitored over more than 40 km (de Pury and Dupasquier, 1995). The major job of the specialists was to define the depth of the materials to be stripped and to be salvaged in the zone that would be reserved for the road and its numerous links: access roads, areas for temporary storage of materials and vehicles, various manufacturing zones and zones for dumping unsalvageable materials. The zone of influence of the work and its annexes was on average 170-m wide during construction (700 ha for 40 km). Under these conditions, the soil to be stripped and stored represents a considerable volume, almost 12,000 m3 for each hectometre length of road, if 70-cm thickness is removed on average. In general, the A and B horizons of the soil were put in separate heaps. The blueprint for stripping was provided by the soils specialists in the form of a map indicating for each sector the depth to which cutting was necessary. The document was presented on scale 1/1000 because that was the scale for planning the road works (Fig. 9.2).

The fertile soil material recovered serves to repair at the end of the construction the various areas worked in and also for vegetating the area around the motorway. But other uses can be thought of. Customers have the chance to appear at a 'soil exchange' such as the one set up by the canton of Geneva several years ago.

Lessons drawn

Reconstitution of soils is a delicate operation. the studies conducted with the cooperation of EPFL showed that some essential precautions must be observed:
- Storage time must be as short as possible to avoid excessively compacting the soil and depriving it of oxygen. For the same reasons the soil should not be piled up in very high heaps, though it may be

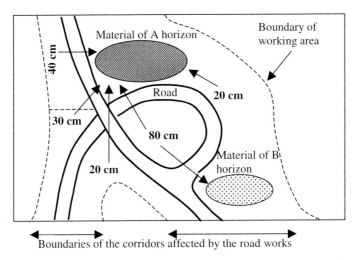

Fig. 9.2 Example of map of soil stripping for road development (from de Pury and Dupasquier, 1993). The numbers indicate the soil thickness to be stripped and stored temporarily. The arrows indicate in which heaps the extracted materials will be stored.

clever to use them as soundproofing mounds during work (technique of the Swiss Federal Railways). Turfing the mound is a good precaution for avoiding erosion and also invasion by weeds (it is necessary to think of future agricultural use). Also, the roots can contribute internal drainage and aeration.
- The restoration phase is delicate. The B horizon is first put back without much tamping. If necessary it is scratched before putting the A horizon on it. The work is done in such a way that machines do not run on the A horizon. In any event, use of tracked vehicles is preferable. The reconstituted soil must be grassed immediately so that its porosity is not reduced by compaction. After three years of this pioneering use, it can be put to cropping.
- All the work is done when the soil is suitably moist. Works must be stopped in bad weather.

When the work is properly done it is particularly pleasing to see the results of such developments. The land regains its original aesthetic quality and its full production capacity. In the case of construction of the Grauholz tunnel near Berne, it is spectacular to compare the photographs of the site taken during and after the work (Krebs, 1995). On the other hand, if the soil is excessively compacted, several years would certainly be needed, in the best of cases, to restore suitable physical properties. In some situations, it has been necessary to decide to strip the soil again and start all over!

9.2.2 Economic development of material from tunnel digging

Objectives
In the canton of Tessin, digging of the Mappo-Morettina road tunnel resulted in extraction of sandy materials from crushed gneiss. This represented hundreds of thousands of cubic metres that had to be disposed of. Four kilometres from there, the farming zone of Terreni Carcale, located at the Swiss end of Lac Majeur, was periodically flooded on a hundred hectares. The situation slowly deteriorated to the extent that the level of the lake, fixed by the Italians, rose gradually. In 1995, the loss of income was estimated to be Sfr 1400 ha^{-1} y^{-1} (low efficiency of fertilizers and plant protection, loss of seed). It was then planned to raise the level of the land by using the material from the tunnel. Considering the earth available, it was possible to treat at least 60 ha. In principle, the A and B horizons were first stripped and their material stored in two separate heaps. The earth from the tunnel was then spread in a layer 1-m thick, then the soil was put back in place. Such raising by one metre was a little insufficient for avoiding all risk of flooding, but it was difficult to go beyond that considering the presence in the zone to be treated of substructures and houses that could not be buried.

Method (Prélaz-Droux and Musy, 1991, 1992)
An operation of this sort poses problems of mapping, experimentation and organization of the construction.
- *Mapping* consisted of first choosing exactly the zone to be treated, then studying the soils through 400 auger holes. The soils were composed of sandy or loamy sand alluvium, at places with intercalated peat lenses (hydraulic conductivity 10^{-4} to 10^{-2} cm s^{-1}).
- *Experiments* had the objective of verifying that the reconstituted profile will have appropriate hydric properties. Experiments were first done in lysimeters and confirmed that permeability of the reconstituted soil will be excellent and that compaction could be within acceptable limits. Later, truly large-scale trials were done on 2 ha. Various determinations were then done with infiltrometers, penetrometers, tensiometers, neutron probes, etc. The hydraulic conductivities of the reconstituted soils were found to be similar to those of the soils *in situ*.
- *The principal aim of organizing construction* was to avoid repeated handling of large quantities of the materials in question. For this, the procedure was something like what is done while ploughing. The soil that was removed was not stored while waiting to be put back in the same place. It was immediately poured on the adjacent strip. Actually things were more complicated because it was necessary to insert the material from the tunnel and treat the A and B horizons separately. Figure 9.3 shows how one proceeded in three stages, treating three

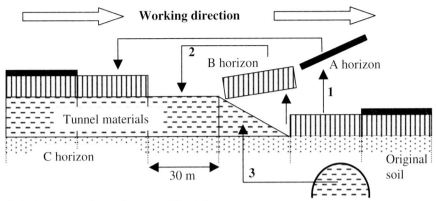

1: Stripping and immediate use of the A horizon
2: Stripping and immediate use of the B horizon
3: Replacing the newly stripped B horizon by material from the tunnel

Fig. 9.3 Raising the soils of Terreni Carcale (from Prélaz-Droux and Musy, 1992).

strips of land at the same time. The reader will notice that we find ourselves again in the same situation as at the start, one strip further in the direction of the right of the diagram, when the operation is completed. Only the soil volumes corresponding to three A horizons and one B horizon over the width of one band must be kept in heaps during the construction.

In the given case, the strips measured 30 m in width. They were moved by means of scrapers.

Results
Although the areas already raised gave satisfaction, it was still too soon to make an overall judgement on this kind of programme. Specifically, one could be afraid of rapid weathering of micas that were abundant in the earth from the tunnel. Such weathering could result in the bottom of the profile becoming impermeable in a few years. Be that as it may, this investigation is exemplary. Firstly, it showed that it is not always impossible to reconcile the interests of farming and those of building roads or railways. Secondly, considering present-day developments, it will certainly be a standard the results of which will be useful for other projects in the future. Lastly, it is necessary to underline the good cooperation in this project of administrative authorities, developers, scientists and farmers. In such a situation, understanding is established and good ideas are born.

9.3 SOIL MAPPING AND ECOLOGY

When we proceed from large-scale maps to small-scale maps, the position of soil gradually diminishes and purely pedological mapping gives way to physiographic mapping (Chap. 2). Thus the proposed syntheses at the scale of large countries or geographical regions (USA, Canada, Europe…) are of ecological as well as pedological use. For example, the soil map of Europe planned on 1/250,000 (Chap 10, § 10.1.3) will have to present 'Soil Regions' broadly taking into account the climate and geological materials. Ecological mapping thus deserves attention. It is not possible to consider it here in detail. The key words 'ecological mapping' result in more than 384,000 references on the Web. Above all, let us see what has been achieved in Canada. This country, with extent, relief and geographical position favouring great ecological diversity has made many efforts in the domain that concerns us here.

9.3.1 Methodology

Definition of objectives
Ecological partitioning of territory is aimed at recognizing natural entities of rather large size. These entities represent a kind of hierarchy and are exactly fitted together in that each element of a given level contains several elements of the next lower rank (Fig. 9.4).

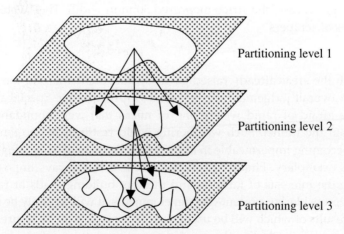

Fig. 9.4 Principle of ecological partitioning of a territory.

We had, in the first edition of this book, presented in detail the efforts made in Quebec for ecological partitioning (Robitaille, 1989, 1992; Ducruc, 1991a, 1991b; Bergeron et al., 1992; Robitaille and Grondin, 1992; Bissonnette et al., 1995). The objective being to separate well-defined entities on the

basis of climate, soils, vegetation and parent material, it was necessary to rationalize the procedure. For this, the authors considered the following criteria:
- Principal determinants for a given level of the hierarchy (soils OR vegetation OR parent material, etc.),
- Order of magnitude of the size of the entities delimited as a function of their place in the hierarchy, particularly the scale of the map that enables legible display of the boundaries of these entities),
- Number of soil observations necessary for establishing the delineations at various levels of the hierarchy.

Similar attempts were made in the other provinces of Canada. The period 1985-1995 saw a profusion of attempts and achievements (ESWG, 1995).

The methodological foundations being defined for practically doing the delimitations, there are two broad types of approaches possible. They are not totally exclusive, rather are complementary and are presented below.

Empirical approach
In Canada the general synthesis of local efforts and works were done in a pragmatic way with case-by-case examination. Working groups were formed. People with practical experience were relied on: each one could share his experience and suggest the boundaries relevant to his sector. Actually it is necessary to consider the local peculiarities and land features. What is of great importance in one place is of no interest in another. For example, 'Rougier de Camarès' in France is a Small Natural Region very well characterized and differentiated by the red colour of the soils related to Permian continental deposits. Everything is indirectly related to this colour: types of soil, types of agriculture, specific types of environmental risks. Thus colour is, in the Rougier, an excellent criterion for delimitation. But 50 km away, colour has no significance for the soils on 'variegated' marls of various mixed colours. So a general system of weighting of criteria cannot be applied. However, it is surprising to do a partitioning with criteria that have variable weights within the same application. This is using an unstable method but we do not see how that can be avoided! Under these circumstances the scientific foundations of the partitioning seem to be rather weak. Though technical reports are many, papers in scientific journals are fewer....

Approach through map overlay
Dokuchaev demonstrated that the soil is bound to the climate, relief, parent material, biological agents and time. This is most certainly correct. But this is to take a pedological view of things. Many of these elements interact directly even if the soil can play the intermediary role (Fig. 9.5).

It is even possible that more arrows could be added in the right-hand part of the diagram. This means, very practically, that by superposing through map overlay the topography, vegetation, etc., we have every chance of

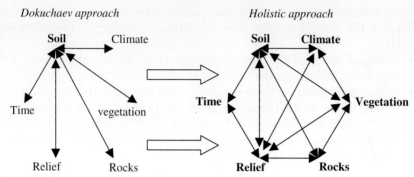

Fig. 9.5 From Dokuchaev's view to a more holistic approach.

arriving at a partitioning that is at the same time significant and reasonable: the various units will more or less correspond and we will not get a fragmentation into very small polygons. This was well demonstrated in New Zealand (Harding and Winterbourn, 1997).

It is thus desirable to combine the two approaches presented. Map overlay serves to get a general idea of the problem and to suggest a preliminary partitioning of the area. The work of experts is to discuss, validate or challenge it based on their knowledge of the setting.

9.3.2 Results

The Canadian scientists recognized the following four entities, going from the largest to the smallest:
- *Ecozones*; they are 15 in number for the whole of Canada and gave rise to the preparation of a map at 1/7,500,000; they correspond to the broad climatic and vegetation groups [(mixed-wood plains, taiga, polar domain...); see Fig. 9.6],
- *Ecoprovinces*; there are 53 of them,
- *Ecoregions*; they number 194 but some of them are split up so that the corresponding map comprises 217 polygons; at this level, climate shows some homogeneity and can be characterized by average values (temperature, rainfall...); the concept of ecoregion was introduced subsequent to the ecological partitioning organized at the same time in USA (Lammers and Johnson, 1991).
- *Ecodistricts*: they are 1021 in number, but represent nearly 17,000 polygons. At this level, territorial continuity of the entity is not an essential criterion. It is based on topography and on subgroups of soils (Canadian classification), the typical combinations of which can be found in many areas of the country. Within an ecodistrict, the potential for forest production or crop production is theoretically uniform.

To these four levels of ecological partitioning must no doubt be added a fifth: *Soil Landscape*. The Soil Landscapes of Canada (SLC) constitute the

SOIL MAPPING AND MULTIDISCIPLINARY APPROACH 323

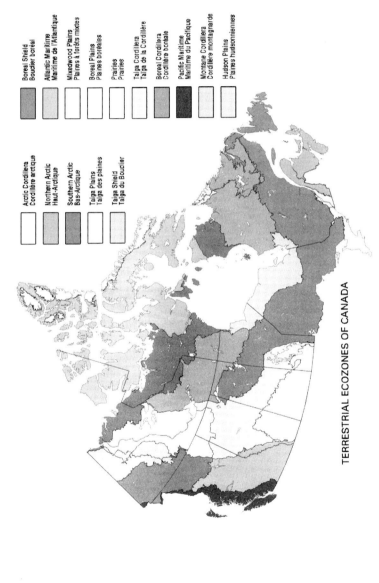

Fig. 9.6 Map of terrestrial ecozones of canada (with permission)

basic units and, at the same time, the polygons that can be related to various Databases (soils, cropping practices) for various purposes. These SLCs can be represented on a map at scale 1/1,000,000. This may shock Europeans who see in their countries smaller Landscape Units. But then, Canada is a vast country.

9.3.3 Discussion

The essential thing in this kind of ecological mapping is the recognition of entities with uniform appearance. One is not interested in details. Under these conditions, photointerpretation is mostly used. Studies of soil are limited to point observations and verification. Thus the approach is rapid and does not cost much. It is quite suited to regions with limited or no agricultural interest and where detailed studies will not be justified. It is particularly suitable in the case of well-forested areas because vegetation can be observed in satellite photographs and images. Thus the study is useful from the forestry point of view. Moreover, it is foresters who developed these approaches. It enables the following applications for them (Ducruc, 1991a):
- estimation of the potential for forest production;
- searching for materials for building forest roads;
- definition of the capacity of soils to bear heavy machinery (trafficability);
- estimation of risk of soil erosion after deforestation;
- choice of species for reforestation;
- searching for favourable sites for burying refuse, considering the vulnerability of ground water to pollution;
- corridors for power-transmission lines;
- definition of the potential habitat for the black bear!

On the other hand, the place of the soil is at times reduced to very little in the system of ecological characterization. Some authors even suggest that they are not interested in the one metre of soil that overlies the materials to be characterized. This seems hard to accept when it is known that 95% of roots of trees draw their water and nutrients from this layer. In truth, this attitude of defiance concerning soil science is the expression of the unwillingness to go down to the identifiers that will not be directly visible on aerial photographs. On the contrary, the parent materials can be spotted on aerial photographs with a little practice: scree, moraines, subsurface rock, etc. But it is only fair to add that some parameters recommended by a majority of authors and included under the category 'geomorphology' actually pertain to the soil (texture, stoniness, drainage).

9.3.4 Other experiences, conclusion

In USA, scientists proceeded pragmatically. The map of *Common Ecological Regions of the Conterminous United States* prepared by the *National Interagency*

Technical Team comprises boundaries still under discussion. Also, in the same country, they have progressively defined, over the past 60 years, the *Major Land Resource Areas* (MLRAs) of the country. There are hundreds of them. The contents of each have been progressively refined by processing the data through STATSGO (Chap. 6, § 6.5.1). Each MLRA of USA is denoted by a number. The soil maps that, in USA, were partitioned and processed on county basis are, since 1995, entrusted to 18 MLRA regions and offices (Anderson, 2002). Each office is responsible for a certain number of MLRAs within which correlation teams have been placed. In this situation, the *National Cooperative Soil Survey* has been totally reorganized. This shows the importance given to naturalist (versus administrative) partitioning of the territory in USA.

In Italy, there has been since 1997 a project to prepare an Ecopedological Map of the entire national territory on scale 1/250,000. A methodology is being drawn up with the support of the *European Soil Bureau*.

In sum, this kind of multidisciplinary approach is growing. It allows preparation of a single map for treating a problem of conservation or development of the environment, while in many countries all sorts of disciplines suggest, in competition, preparation of uncoordinated maps: geomorphological map, soil map, forest-vegetation map, etc. Ecological mapping thus represents great economy of means. The experience is worth thinking about. In some countries that cannot or do not wish to allocate the necessary funds for establishment of numerous varied coverages of the territory, a pedo-geomorpho-ecological zoning could represent a valid alternative to often unsystematic efforts not always taken to completion.

9.4 SOIL MAPS AND CLIMATIC DATA

9.4.1 The requirements

It is necessary in all sorts of applications to use soil data and climatic data together. In certain cases, at very small scale, this means ecological zoning (§ 9.3). One can then superpose soil maps and climatic maps. But at large scale, map overlay is unsuitable: firstly, climatic entities have no great significance (for example, climate of a pixel); secondly, polygons are multiplied with this superposition. The resulting map risks losing clarity. Then, processing one by one all the polygons contained in the map can be a great volume of calculation be it for modelling erosion or modelling runoff, for predicting crop yields or for calculating requirements of irrigation water at regional scale, etc. Some studies pertaining to global change also fit in this context of simultaneous manipulation of large volumes of soil data and climatic data.

We came up against this kind of problem and the necessity of properly organizing and manipulating data for a European programme conducted collaboratively by seven laboratories belonging to six countries (England, Spain, France, Hungary, Poland and Romania). The objective was to determine, at the level of Europe, the agricultural consequences of predicted climatic changes. To be more exact, we had to answer questions of the type: if the temperature and rainfall are changed in the proportions calculated by the specialists, what will be the agricultural consequences? For this, we were led to develop and use the crop models EuroAccessII, then IMPEL, mentioned in the preceding chapter (Chap. 8, § 8.6.3) (Loveland et al., 1994; Legros et al., 1996a; Henric, 1995; Rounsevell, 1999). The models were validated by using agronomic experiments conducted in the participants' countries, that is, under very varied conditions. We chose Languedoc as test zone. This meant that it was necessary to take charge first of a soil map at 1/250,000 comprising 2710 polygons and then of 64 meteorological stations represented by points and data files. How, under these conditions, are the data to be organized suitably for easy manipulation?

9.4.2 The rejected solutions

We explored unsatisfactory solutions first. It is useful to point them out so that others do not face similar problems.

We first thought of reconstructing the climatic data by interpolation at the centroid of each pixel of soils. This was complicated because the data to be interpolated were five in number (minimum temperature, maximum temperature, rainfall, potential evapotranspiration, solar energy) and each one had to be available for each day over a period of at least ten years. This method amounted to reconstructing an entire meteorological station for each pixel! It involved manipulation of very many digital files. Furthermore, interpolation of daily rainfall from one station to another has no scientific value.

We also tried to simplify the crop model so that it could run faster and thus process all the 'soil polygon + meteorological station' pairs. This was a poor idea. Firstly, it is hardly scientific. Then, progress in computer technology is rapid. A program that runs slowly on a machine one year appears efficient the next year. Many of the efforts at optimizing and simplifying models are thus wiped out by technological progress... .

The best idea was undoubtedly to make the model work at a few reference sites and to find out the means to generalize the findings. But in spite of there being a few ways of thinking still to be researched, the method for this has not yet been established in totality.

9.4.3 The solution retained

Basic principles (Jung, 1994; Legros et al., 1999).
The first principle is: the boundaries of map delineations are not questioned and the climate determined for the centroid of the delineation is applied to the whole of the latter. In the rare cases where the delineation is clearly too large for the hypothesis of climatic uniformity to be credible, it is subdivided manually into smaller delineations. This way of proceeding avoids overlaying a soil map and a climatic map, which will generate new polygons. It also avoids giving importance to climatic boundaries which, more than soil boundaries, are artificial.

The second principle is: the climate at the centroid of the delineation is the climate of the nearest meteorological station, which amounts to saying that the method of Thiessen polygons or Voronoi polygons is used. Very practically, each soil polygon is assigned a number indicating the meteorological station to which it corresponds. In processing, meteorological data of that station enter the calculation (Fig. 9.7).

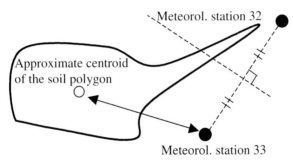

Fig. 9.7 Allocation of a soil polygon to a meteorological station. This polygon is presumed to correspond to the climate of station 33 although it can be claimed that it is nearer in a way to station 32.

Consideration of wind
Working in Languedoc (the South of France) we realized that there often were several meteorological stations at nearly the same distance from a soil polygon. Is the nearest which will act from just a few hectometres, the most representative? How are we to choose the correct one?

We applied the principle that it would be necessary to consider the wind direction. For example, with a west wind, the meteorological station to the west or to the east will see the same clouds as the soil polygon before or after it sees them. It is thus probably more representative than a station located at the same distance from the polygon but to the north or to the south of it.

For simplifying the problem, we reduced the true direction of the wind to the four main directions (Fig. 9.8), the direction of flow not having any

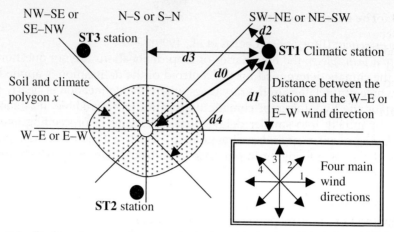

Fig. 9.8 Finding the nearest meteorological station by considering wind direction

significance. If the wind comes from the west or from the east, for example, the meteorological station ST1 is at a distance $d1$ from the line of flow passing over the polygon. If the wind is from the north, the corresponding distance is $d3$... .

Actually it is also necessary to consider the true geographical distance between the polygon and the stations ($d0$). Otherwise a very distant station located exactly on a wind current will be considered representative on the day when the wind is in that direction. After all is said and done, the station i considered nearest to the polygon in question, with wind from the east, say (direction 1), is that which minimizes the quantity:

$$d_i = \alpha \cdot d0_i + \beta \cdot d1_i$$

where $d0_i$ is the true distance between station ST_i and the polygon considered; $d1_i$ the distance from ST_i to the E-W axis passing through the polygon; α and β are coefficients related by $\alpha + \beta = 1$.

It was determined empirically that in the general case α must be between 1/3 and 1/2. These parameters are fixed once for all in the application considered.

But when there is no wind or the wind is weak, α is taken to be equal to 1; this means that only the geographical distance is considered. The estimate of the limiting value of wind speed $V_{\lim}$ below which wind speed is not taken into account depends on the local climate and density of meteorological stations in the region considered. It is a regional characteristic. For establishing the value, the average station-polygon distance D is calculated using a GIS, by an approach of the Voronoi type. Then $V_{\lim} = D/T_{\lim}$. The value of $T\lim$ is to be chosen. It depends on the climate of the study area. For example, we can take 120 minutes or even less if the area has unsettled weather.

To establish these ideas, let us go back to Fig. 9.8. Even without calculation we can well see that for characterizing the climate of polygon x shown, station ST1 will probably be retained for wind from the NE or SW, station

ST3 for wind from the NW or SE, and station ST2 for wind from the South or from the North and also when there is no wind. The calculations are done once for all. A table stores the references to the stations that will be used for characterizing the weather of the polygon according to the day's wind (Table 9.1).

Table 9.1 Table giving the code of the meteorological station for each soil delineation in relation to wind direction

Polygon	No wind	N-S	E-W	SW-NE	NW-SE
x	ST2	ST2	ST2	ST3	ST1
y	ST31	ST31	ST31	ST31	ST31

The wind speed and wind direction are known regionally. They are not needed for each meteorological station! Overall, using meteorological data according to the suggested method will not be more difficult than using the Voronoi method. The latter would come up for consideration in just one column, that of 'no wind' in Table 9.1.

In the test done in Languedoc by giving α the value 1/3, we obtained the following results for the 2710 polygons of the study area, when the 4 wind directions were considered:
- attributed to only one station whatever the wind direction: 987 polygons,
- attributed to two stations: 675,
- attributed to three stations: 643,
- attributed to four stations: 379,
- attributed to five different stations according to the winds: 26 (actually the station retained when there is no wind can be different from the other four, which case is more rare the larger the value of α).

Figure 9.9 represents the central part of Languedoc with its meteorological stations (marked by sunbursts). The sun's rays correspond to the segments that link the stations to the centroids of the soil polygons, for wind from the southwest. It is clearly seen that some polygons are attributed to stations that are not geographically the closest ($\alpha = 1/3$).

Other corrections
Climatic data from the reference stations are not directly used. A correction is applied in order to compensate for the difference in altitude between the station and the polygon considered. Languedoc, where we worked, is a region where the variations in temperature and rainfall with altitude are large. It was appropriate to take them into account. For this, the correlations between climatic parameters and altitude were calculated on the scale of the year, for the entire region. But the corresponding corrections were applied at the scale of the day (adding or subtracting degrees; adding or subtracting mm for rain).

330   MAPPING OF THE SOIL

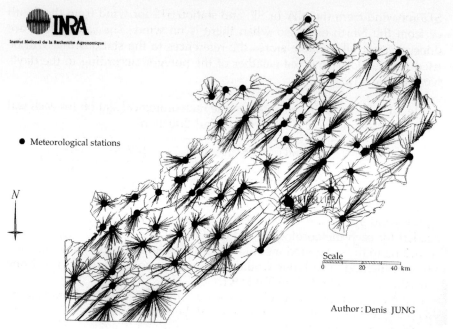

Fig. 9.9 Diagram of the links established by the modified Thiessen method (Jung, 1994; André et al., 1995) between the centroids of the map delineations and corresponding meteorological stations.

The method as suggested has still to be improved to take into account the fact that a geographically close station is actually hardly representative if it is separated from the polygon by a high-relief feature. But this sort of correction has not yet been introduced.

9.4.4 Some results

We have verified the suitability and validity of the method by predicting the weather at points corresponding exactly to 64 meteorological stations but taken out of the test data, one after the other, and then used for the validation. We could thus calculate the mean prediction error. The mean error in prediction of annual rainfall was 9.5%. Taking into account wind direction and difference in altitude reduced the error by 37.5% compared to that given by the unmodified Voronoi method (that is, the polygon is presumed to have the weather of the meteorological station closest to it). With kriging, one can certainly do better, but as much interpolation would be necessary as there are days to be processed. And yet we often work day after day for ten years or more… .

With this method, we have highlighted the complexity of agricultural consequences of global change and the impossibility of simplifying the results through simplistic evaluations of the type: 'this will be good' or 'this will be bad'. For example, for wheat in the Mediterranean region, if the soil is deep enough to ensure good water supply in spring, rise in temperature will be favourable for the yield. The situation will be the opposite if the soil is shallow. The resulting economic assessment will thus depend on the proportions of shallow soils and deep soils in the region in question. But for maize, which is a summer crop, water deficit will be seen in all soils, even the deepest. It would be necessary, for this crop, to accept increase in irrigation or diminution in yield.

In sum, the relatively simple solutions that we have suggested must enable use of soil data and meteorological data jointly, without neglecting any information but without also artificially creating data through overlays and interpolations. Thus there is no reason to ignore spatial data, provided by the soil maps, for treating a number of ecological and agronomic problems that are sometimes visualized in a point-based manner.

Conclusion and perspectives

The examples reviewed above clearly show that soil mappers must pause to think whether it could occur to them some day that their discipline is, if not at the centre of the world, at least at the centre of the concept of regional development. Nothing like that! Canadian foresters, Belgian agronomists, Swiss developers and European 'global-change' specialists will much like to be able to simplify the problems, restrict the cost of their investigations and do totally without knowledge of soil geography. It is no use the mappers of our discipline grabbing their pilgrim's staff and with that go to convert them. They will not be heard. But it is essential that they organize their pedological data into databases and in Geographical Information Systems so that they will be easy to use. It is also necessary to indicate to them how to take the data into account; this is what we have tried to do above. It will be for the survival of our discipline. Who can think that the authorities and parties involved in economic life will support investigations in soil science, even the most remarkable, if they do not result in development or protection of the territory at regional scale?

CHAPTER 10

SOIL MAPS OF THE WORLD AND FRENCH-SPEAKING COUNTRIES

In this chapter, the last in the book, we begin by summarizing international efforts in publishing synthesized small-scale maps. Many of these maps are didactic and useful for teaching courses in Pedology, Physical Geology or Ecology. Also, most importantly, some of them can be used for studies pertaining to large areas: role of the soil in climatic change, continental planning, protection of natural environments, etc.

In the succeeding parts of the chapter are reviewed, through examples, the case of certain countries and their efforts in soil mapping: Canada, USA, Belgium, England and Wales, France, Switzerland, Africa and Australia. Obviously an exhaustive summary at global level cannot be presented here! In the case of the countries considered, it is only rarely necessary to give details of the maps available, for this information is found on the Web, where it is regularly updated. Consequently, we will prefer to give historical reminders before examining the volume of map data collected and their accessibility.

10.1 INTERNATIONAL MAPPING EFFORTS

10.1.1 Soil maps of the world

According to Purnell (1994), the principal international organizations that demand or financially support soil-mapping programmes are the following (Table 10.1):

Table 10.1 International organizations involved in the financing of soils mapping programmes

ACRONYM	NOM
FAO	Food and Agriculture Organization (headquarters in Rome)
UNEP	United Nations Environment Programme (headquarters in Nairobi)
UNESCO	United Nations Educational Scientific and Cultural Organization (headquarters in Paris)
IBSRAM	International Board for Soil Research and Management (headquarters in Bangkok; integrated with IWMI April 2001)
CGIAR	Consultative Group on International Agricultural Research (45 donors including FAO, World Bank, United Nations; the Group supports 16 international agricultural research centres)
ISRIC	International Soil Reference and Information Centre (headquarters in Wageningen, The Netherlands)

Let us examine the work done at world level, from the most schematic to the finest.

There is a soil map of the world on scale 1/30,000,000, prepared by the Soil Conservation Service (now Natural Resources Conservation Service) of the U.S. Department of Agriculture. Another, drawn by the FAO is on scale 1/25,000,000.

The ISRIC and UNEP have published a soil-degradation map on 1/10,000,000, the *World Map of the Status of Human-induced Soil Degradation* (Oldeman et al., 1990). Water erosion, wind erosion, physical degradation and chemical degradation are distinguished on the map. The severity of damage is expressed based on a four-level classification. Deserts and other uncultivable lands are grouped separately.

The major pedological synthesis available at the world level is the *Soil Map of the World* prepared in 1981 by the FAO (see Table 10.1) and UNESCO (see Table 10.1). It comprises 106 Mapping Units that are soil classes or broad soil types. The legend is essentially pedogenetic, developed under the guidance of R. Dudal. The map is published on scale 1/5,000,000 in 19 sheets and in four languages.

The map has been digitized in vector mode with the ArcInfo GIS; it may be obtained on CD-ROM corresponding to 10 sheets (search for site FAO ROME WORLD SOIL MAP). Its legend has been revised (FAO-UNESCO, 1988), leading to improvement of the system of grouping of soils of the world. This system is not detailed but it has the great advantage of being a preliminary partitioning of the soil continuum of our planet into clearly identified sub-assemblages using bases known in all countries and quite well accepted. It has since been revised to give a more complete classification of soils: the *World Reference Base* or WRB (search for site FAO WRB IUSS).

In addition, this soil map of the world has been digitized in raster mode by the UNEP in Geneva using the ELAS software. This map shows pixels of two minutes latitude by two minutes longitude or approximately 3.7 km on the side or 13.7 km². This second version of the map offers a slightly more detailed and more uniform legend comprising 132 soil types. It is available free on the internet (search for site GRID UNEP GENEVA).

In the international project SOTER (*World SOil and TERrain digital database*; also see Chap. 6, § 6.5.3) set up by the ISRIC (see Table 10.1), it was hoped to prepared a new world map, this time on 1/1,000,000. The major declared objective was to quantify degradation of soils by man. But progress has been slow and the work effectively limited to test-sectors identified in Argentina, Brazil, Uruguay, USA and Canada and eastern Europe. At the same time, the specialists involved in SOTER achieved the methodological developments necessary for the project: design of data storage, drawing up a legend and publication of a manual of procedure for soil characterization (FAO, 1995). A whole literature is available (Oldeman and van Engelen, 1993; Baumgardner, 1994). At present, it seems that the SOTER project has evolved towards construction of a database with the objective of recording non-graphic data corresponding to the FAO-UNESCO map on 1/5,000,000 (see below).

According to the FAO (Purnell, 1994), less than half of the world is covered by soil maps on scale 1/1,000,000 or larger. This half-coverage available represents the work of 109 countries, but they often are the smallest.

10.1.2 International soil data set

The IBGP (*International Geosphere-Biosphere Programme*) hoped to assemble soil data to support studies concerning the global changes undergone by the planet (Scholes et al., 1994). In this context, the ISRIC worked and assembled a collection of profiles, the Global Pedon Data Base (GPDB), based on profiles from the ISRIC, profiles of the FAO and profiles of the NRCS (*National Resources Conservation Service*) of the USDA. The collection has grown over time and it seems that a big sorting was done of the 20,920 profiles initially collected. At present the collection comprises 1125 profiles (search for site IBGP DIS ISRIC and for site BATJES ISRIC WISE GPDB). The selected profiles certainly do not have great global representativeness. But these data are designed to develop pedotransfer functions (Chap. 8, § 8.1.2), for which geographic representativeness is not essential. On the other hand, the samples have been characterized rather well; this was a criterion for their selection: 85% for particle-size distribution, 85% for organic carbon, 30% for bulk density, 32% for moisture content at pF 2.3, etc. The data can be freely accessed on the Internet.

This short review of international work is not exhaustive. Here and there various organizations hope for the prestige and financial windfalls conferred by the responsibility of a world programme of inventory of soils with respect to potentials, pollution risks, degradation or any other subject in vogue. To put it briefly, although the efforts are all directed to a better understanding of the natural environment, they are far from being perfectly coordinated. Perhaps we must not regret it. Competitiveness is also a factor in progress.

10.1.3 Soil maps of Europe and applications

Mapping
In 1965 the FAO published a soil map of Europe on 1/2,500,000. This was the outcome of an idea that dated back to 1959 (FAO, 1966). But later this work was stopped for financial reasons. Revived by the European Economic Community, work led to publication of a soil map of Europe (CEC, 1985) under the title *Soil Map of the European Communities*. This map only pertains to the twelve founding members of the Community. The scale is 1/1,000,000 and 7 sheets are covered. The legend is identical to that used by the FAO and UNESCO for the soil map of the world, which was published and then revised with simultaneous editions in English and in French (FAO-UNESCO, 1988).

The map was digitized in Denmark in 1986 by the then available means (Platou et al., 1989). This obviously was a huge job. Later it was necessary to verify the data, suppress some errors and bring the data together in a GIS (King et al., 1995). At the same time the non-graphic data were completed. Actually, they were at first almost limited to the name of the soil type in question. This of course was insufficient for proceeding to practical applications. Then attributes were added, particularly:

- altitude
- land use
- parent material
- texture
- redoximorphic features
- water regime
- obstacles to root penetration

For this, it was necessary to go back to the original documents that had very fortunately been preserved in the archive by R. Tavernier at Ghent in Belgium. As always in these domains, the principal difficulty was to correctly define the data structure in the data bank to be created, that is, the relational data model or, if preferred, the Soil Organization Model (Jamagne et al., 1993). For the twelve countries constituting Europe in 1985, the map comprised about 16,000 polygons and 312 Map Units each consisting of up to 4 or 5 different soils in association. Efforts were made to include the new members of the European Union (Jamagne, 1993; Jamagne et al., 1994, 1995b; Meyer-Roux and Montanarella, 1998) and thus extend the area of the map to eastern Europe and northern Europe. Here the view was more geographic than political. For example, Switzerland was included. In principle the synthesis was achieved in 1999 (King and Montanarella, 1998, 2002). This was a hard task. It was facilitated by the fact that the countries of eastern Europe had already surveyed their soil mantle, often on detailed scales. However, improvements and later editions of the map will be necessary because, if carefully seen, certain countries of the north have not been characterized in detail (Sweden, Norway, Finland).

Now, extension is envisaged to Russia and to the southern and eastern shores of the Mediterranean, to Morocco and Turkey, passing through Egypt.

The work pertaining to the map of Europe has been coordinated by a *Soil and GIS Support Group*, which became a *Soil Information Focal Point* in 1994 and was transformed in 1996 to the *European Soil Bureau* (ESB).

In its current form the map can be viewed on the Internet but not exported to potential users. According to the ESB (Jones et al., 1998), *'There has been a radical change in attitude toward the availability of data in recent years ... from ... all data should be available free to everybody ... to ... no data available without charge to anybody'*...

Around 1993, scientists started examining the possibility of constructing a soil map on scale 1/250,000 by first covering some pilot regions (Dudal et al., 1995). The work was placed under the aegis of DG XI of Europe (Directorate-General of the Environment) and organized within the ESB from 1996. A manual of procedure was first published in English and then translated into French (European Union, 2001). In it are found the concepts examined in Chapter 6, in particular an organization into hierarchical levels comprising, from the lowest to the highest:
- *Horizon*,
- *Soil body*, equivalent to 'profile' in the instance considered,
- *Soilscape*, a kind of 'Map Unit',
- *Soil region*; this soil region is linked at the same time to geographical location, climate, geological material and dominant soil type. For example, Podzol-Andosol region of northeastern Iceland. Altogether, there will be 176 soil regions of this kind in Europe.

At present, what is mainly envisaged is a *nested multiscale soil information system for Europe* (EUSIS). It will link map data pertaining to this part of the world: international data, data of instances in the European Union and national data.

Profile data base

It seems logical to illustrate the Map Units of the map of Europe on 1/1,000,000 by some representative profiles. This is the *EU Soil Profile Analytical Database*. On the map, a Map Unit is accurately characterized by:
- a dominant Soil Unit,
- one or several sub-dominant Soil Units each representing between 10% and 50% of the area of the Map Unit,
- inclusions that represent less than 10% of the area.

Under these conditions, it was desired to build first the collection of profiles corresponding to the *dominant Soil Units*. This represented between 100 and 300 profiles according to the number of countries considered. In a way this is a small number. But the work is not finished because coordination is difficult and the persons in charge are not fully convinced of the representativeness of the profiles provided to them (Gardiner, 1989; Madsen, 1989; Madsen and Jones, 1995; Breuning-Madsen and Jones, 1998). The second stage (characterization of the *sub-dominant soil units*, thus of some hundreds of additional profiles) and the third stage (characterization of the *inclusions* through some thousands of profiles) have not been completed.

Naturally, whatever the effort put in might be, characterization of the soils of a territory as vast as that of the European Union will necessarily remain cursory. At regional scales, it will be necessary in many cases to reconstruct the properties of the soils to be considered. Pedotransfer functions (Chap. 8, § 8.1.2; also see § 10.1.2) must be used for this.

Also, the soil map of Europe represents only one of the data layers of the ambitious (too ambitious?) European programme, CORINE (*COoRdination of INformation on the Environment*). This programme is experimental. It consists of regrouping and coordinating the essential data on the state of the environment and natural resources in the European Community. The soil map and other data recorded in CORINE have already given rise to two major applications in the domain of our interest. Let us examine them.

CORINE erosion project

The CORINE erosion project, launched by DG XI, has for objective the evaluation and mapping of erosion risks in the southern regions of the European Community. Started in 1985 (Briggs and Martin, 1988), this programme resulted in a publication some years later (CORINE, 1992). The method is the one already mentioned (Chap. 8, § 8.6.6). It requires consideration of soil erodibility, erosivity of the climate, crop type and slope (Bonfils, 1989). Nature of the soil is given by the map presented above. Climatic characteristics are provided by 160 meteorological stations distributed quite regularly all over Europe. The slope according to region is calculated manually or obtained from a DEM (Chap. 8, § 8.6.1). One single average value is given per km^2. If the published map results are to be believed, erosion risks are particularly high in Spain and in Sicily. As for good agricultural areas, they are unfortunately very small and greatly scattered in southern Europe. However, two exceptions are seen, both Italian: firstly, the plain of the river Po and secondly, Apulia (the heel of the boot, near the cities of Foggia, Bari and Taranto).

The MARS project

The MARS project of the *Directorate-General of Agriculture* (DG VI) is so named for *Monitoring Agriculture by Remote Sensing*. It consisted of using remote-sensing resources to predict agricultural production in and outside the European Union. It is not necessary to emphasize the strategic significance of the question. For example, knowing that Europe is going to be surplus in such or such produce can result in anticipating the sales before international prices collapse following the presence of surpluses... . In reality, prediction of agricultural production is particularly difficult. We can even say that the question has not yet been satisfactorily resolved. It is very complex and simultaneously involves finding and using (Meyer-Roux, 1987; Vossen, 1992):

- data on the state and location of crops (use of satellites);
- agronomic data on the crops considered: average yields obtained and production zones (agricultural statistics), reactions of the crop in question to climatic conditions (experimental results);
- data pertaining to the natural environment (soil map of Europe on 1/1,000,000);
- climatic data obtained in real time;
- plant-growth models of the type mentioned earlier in Chapter 8, § 8.6.3 and in Chapter 9, § 9.4.

The methods and papers published have been summarized by Vossen and Meyer-Roux (1995).

Also, the ISRIC has initiated the SOVEUR project (*SOil Vulnerability mapping project for EURope*), which deals with the sensitivity of soils to different types of stresses.

To sum up, the various European programmes discussed complement one another. For example, DG VI financed preparation of the soil map of Europe, while DG XI took charge of its digitization. The administrators of the European Union and the scientists involved endeavoured, year after year, to expand in a coordinated way the range of tools available for managing their territory.

10.1.4 Progress of national studies in Europe

Table 10.2 summarizes the detailed statistics published by European officials (King and Montanarella, 2002). It distinguishes the countries according to progress in their soil mapping on scales larger than 1/250,000 taken together. This table is approximate and for certain countries one could question the grouping according to the efforts made, scale by scale.

Table 10.2 Progress of mapping in the various countries of Europe

Soil coverage completed	More than 50% of the territory covered	Less than 50% of the territory covered
Austria (a) Belgium Denmark (a) Hungary Iceland Lithuania Netherlands Poland (a) Romania Slovakia Czech Republic **United Kingdom**	Finland (a) Germany Portugal **Switzerland**	France Greece (a) Ireland Italy Norway (a) Spain Sweden

(a) indicates that the percentage is calculated as proportion of the cultivable area

We shall present later, by way of example, the studies done in one country in each category (countries in bold).

10.2 SOIL MAPPING IN CANADA

10.2.1 Historical

The Canadian 'Soil Survey' was created before 1920, but it started preparing maps only after 1953. The country being vast, all the means were good for reaching the remote regions and mapping the soils. Thus the scientists used, turn by turn, horses, canoes, seaplanes and, more recently, helicopters (McKeague, 1995). One may also find on the Internet an older, more expanded text, *History of Soil Survey in Canada* (McKeague and Stobbe, 1978).

Progressively built up and improved, the Canadian soil classification was published in 1970 under the name *The System of Soil Classification for Canada*. This gave the conceptual framework necessary for delimiting the land. Later on there was a revision in 1978 and another in 1987. It should be noted that Canada did not adopt the American classification in spite of the fact that many young scientists trained in USA had been invited to use and evaluate it. There was a democratic vote on this subject in a *Soil Survey Committee*, following which everyone rallied behind the majority opinion, which was to have a special classification. In fact, considering the characteristics of their country, Canadians were in a very good position to advance knowledge of

Podzolic soils that cover 1/6 of their enormous country and soils with permafrost (*Cryosolic soils*) that represent 1/3 of the total area. They converted the Americans who, in 1994, introduced in their own system *Gelisols*, equivalent to Cryosolic soils.

Canadians were among the first to design computer systems (GIS) for managing geographic data (Tomlinson, 1987). In the area of management of soil data too, Canadian progress has been very great. Under the direction of Julian Dumanski, the 'CanSIS' (*Canadian Soil Information System*) was developed starting from 1972. The aim was to combine all sorts of data, including economic and social, to evaluate soils and define their best use. The Canadian GIS was specific from 1975 to 1986; the centre in Ottawa was a world leader in this field of activity. The CanSIS map sheets pertaining to soil served as model for the design of the French model STIPA-1979 (Chap. 4, § 4.4.1). Since 1986, for ensuring compatibility with other mapping centres, CanSIS adopted the ArcInfo software. The information has been put on the Internet since 1994.

It was also sought in Canada, as in other countries, to develop a system of land evaluation for different uses (agriculture, forest, tourism, wildlife) without being encumbered with detailed characterization of the soils, considered too time-consuming to implement and too costly. Thus was born the *Canada Land Inventory* (CLI) based on mapping at 1/250,000 of test sectors, accompanied by many varied extrapolations. Hundreds of *capability maps* were thus produced. Depending on point of view, this approach can be considered very useful (coverage of the country was completed in 1975 and many problems of land use were solved efficiently and simply in this way) or unfavourable (the procedure is too empirical at the level of combination of data and this kind of simplified mapping used up the funds that otherwise would have devolved to soil mapping).

In Quebec, as elsewhere, soil science followed agricultural geology. The Division of Soils was created in the 1940s by the Ministry of Agriculture. This Division was moved in 1971 to the science complex of Quebec at Sainte-Foy; it then became the *Service des Sols* of the *Direction de la Recherche et du Développement*. Soil mapping and various studies pertaining to soil-plant relationships and fertility were done under the instigation of Director Marton Tabi.

Soil mapping was done on scales ranging from 1/20,000 to 1/126,720. The administrative divisions were counties. In a country where roads and highways are not always numerous, the topographic base is sometimes substituted, on large scale, by a reproduction of aerial photographs, as in neighbouring USA. This is practical for immediately seeing lakes, woods and parcels if any. See, for example, the map of Richelieu county (Nolin and Lamontagne, 1990). The photographic base is in brown and the soil boundaries in black. There are no colours on the map on which only the soil unit numbers are added. The mapping was done at series level (Chap. 1, § 1.4.3) for scales equal to or larger than 1/50,000. The series name is that of the place where the soil was observed, for example *Francoeur series, Séraphine series, Saint-Samuel series*, etc. In detailed studies, the *phase* is added to indicate texture, for example, *Comtois, clay loam* (Laflamme et al., 1989). In the legend, these series are grouped according to broad types of surface geological formations. In Canada, soil mappers and ecologists (Chap. 9, § 9.3) accord an essential role in differentiation of soils and landscapes to these formations.

In the 'belle Province' as in the entire country, there is great interest in the suitability of soils under the name ARDA inventory, *Inventaire Systématique des Ressources Revouvelables au Canada*. For this, 32.6 million ha have been inventoried and 26 maps on scale 1/250,000 published. This only represents 21% of Quebec, but it is the area south of the 49th parallel and thus encompasses most of the soils of high agricultural potential. In fact, according to Marton Tabi, Quebec has only 2.4 million ha suitable for crops, with no major problem for mechanized agriculture. This enormous country must therefore properly manage its relatively rare cultivable soils. The inventory programme mentioned above reminded people of the problem. The Land Protection Act, voted into law in 1978, has the objective of curbing anarchic urbanization and readjusting the market in land.

10.2.2 Data and their accessibility

Data on the soils of Canada are currently assembled in the *National Soil Data base* (NSDB) maintained under the Ministry of Agriculture by the CanSIS staff (search for site CANSIS CANADA NSDB).

In the south, that is along the border with America, soil mapping is nearly complete. There is no soil inventory north of the 60th parallel, except for some scientific studies at selected sites.

In Quebec, the overall balance sheet of soil mapping is as follows (Laflamme and Rompre, 1993; Tabi and Carrier, 1993): 60 soil surveys have been published. Nearly 8 million ha have been covered. Still this only represents 5% of the area of the province. Thus one could think that this is very little, but it is nearly three times as much as the area of Belgium! Some maps have been digitized and are available on the Internet in the ArcInfo format, compressed. An index map enables us locate all the surveys and distinguish the digitized maps from the printed maps. A querying system provides facility for finding the desired information by using various search criteria. The maps are free for Canadian collaborators in the system. Other interested persons can be made to pay for the time required to extract the data asked for, according to a very modest tariff.

It should be noted that there also is, on the Web site, a map on very small scale, at the same time schematic and didactic, which shows the broad soil types of Canada. This is free of charge and can easily be downloaded. Maps of *Ecozones, Ecoregions* and *Ecodistricts* can also be obtained.

10.3 SOIL MAPPING IN USA

10.3.1 Historical

In USA, Eugene W. Hilgard (1833–1916) and his boss Major John Wesley Powell played an important role in soil mapping (Amundson and Yaalon, 1995). Hilgard was an immigrant of German origin. He left Heidelberg to accept a post of Geological Assistant to the state of Mississippi in 1855. In 1858 he was already in the rank of Geologist, then he pursued a career as University Professor. He very quickly realized the importance of observation and analysis of soils and the study of vegetation. All his life, with the support of Powell, Hilgard fought to have an *Agricultural Survey* created within the Geological Service. This was not realized because of opposition from senators alarmed by the duration and cost of a programme of mapping USA. But Hilgard's attitude and work showed that he was one of the first to understand the importance of the soil mantle and the need for its spatial study. For him the soil was not to considered only through geology… .

What Hilgard and Powell had planned, Dabney and Whitney would achieve. In 1889 Charles Dabney was appointed *Assistant Secretary of the U.S. Department of Agriculture*. This was a very important post, which allowed him to regularly meet the President of USA, at that time S.G. Cleveland. In 1894 Dabney appointed a chemist Milton Whitney (1860–1927) Chief of the new *Division of Agricultural Soils*, which he created within the U.S. Department of Agriculture (McCracken, 1990). Milton Whitney had a difficult start to his career and found himself posted to agricultural experiment stations successively in North Carolina, Connecticut, South Carolina and then Maryland. Undoubtedly he was struck by the diversity of the soils. He became extremist in the physical and morphological approach (Boulaine, 1989). He stated that that the soil had almost infinite fertility and that it was therefore enough to study it from the water economy point of view. Also he was a colourful personality difficult to interact with. He drove everybody away from himself and was not on good terms with Hilgard.

However, we owe many things to Whitney. At the Chicago World's Fair in 1893 he presented the American achievements: 265 monoliths (defined in Chap. 1, § 1.3.3) were collected from 24 states of the Union. The same exhibition allowed him to see the progress of work of the Russian school of pedology (Chap. 1, § 1.3). He appeared interested. Later he gave direction for launching the first soil maps. Field surveys started in 1899 and the first publications date back to 1900. They pertained to North Carolina, California and Texas. Lastly, it was Whitney who, in 1904, discarded the mapping of soils solely on the basis of texture and introduced the concept of series (Chap. 1, § 1.4).

All in all, this man played a big role in the start of soil mapping in USA. One can learn much more by referring to the fine book *Profiles in the History of the U.S. Soil Survey* (Helms et al., 2002).

10.3.2 Data and their accessibility

Detailed maps published after 1957 are printed on a photomosaic base on scales of 1/12,000, 1/15,840 or 1/24,000.

The organization of soils data in USA has already been presented (Chap. 6, § 6.5.1). The information is thus dispersed in various systems, not all of them interconnected. The key words to be used for finding information are many: *nasis* (**Na**tional **S**oil **I**nformation **S**ystem), *lims* (**L**aboratory **I**nformation **M**anagement **S**ystem), *ssurgo* (**S**oil **SUR**vey **G**e**O**graphic Data Base), *statsgo* (**STAT**e **S**oil **G**e**O**graphic Data Base), *fotg* (**F**ield **O**ffice **T**echnical **G**uide), *sdv* (**S**oil **D**ata **V**iewer), *cst* (**C**ustomer **S**ervice **T**oolkit), *published soil surveys, nri, pedon descriptions, sir, muir* (National **M**ap **U**nit **I**nterpretation **R**ecord), *osd* (**O**fficial **S**eries **D**escriptions), taxonomy, *nscd* (**N**ational **S**oil **C**haracterization **D**atabase)… . Our American colleagues are forced to tidy up this abundance of information. Mainly one must search for *nrcs usda soil survey*.

One can also find a 'map of the maps' pertaining to the whole country (Fig. 10.1). This is provided here by way of illustration with permission of the Natural Resources Conservation Service, but it can be found on the Web, where it is updated regularly.

One is struck by the fact that in USA, unlike what happens in Europe, mapping efforts seem to be pursued with sustained speed. One-half of the area is covered and published or soon will be. One-third is in the process of survey. Mountain regions and Alaska are practically the only *non-project* areas or only 18% of the area of the country.

A list of *Published Soil Surveys by State* is put on the Web. Certain reports are available on line. They can be downloaded free of charge. For the American scientists, '*The soil survey information is public information*'. The printed maps can be obtained by post. They are free for decision-makers and organizations in the area the map pertains to. Some documents can be bought from libraries.

Some particularly attractive teaching materials such as *Dominant Soil Orders, Ecological Regions,* etc., are available on the Internet.

10.4 SOIL MAPPING IN BELGIUM

10.4.1 Historical

As in all the countries of old Europe, efforts at characterizing soils began in the 19th century generally associated with agricultural geology (§ 10.6.1).

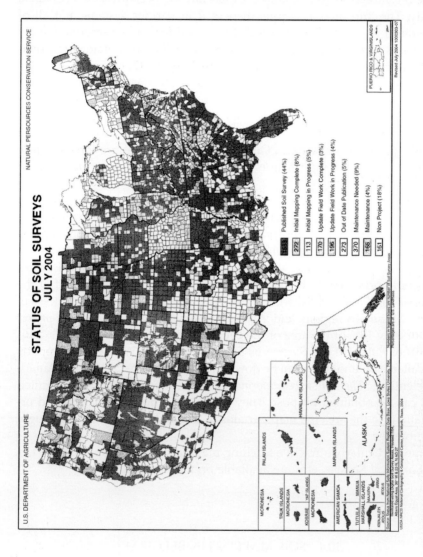

Fig. 10.1 Status of soil surveys in USA in 2004; source. Natural Resources Conservation Service (NRCS). Most of the country has been covered.

Various maps were then published. But it took till 1947 for the systematic study of the country to begin on scale 1/5000 for publication at 1/20,000. The Secretary of the *Comité pour l'Établissement de la Carte des Sols et de la Végétation*, R. Tavernier, had from that time on appreciated the issues and the essential role of soil mapping in properly conducting investigations on fertility, agronomic potential, competition for space and in understanding the evolution of rocks on the surface of the earth (Tavernier, 1950).

The work was done by the cartography centres associated with the Agronomy Faculties of the Universities of Gembloux, Louvain and Ghent. The last city, where R. Tavernier worked, was in charge of general coordination. Surveys were conducted on 1/5000. The way the Belgian mappers work is drawn from geological methods: respect for the same geographic divisions, drafting deliberately brief reports, parallel publication of monographs on the soils of the regions studied. The delimitations were made on the basis of recognition of series, on the American model. The three major soil characteristics considered were:
- texture,
- natural drainage class,
- degree of profile differentiation.

Profile studies were completed in 1971. The density of observations was 1 per 200 ha to 1 per 800 ha. Almost 18,000 soil pits were opened, described and analysed. Then, after verification, the data corresponding to 13,000 pits were stored in the data base (van Orshoven et al., 1991).

Auger holes were dug following a regular grid about 75 m on the side with an average density of 1.8 per ha.

Compared to the standards given earlier (Chap. 3, § 3.2), the profile density is slightly low and the density of auger holes, on the contrary, very high for a scale of 1/20,000. If the Belgian mappers are given a mapping efficiency of 10 (Chap. 2, § 2.4.3), the virtual scale at which they made their map is 1/4714. But our standards are only approximations!

Today coverage of the entire country is almost complete. Soil surveys have ceased. There are only 59 sheets left unprinted out of a total of 457 (status in June 1995, still valid in 2002). The published maps are in superb colour, which shows the technical and artistic mastery of the designers since fifty years ago. In western Europe, Belgium and Netherlands are the only countries that have done their soil coverage on large scale.

In addition, to overcome the lack of certain data necessary for solving current problems, pedotransfer functions have been developed (Hubrechts et al., 1998).

10.4.2 Data and their accessibility

Figure 10.2 gives a picture of soil mapping in Belgium. It is not yet available on the Web.

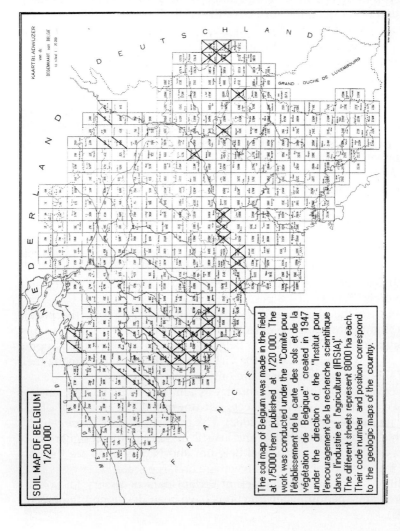

Fig. 10.2 Progress of soil mapping on 1/20,000 in Belgium. The sheets marked 'x' are not available yet; those marked with a line can be obtained in the form of photocopies. Source of information: Geologisch Instituut, Universiteit Gent.

The maps on 1/20,000 can be referred to. They are sold with their explanatory booklets. Produced on an old topographic base, they sometimes lack accuracy (locational error). Correction procedures have been devised in the PC-ArcInfo environment.

In Flanders, all the maps have already been digitized (A. Grillet, pers. comm.). They can be purchased. (Search for site, in Dutch: BODEMKAART, SPIDI, GIRAF, BELGIE).

In the French-speaking region, the Faculté Universitaire des Sciences Agronomiques de Gembloux is developing the *Project on Digital Mapping of the Soils of the Walloon Area* (PCNSW in French). The maps already published are scanned in order to facilitate removal of distortion. Then these maps are digitized. Maps still in the draft stage are directly digitized (L. Bock and C. Bracke, pers. comm.). Quality of the maps is checked: edge matching between maps, construction of a uniform general legend, etc. There are 270 maps of 8000 ha each, of which 80% were printed earlier.

10.5 SOIL MAPPING IN GREAT BRITAIN

10.5.1 Historical

The first maps in England with a legend specifically pertaining to the soil appeared in the years 1790–1800 at the request of the *British Board of Agriculture*. They are now considered the oldest soil maps known (Yaalon, 1989). The maps were accompanied by summaries of the agriculture in each county. Nearly a hundred of them were published before 1814. They were done on scale 1/63,360; their legend presented four to ten soil types based on texture (clay, sand), geological materials (chalk, igneous rocks) and wetness (bogs, peats...). Their accuracy was very variable and was much criticized even at that time. The most relevant were used 150 years later to draw the modern soil maps. Also, Great Britain was quite early with computer processing of soils data (Ragg, 1977).

Most of the following information is extracted from the Soil Survey and Land Research Centre (SSLRC) website: the SSLRC was established in 1939 as the Soil Survey of England and Wales. This was before the War of 1939–1945. Only Portugal is in the same situation in Western Europe. This Soil Survey was funded by the U.K. government between 1946 and 1987. Since 1987, SSLRC has been part of the Cranfield University at Silsoe and now operates as a research and consultancy organization.

10.5.2 Data and their accessibility

The *Land Information System for the UK* (LandIS) was developed over nearly fifteen years. It contains the following soil data (Proctor et al., 1998):
- A National Soil Map on 1/250,000 (NATMAP); it was published in six sheets around 1983, digitized in raster mode in 1986 and comprises 15.5 million pixels,
- A National Soil Inventory (NSI), which is a soil-monitoring network covering England and Wales. It contains site information, soil-profile characteristics and topsoil analytical data for 5691 sites. These sites are located at the intersects of a 5-km orthogonal grid. Up to 127 attributes are available at each site. The NSI was built up between 1978 and 1981. Then, between 1994 and 1996, one-third of the sites were resampled and many of the determinations repeated,
- Auger-hole records (150,000 holes),
- Representative profile records (PROFILES), more than 2400,
- A National Catalogue of nationally important soil types (NATCAT) each identified by a *series* name; there are about 720 of them,
- Additionally, *agroclimatic data* (AGROCLIM) that can be made use of for various agronomic or environmental applications.

The work was funded by the Ministry of Agriculture, Fisheries and Food (MAFF). Data are available for lease (search for site SSLRC UK SILSOE SOIL MAPPING).

10.6 SOIL MAPPING IN FRANCE

10.6.1 The beginnings of soil mapping in France (Boulaine, 1989; Boulaine and Aubert, 1989)

In 1792, just after the French Revolution, the Englishman Arthur Young, a sort of gentleman farmer and also an agricultural journalist, summarized his observations made on three trips to France in 1787, 1788 and 1789. He had kind words for the French Revolution so much so his complete works were translated into French with the help of the authorities in Paris. In this context, he drew a map of France in which he distinguished seven broad zones: mountains, loams, chalk, sands, rocks, rich loams and heaths and moors. The validity of the division is almost non-existent, but it was the first attempt at mapping pertaining to the whole country.

In the middle of the 19th century, great efforts were made to characterize the soils chemically and to understand the fundamentals of mineral nutrition of plants. Thus it seemed to be a guiding principle for characterization of soils (calcareous soils, siliceous soils, etc.). But it would be revealed that this was not enough to account for fertility and it was also necessary to consider physical characteristics: *hard soils, open soils, light soils,* etc. (These old terms are no longer used!). Also, certain rocks imprint important features on the soil. One cannot decide and, in sum, all sorts of ideas are mixed up to present classifications of heterogeneous appearance. Actually, agronomists of that age found themselves trapped in a sort of vicious cycle: firstly, for lack of suitable concepts they were unable to delimit and group soils; secondly, for lack of soil mapping and environmental characterization, they could not bring out ideas in classification of soils.

Progress will come from those who traverse the land: geologists and botanists. Geologists, most of all, observe that there is a close link between rocks and the soils overlying them. They are right in certain cases: dry regions subject to intense erosion where the soils scarcely have the time to develop before disappearing, regions (such as Switzerland) that were scraped by glaciers in the Quaternary and where the soils are very young (less than 14,000 years) and strongly marked by features of the underlying parent materials. Eugène Risler, a Frenchman who owned a property in Switzerland and married a native of Geneva, joined in this current of thought. He published his *Géologie Agricole* in four volumes between 1884 and 1897. Taking as guide the geological stages and periods from the Precambrian to the Quaternary, he presented the natural environments not forgetting the soil but giving it much importance in agricultural land use observed through the functioning of a few model cropping systems. On the one hand this was therefore the first attempt in France to give a serious and regionalized general overview of the soils of the country. On the other hand, the relevance of the geological model as a tool for spatial recognition of soil was not questioned. It was axiomatic. The contribution of the Russian school was necessary to stop pedology from being reduced to a minor branch of geology.

Dokuchaev's student Agafonoff, who emigrated to Paris, constructed a map of *pedological zones of France* at scale 1/10,000,000. This document was published in the *Annales de la Science Agronomique* (April-May 1928 issue). Then the same author refined the first outline and published the schematic soil map of France on 1/2,500,000 in his book *Les Sols de France au Point de Vue Pédologique* (1936). Unfortunately, Agafonoff did not know how to get rid of the idea of zonality, scarcely useful for the small countries of western Europe where climatic variations are limited whereas the parent materials are very different and play an important part in the typology of soils. Thus the legend to the map is quite clumsy.

10.6.2 Work on small-scale mapping

At the beginning of the 1950's, an attempt was made to prepare a soil map of France on 1/1,000,000. The first one-fourth of this map, the sheet of the north-east, was presented in 1952 at the *Association Française pour l'Étude du Sol* (A.F.E.S.), a society created in 1934 and very active even today. The south-west sheet was completed in 1956. The two other sheets were prepared later. The provisional drafts are published in black and white (Dupuis, 1960). They were circulated to the French soil science community for critical comments and improvement. The final map, in colour, was published in 1967 (Pedro, 1984b).

In the beginning, it was planned to use the FAO system of classification as developed until that time. But a French school of pedology had already appeared and was going to draw up a preliminary soil classification (Duchaufour, 1956). Then in 1962 came the Aubert-Duchaufour classification; lastly the French soil classification under the name *CPCS* (Commission de Pédologie et de Cartographie des Sols, CPCS, 1967). The latter was a revision of the former on the basis of the ideas of the entire French scientific community assembled by Jean Boulaine at the request of G. Drouineau, *Inspecteur général* of the INRA. The final soil map on 1/1,000,000 was published the same year. This was no coincidence: the inventorying had promoted the emergence of new concepts pertaining to classification. Classification in turn facilitated naming the soils on the map.

The data of this map on 1/1,000,000 were completed and later revised for helping the construction of the map of Europe mentioned earlier.

Other syntheses were done on the national scale. There is a *Carte de France de l'Hydromorphie à l'Échelle des Petits Régions Naturelles* (Map of hydromorphy in France at the scale of small natural regions) that was presented at the Académie d'Agriculture de France (Pedro, 1989b).

10.6.3 Medium-scale mapping and regional development

Then came a fruitful period for soil mapping in France (1960–1975). It was necessary to develop the territory. For this, development corporations were set up: Compagnie nationale d'Aménagement du Bas-Rhône-Languedoc, Société de Mise en Valeur de la Corse, Compagnie d'Aménagement des Coteaux de Gascogne, Société du Canal de Provence, etc. One of the first worries of these corporations was to prepare soil maps of their area of function. This was the best way for them to know the environment in which they were going to work. These maps were used particularly for planning irrigation. The corporations recruited soil scientist-mappers in large numbers. Very many maps appeared, scales of 1/25,000 and 1/50,000 often being preferred. These studies rapidly led to advance in knowledge of the soils of the region. But each organization remained entirely its own master. Also the surveys lacked coordination and their presentation was not uniform. However, thanks to them, very many development problems could be dealt with: study of crop suitability of the soils, drainage, erosion, regrouping of parcels, reclamation and, of course, irrigation.

Similarly, the Farmers' Association of the Aisne and the INRA joined forces to form a team of mappers for fully covering the department of the Aisne on scale 1/5000 with map publication on 1/25,000. The project started in 1958 and was completed in 1979. The department of the Aisne is the only one in France to benefit from a total coverage of its soils on large scale. The original presentation of this work will be remembered in the context of the 'analytical' method, the principle of which was presented earlier (Chap. 2, § 2.2.3).

Among the departments that have put in the greatest effort in mapping their soils must be mentioned those of the Centre Region. Under the direction of R. Studer of the INRA, systematic mapping on scale 1/50,000 was started in 1975. Since then, it has progressed considerably and most of the area of the department has been covered to date.

Other departments have put in much effort in mapping and deserve mention, particularly Ille-et-Vilaine, Mayenne, the Yonne and the Sarthe.

A *Service d'Étude des Sols* (SES) was similarly formed in 1960 at Montpellier by Professor E. Servat, in a laboratory attached to ENSAM and INRA. From 1965, this service published its soil maps in colour and suggested accurate, non-genetic detailed legends that were therefore easy to understand. In ten years, the service surveyed more than 2 million ha, all scales taken together. These exemplary studies were imitated and thus brought out quality standards for publishing maps and compiling reports. The various private firms had to comply and the quality of mapping generally improved. The major achievements of this service pertained to the following regions:

- Valley of the Allier where more than 600,000 ha were surveyed on scale 1/100,000 (Bornand et al., 1968);
- Middle valley of the Rhône partitioned into several sheets on scale 1/25,000 with a final synthesis on 1/100,000 (Bornand, 1972);
- The d'Ouche region on scale 1/10,000 where 13 sheets cover a total of 40,000 ha (Favrot and Bouzigues, 1975).

10.6.4 Soil information of France on 1/1,000,000 (CPF project)

The INRA wanted to organize a national mapping service. The Montpellier team could have formed the nucleus for it, but its location seemed too distant from the centre of the national territory. Therefore, in January 1968, the *Service d'Étude des Sols et de la Carte Pédologique de France* (SESCPF) was created within the Versailles centre of the INRA. It was transferred to the Orleans centre of the INRA in 1983. Marcel Jamagne, who was earlier one of the principal architects of soil mapping in the Aisne, was named Director of this service.

It was decided to cover France uniformly on scale 1/100,000. This represented 293 sheets of about 220,000 ha each. Thus the effort would, in principle, have been of reasonable magnitude while ensuring characterization of the country with adequate accuracy. But the work progressed slowly, linked to lack of sufficient financial support from the decision-makers (Jamagne et al., 1995a). No more than 25 sheets have been published since the start of the operation. Another 15 are under completion. The total represents about 15% of the whole of France. The speed of completion has been about a sheet and a half per year (Jamagne et al., 1989). At this rate, the programme should end around the year 2170! Presently the programme as been abandoned but there is an attempt to digitize all the sheets already done. The surveys, often done by researchers to accompany theses, are of excellent quality.

For further information, an article by King and Saby (2001) that summarizes the issue may be consulted.

10.6.5 Department maps of agricultural lands on 1/50,000

Towards the end of the 1960's, soil maps no longer appeared to be the most suitable means to approach the problem of land use because of their cost and slowness of completion, and because they did not take socio-economic aspects into account. This was a development of ideas quite similar to what had taken place in Canada a few years earlier (§ 10.4.1). The officials of the Ministry of Agriculture then thought of another way: establish *Cartes Départementals des Terres Agricoles* (CDTA) (Department maps of agricultural lands) that took into account characteristics of the soils along with the economic conditions of farming them to provide decision-makers with a direct estimate of the agricultural value of the lands. The Agricultural Direction Law of 1980 provided in its Article 73 the aims of the operation (APCA, 1984). In practical terms, this involved proceeding very fast, without getting mired in details of soil characterization, and covering France in 5 years. The scale 1/50,000 was chosen, which resulted in the country being divided into 1103 sheets, each representing about 60,000 ha. With funding of 20 million francs, the operation effectively started in 1982. Thirty seven sheets were published the following year. Different organizations contributed to their preparation (Duclos, 1985).

But in the early 1980's the technical tools for doing map overlays were not available in France. Also, multicriteria analysis was still not used. It was difficult to simultaneously use soil data and related economic data, each one belonging to different geographic entities (Guyot and Bornand, 1987). Under these conditions, it quickly appeared that the delimitations left much to intuition! Also all those responsible for the study did not do the work expected of them. Almost one half of the CDTA were not new surveys. Actually certain organizations demanded and obtained public credit for preparing the maps the soils part of which already existed. The CDTA programme thus became a means of granting hidden subsidies to organizations some of which were already on the verge of bankruptcy and risked laying off their personnel. Lastly, government subsidies, which should have been granted with proper judgement and over half a century, disappeared as they had come, that is, abruptly.

The final balance sheet is as follows:

Sheets to be done	1103
Sheets published	about 132
Redundant sheets published	57 (soils already mapped).

The best CDTA, which corresponded to new soil surveys seriously done, are now considered to facilitate preparation of the systematic map of the territory on 1/250,000.

10.6.6 The 'reference drainage sectors' campaign

Basic concepts (Favrot, 1987, 1989)

All of this showed that it was necessary to give up, at least in the short term, the ambition of systematic uniform coverage of France done on large or medium scale. Under these conditions, INRA-Montpellier put forth the idea (Favrot, 1981) of a limited mapping of Reference Areas (RAs) representative of Small Natural Regions (SNRs) that include them. The SNRs are territorial entities of area a few tens or thousands of hectares identifiable by their geology, their pattern and the unity of their soil cover (finite number of soil types, spatially organized according to specific distribution laws). The variability found in an RA, in principle, gives an account of all the variability found in the SNR. Just like for wallpaper, the RA corresponds to a 'motif' as the SNR corresponds to the whole tapestry. In practical terms, the experience gained at the level of the RA is used in the SNR, in the fields or in the parcels where there is no soil map but problems are faced. This is what the author of the method calls 'going back to the parcel'. Returning to the parcel is actually going down again from the synthesis (the map) to the observation (parcel to be examined). This, on the contrary, does not mean that the parcel in question had been visited before. Some soil observations are necessary at the level of that parcel but, considering the experience gained, these new investigations are reduced to the minimum. Also, the 'going back' is the opportunity beyond the application that leads to defining correctly, at least locally, the validity of the procedure. Actually, in every parcel or group of parcels of the SNR the soils of the RA must be found, with:

- Less than 25% *new Map Units* by number;
- Less than 10% of *new Map Units* by area.

Extrapolation outside the *Reference Area* naturally gives rise to various methodological problems. This is why a thesis was written on the subject (Chap. 7, § 7.3.2). A general summary of the papers has also been published (Favrot and Lagacherie, 1993).

Implementation

A project implementing the RA and SNR concepts was launched in 1980 jointly by the Ministry of Agriculture and *l'Office National Interprofessionel des Céréales* (ONIC). It was designed for aiding drainage of cultivated lands showing water in excess. Scientific coordination was provided by scientists of CEMAGREF and INRA, brought together to form a national steering committee. Three kinds of data and knowledge are necessary for applying knowledge and experience gained in the RA to the level of the SNR:

- knowledge of the nature and distribution of soils in the RA (mapping),
- availability in the RA of measured data on the hydric regime of the soils and on the response of the soils and crops to drainage (surveys, observations, hydrodynamic measurements…),
- availability of a written synthesis giving the practical procedures applicable in every parcel of the SNR where some quick supplementary observations will be made before installing drainage. For example: necessity in such or such soil of lining the drains to avoid their being getting sealed by iron, or the need for augmenting drain density.

Results

The studies were conducted between 1980 and 1985. They resulted in the creation of 70 Reference Areas that characterized as many Small Natural Regions representing totally 2 million ha. Experience showed that an RA must cover around 1000 ha. The scale of 1/10,000 was found to be particularly suitable. According to the studies conducted, there were on average 28 different Map Units per RA. The SNRs were obviously of very variable size; the average area was about 30,000 ha. The project scientists did one auger hole per 1.4 ha on average.

These mapping efforts gave work for 5 years to about 60 mappers belonging to 44 research departments of private or government services. Lastly, the RA method was exported outside France.

It is not known exactly how many millions of hectares were drained in France using the method developed. In 1980, it was said that there were about 6 million ha of wet agricultural lands out of the total of 12 million ha of wet lands and marshes. With passage of time, the latter figure seems too large. Surely it was not necessary to alarm the ecologists! They now state that there are less than 2 million ha of wet lands and marshes in France. Let us hope they are exaggerating, in the other direction this time… .

10.6.7 Project on inventory of management and conservation of soils (IGCS)

The time had come to salvage all the mapping documents available in France for reorganizing them and making them uniform in a projected national synthesis on scale 1/250,000. This is the *Inventaire Gestion et*

Conservation des Sols (IGCS), launched in April 1990 under a contract between the Ministry of Agriculture and the INRA. Actually, additional important, not to say essential, partners were invited to participate in the project. These were the 22 administrative Regions of the country. A 50-50 funding was anticipated: 50% by the Ministry of Agriculture and 50% by the Regions. In reality, the Regions assured a large part of the financial cost of the project and at times turned to the European Union for aid. The limits adopted for the soils sheets were those of these Regions. It is necessary to motivate the backers.... These maps were called 'Regional Soil Reference Bases'.

Scientists of the INRA appeared at several levels in this operation: local surveys, expertise in quality of the work, participation in the Steering Committee.

Preparation of regional soil reference bases on scale 1/250,000 posed new problems. Firstly, it was necessary to establish a method suited to this scale and to the delineations relying considerably on the landscape (the *soilscape* concept, Chap. 1, § 1.4.3): definition of rates of survey, observation densities, methods of synthesis, etc. Secondly, in this programme, construction of graphic databases appeared to be something essential for safeguarding the data and making them available to users. It was therefore necessary to develop methods of data storage, updating of data, processing, graphic presentation, etc. Going from conventional maps on paper to digital databases is obviously an important step. Standards (DONESOL system, STIPA-2000; see Chap. 6, § 6.3 and Chap. 4, § 4.4.2) pertaining to the volume and type of data to be gathered were defined.

Above all, the rules of the game, that is, the conventions to be followed by the different partners, were clearly established (Urbano, 1994). A three-tier quality label was proposed for application to the results of the studies (Chap. 3, § 3.1.3). A very active Steering Committee is at work to provide liaison between the partners, to do adequate quality checks, to promote the project amongst users and to motivate the Regions that are not yet collaborators in the IGCS. At present there is every chance of completing coverage of soils of the territory in a reasonable span of time. This will be very desirable at a time when certain specialists are already thinking of covering all of Europe by maps of this type (Dudal et al., 1995). This chance for mapping is also the last and the mappers of France must realize it: one cannot shamelessly start all sorts of programmes that no one ever completes.

Languedoc-Roussillon, the map of which was the first one completed including digitization, played the role of pilot Region (Bornand et al., 1994).

10.6.8 Data and their accessibility

The 'Service d'Étude des Sols et de la Carte Pédologique de France' (SESCPF, INRA-Orleans) is, as the name shows, in charge of soil mapping in France.

It organizes the work, keeps and distributes the data, particularly those pertaining to the systematic inventory programmes (1/250,000, 1/100,000 and CDTA at 1/50,000).

For some years, as an extension of the initial efforts made by INRA-Montpellier (Favrot, 1994), the SESCPF has at its disposal a database termed 'RÉFERSOLS' (*Répertoire pour la France des études pédologiques avec cartographies détaillées des sols*) that lists the maps done in France on scale 1/25,000 or larger and covering more than 500 ha. More than 1500 files of soil studies on large scale have been found (search for site INRA INFOSOL – IGCS CARTE PÉDOLOGIQUE DE FRANCE).

In all, about two-thirds of France have been covered, including all scales 1/250,000 or larger. Progress is reported regularly (Bornand and Lehman, 1997; King et al., 1999).

Furthermore, perhaps aided by the allure of the islands, French mappers of ORSTOM (now IRD, *Institut pour la Recherche en Développement*) have fully covered these Departments, Territories and former colonies: the Comoros (on 1/20,000, when they were French), Guadeloupe and Marie-Galante (1/20,000), Martinique (1/100,000), Mauritius (1/50,000), New Caledonia (1/300,000), New Hebrides now Vanuatu (1/100,000), Reunion (1/100,000), and of course not to forget Tahiti (1/40,000) (search for site IRD FRANCE CARTE PEDOLOGIQUE).

10.7 SOIL MAPPING IN SWITZERLAND

In spite of the scarcity of good soils in their country and great attention being paid to protecting their natural environment, the Swiss have not put in much effort in detailed soil mapping. The few surveys done pertain to the Swiss Plateau, the Jura mountains and a few valleys in the Alps (Bonnard, 1989).

10.7.1 Small-scale maps

The major achievement is the map on 1/500,000 that covers the entire country.

Besides this, the four sheets of the *Capability Map of the Soils of Switzerland* (1980) on 1/200,000 and the *Crop-suitability Map of the Soils of Switzerland* on 1/300,000 have been prepared. These documents have very practical objectives. They directly present the kinds of soils and their suitability/capability. Nature of the soil is only indicated by a code referring to the FAO legend.

10.7.2 Systematic mapping on scale 1/25,000

Switzerland started systematic coverage of its territory on scale 1/25,000 in 1977. The department charged with coordination of this programme is the *Station Fédérale des Recherches Agronomiques de Reckenholz* near Zurich. About 247 sheets were to have been surveyed, completely or partly at the country's borders. Each sheet represents 17.5-km length and 12-km width, or 21,000 ha. The major characteristics used for delimiting and representing soils are usable depth and possible degree of excess water. Unfortunately, very few sheets have been published: not more than about ten (Fig. 10.3). Above all, reorganization of federal agronomic research stations led to cessation in 1997, as an economy measure, of systematic mapping of the soils of Switzerland.

10.7.3 Very large scale mapping in the canton of Vaud

In Switzerland, the greatest interest lies in large scales related to specific problems to be solved: irrigation, drainage, fertilizer application, etc. To take the example of the canton of Vaud, surveys were done on 1/10,000 by the canton's development department. Referred to the partitioning of Switzerland on 1/25,000, this corresponds to the following sheets: *Payerne* (1184, eastern half), *Orbe* (1202), *Yverdon-les-Bains* (1203) and *Cossonay* (1222). Characterization of the wine-growing soils in the canton was also started on 1/10,000. More than 2000 ha have been surveyed by a French research department. Also, there are detailed studies done locally on 1/1000 or 1/5000. They pertain to part of sheets 1183, 1204, 1223, 124, 1243, 1244, 1261, 1281 and more particularly to sheets 1300 and 1301.

Let us also keep in mind that there has been great effort in Switzerland for monitoring soil pollution (the NABO programme). This has resulted in characterization of the soils at 102 sites quite regularly distributed in the country.

10.8 SOIL MAPPING IN AFRICA

Space is lacking for presenting in detail the mapping that has been done in African countries. In the French-speaking zone, for example, we must mention the work of ORSTOM (now IRD), of CIRAD-IRAT and certainly of the countries themselves after attaining independence. A survey has been done

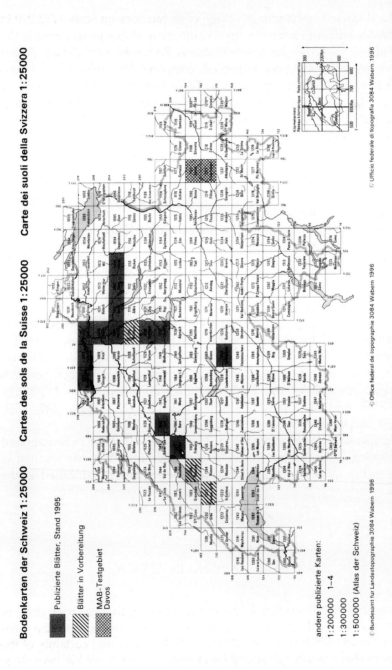

Fig. 10.3 Status of progress of mapping of Switzerland on scale 1/25,000. Source of information: *Station Fédérale de Recherches Agronomiques*, Zurich-Reckenholz.

of the status of progress of mapping in the developing countries (Zinck, 1994), which could be consulted. Table 10.3 was prepared by taking out only the results for the French-speaking countries.

Table 10.3 Progress of soil mapping in the French-speaking countries of Africa (% of area covered on various scales) (Zinck, 1994)

Country	Small-scale: 1/500,000 to 1/100,000	Medium-scale: 1/100,000 to 1/50,000	Large-scale: 1/25,000 and larger
Algeria	0	5	5
Benin	100	10	2
Burkina Faso	100	25	0
Burundi	0	100	0
Cameroon	30	5	1
Congo	10	5	0
Gabon	30	0	0
Mali	50	0	0
Morocco	0	40	20
Togo	80	20	0

Among the French-speaking countries that did not respond to this enquiry, the Central African Republic and especially Senegal have covered a considerable part of their territory on scales from 1/100,000 to 1/200,000. Cameroon is engaged in a programme of systematic mapping on these scales (Bindzi-Tsala, 1994). Completion of coverage is a priority for that country. Also, a deeper investigation has been done in the Sudano-Sahelian zone (Bertrand, 1989). This zone is mostly covered on 1/1,000,000 or 1/500,000 except for Mali, Mauritania and Guinea (Fig. 10.4). Further information can be obtained from *Répertoire des Cartes de l'ORSTOM* (ORSTOM, 1985) and its supplement (ORSTOM, 1995) (Search for site IRD FRANCE CARTES DES SOLS).

10.9 SOIL MAPPING IN SOME OTHER COUNTRIES

10.9.1 Soil mapping in Australia

The *Australian Soil Resources Information System (ASRIS)* is a soil-profile database. It contains 160,000 described and analysed soil profiles pertaining to different States and Territories. Degree of characterization is variable. This database, established through the collaboration of several organizations, has resulted in a very complete description (search for site AUSTRALIA,

MAPPING OF THE SOIL

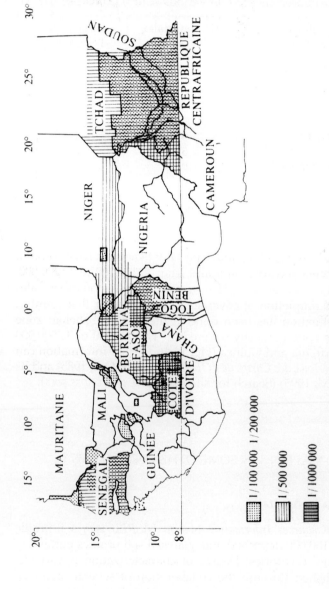

Fig. 10.4 Progress of soil mapping in French-speaking Sudano-Sahelian Africa (Bertrand, 1989).

ASRIS, CSIRO, ACLEP). Much time has been spent in 'cleaning up' the data. The data can be directly downloaded if necessary permission has been obtained.

The *Soil And Land Information System* (SALIS) is another database containing the data corresponding to 48,000 observation points. According to the type of user, the data are given free, provided against reimbursement of the cost of their transfer or sold for a price representing 10% of the actual cost of their acquisition. They can even be exchanged to the extent the supply to the DBMS of data encoded on appropriate sheets authorizes export of an equivalent quantity of data from the database.

10.9.2 Soil mapping in Russia

The former USSR has been completely covered on scale 1/2,500,000 since 1988 but all the sheets have not been published. The database pertaining to Russia was set up by the Dokuchaev Soil Science Institute in Moscow. The soils have been identified in three classification systems: Russian, FAO and USDA (Soil Taxonomy). Printed maps can be obtained against payment (Director-Professor Vyacheslav A. Rojkov, e-mail: rojkov@agro.geonet.ru). A synthesis at scale 1/5,000,000 can be obtained through the Internet (Dr Vladimir Stolbovoi, e-mail: stolbov@iiasa.ac.at).

Coverage on scale 1/1,000,000 is nearly 60% and on scale 1/10,000 about 10%, which corresponds to agricultural areas.

10.9.3 Soil mapping in China

A soil map of China covers the country on scale 1/4,000,000. The legend is broadly genetic. It has been correlated with the FAO-UNESCO revised legend of the *Soil Map of the World*.

Final conclusions and perspectives

Soil-inventory programmes in developed countries have often been reduced and at times abandoned under the effect of economic crisis. The administrators of national mapping centres, who met in the Netherlands some time ago, described the situation as 'desperate' (Zinck, 1994). But though the inventories have been stopped, all work has not ceased. For several years, profiting from the appearance of new computer tools (GIS, RDBMS), scientists of the different countries have put in much effort in gathering available map data, validating them, organizing them in suitable databases and making them available to users. Moreover, the situation could improve. Environmental problems result in soils being properly taken into account. This goes from the greenhouse effect (exchanges of carbon and nitrogen between the earth and the atmosphere) to all kinds of site-specific problems: salinization, metal pollution, etc.

It is noted that USA has largely escaped economic crises and consequent reduction in funding for preparing soil maps.

Mapping efforts appear to be more necessary than ever in developing countries. It is known that at least one-half the potentially cultivable land of the world is already cultivated. It is also known that world population is growing rapidly. A doubling of the demand for agricultural commodities is expected in less than 40 years. One need not be a soothsayer to understand that intensification of agriculture is imperative and in a very short time as well, although birth control must have proved efficient everywhere. One way or another, governments must launch agricultural-development programmes and the studies that precede them. As only half of the area currently cultivated in the world has been mapped, and this on rather small scales, the need for new knowledge concerning soils will be considerable. This has been well understood by administrators. The investigation mentioned above (Table 10.3) shows that mapping efforts have not stopped in developing countries. In 1994, the directors of the concerned departments were mostly optimistic about this. While 36% of them declared their mapping budget was reducing, 54% declared it rising, the rest considering it unchanged. Since that time, unfortunately, political instability in many countries has harmed their development in all areas.

We certainly have put behind us the period of 'mapping depression'. Mappers must ready their methods to be in a position to respond to a very pressing demand that needs to be quickly satisfied. Isn't it a wonderful profession that detects, zones and enhances the ability of land to feed Man?

APPENDICES

Three exercises are presented in these appendices. The first one serves as example for calculating the time and cost of a mapping investigation; the second pertains to the photo-interpretation phase that precedes field work; the last is an attempt at thematic mapping by means of multicriteria analysis, after soil mapping.

A1   PLANNING A FIELD PROGRAMME

A1.1   Statement of the exercise

This exercise will take about an hour and 30 minutes.

The soil survey of Lausanne region in Switzerland is to be done. The scale chosen is 1/50,000 and the area is 100,000 ha. The aim of the work is to plan urban development in a way compatible with agriculture and protection of the environment. For this, it is felt that it is necessary first to prepare a conventional soil map on the basis of free survey (Chap. 2, § 2.1.5) with traversing of the land and a profile-study programme. The company taking charge of the study can utilize the services of four mappers. The problems are:
- (a) to calculate approximately the number of auger holes and profiles to be done;
- (b) to estimate the time required for field work (delimitation of soil units with auger holes and then study of profiles); the survey must start in the month of March;
- (c) to estimate the cost of these studies with the following being shown separately:
    — expenditure on salaries (the four persons are young engineers paid 1600 euros each per month;

— expenditure on travel; the area is located around Lausanne, thus there is expenditure on midday meals (@10 euros per person) but no hotel expenses because the teams return home every day; the distance travelled is estimated assuming that the area is a semicircle with Lausanne as centre and adding 50 km per vehicle per day for the survey proper; the cost of travel is 0.5 euro per km;
— cost of analysis (it is assumed that complete analysis of a profile costs 450 euros on average);
— cost of printing the map and report; it is assumed to be 6000 euros per m^2 of map;

(d) to deduce from all the above the mapping cost per ha;
(e) if the funds available are insufficient, what suggestions could you give for reducing the costs without lowering the quality of the work too much?

A1.2 Answer to the problem

Whatever efforts are put in, it is almost impossible to put a figure to a mapping job if a spreadsheet is not available for enabling review of different items so that none is forgotten. Table A.1 presents one such document. It will be sufficient to fill it by estimating the items of expenditure for which the statement of the exercise does not give precise indication. Following this, only some major points are emphasized.

Table A.1 Spreadsheet for estimating the cost of a soil survey

PLANNING A MAPPING INVESTIGATION		
AREA IN ha	A	
SCALE	E	

STUDY OF STANDARDS (Chap. 3, §. 3.2), CALCULATION OF NUMBER OF OBSERVATIONS		STANDARDS
Number of ha covered per team per day	Nha_j	
Number of ha per profile	Nha_p	
Number of ha per auger hole	Nha_s	
Total number of auger holes to be done	$S = A/Nha_s$	
Number of profiles to be opened	$P = A/Nha_p$	

CALCULATION OF TIME REQUIRED BASED ON A TEAM OF TWO PERSONS		TIME
Number of days for delimiting soils	$J_1 = A/Nha_j$	
Number of days necessary for preparation to open profiles and for informing the owners (Chap. 3, § 3.6.1)	$J_2 = P/20$	
Number of days necessary for opening and closing the profiles (team accompanying the mechanical digger, Chap. 4, § 4.6.4)	$J_3 = 2P/36$	
Number of days for describing the profiles @ 6 profiles/day	$J_4 = P/6$	
Total number of field days	$J_{tot1} = \sum_{J=1}^{J=4} J_i$	
Evaluation of climatic risks ($\beta \cong 1.2$)	$J_{tot2} = \beta \cdot J_{tot1}$	
Number of months (22 days/month), all teams taken together	$Months = J_{tot2}/22$	

ESTIMATION OF km BY VEHICLE		km
km_1 = distance travelled between place of stay and place of work		
Nt = number of trips per day (2 or 4)	$km_1 = Nt(\alpha \cdot D)$	
α = coefficient linked to the road network (Chap. 3, § 3.6.4)		
D = straight-line distance between residence and centre of the survey area		
km_2 = km for survey proper (km_2 = **5** for 1.10,000, **20** for 25,000 and **50** for smaller scales)	km_2	
Total km where J_{tot} is the total number of days for survey (see above)	$km_{tot} = J_{tot1} \cdot (km_1 + km_2)$	

GUIDE FOR ESTIMATION OF COST		COST
SALARIES		
NET		
GROSS		
WITH CHARGES		
PURCHASE OF MATERIAL: geological and topographic maps, map bases and aerial photographs (Chap. 3, § 3.6.5)		
TRANSPORT VEHICLE COST: at the rate of *x per km*	$x \cdot km_{tot}$	
MECHANICAL DIGGER AT SITE: opening/closing of profiles at hourly rate of *y*	$y \cdot 8 \cdot 2P/36$	
TRAVELLING EXPENSES: at daily cost of *z*	$z \cdot J_{tot}$	
COST OF ANALYSES: estimated at the average rate of *l* for each of the profiles *P′* selected for analysis (Chap. 5, § 5.2.3)	$l \cdot P'$	
MAP PRINTING: function of size, number of colours, number of copies and reproduction technique		
REPORT: function of presentation, reproduction in colour and number of copies		
OTHER DOCUMENTS PRESENTING THE DATA: appendixes on paper, CD-ROM, etc.		
GENERAL EXPENSES		
TAXES (if applicable)		

SUMMARY		
TOTAL	**COST**	
EXPECTED DATE OF COMPLETION OF WORK	(Month)	

QUALITY CHECKS		
Number of auger holes to be done per team per day	S/J_1	
Rate per ha (for comparison with standards)	COST/A	

(a) *Estimation of number of auger holes and profiles to be done*

Table 3.2 (Chap. 3, § 3.2) gives the method for calculating the answer to the first question of the exercise.
- Auger holes: about $100{,}000/[(20 + 50) \times 0.5] = 2860$
- Profiles: $100{,}000/[(200 + 500) \times 0.5] = 286$
- Hectares per day: $(200 + 500) \times 0.5 = 375$.

(b) *Estimation of time required for the survey*

The time required for the survey would be about $100{,}000/375 = 267$ days assuming that just one team will work. But the time required for the field programme must not be forgotten:

- informing the owners of the sites where digging will be done (286/20 = 14 days);
- accompanying the mechanical digger for opening and closing the profiles = 286 × 2/36 = 16 days;
- describing the profiles (48 days @ 6 profiles per day).

All this represents 345 days or 172 days for each of two teams. Adding 20% to cover special climatic aberrations (the work starts early in the season), the total comes to 414 days. Reducing this to months (22 working days per month and not 31!), ten months will be needed with two teams of two persons each, not counting the work of synthesis in the office.

(c) *Estimation of distance to be travelled*

The following reasoning can be used. The area is assumed to be approximately semi-circular with a radius of

$$R = \sqrt{2.1000/3.14} \text{ or about 25.2 km.}$$

The place of stay being nearly at the centre of the area, one-half this distance will be crossed on an average to go to work. That is, 25.2/2 = 12.6 km. Actually, we will go to the border of the area in some cases, but be at the gates of Lausanne in others. But this is for travelling one way. We must also calculate for going back. Furthermore, it is necessary to consider the fact that journeys by road are not done in a straight line. Around a town well provided with road network, the distances travelled are not greater than the straight-line distances enhanced by 20%, strange though this might seem. Thus we get 12.6 × 2 × 1.2 ≈ 30 km per vehicle per day without counting the 50 km directly related to survey, for 345 days. The 20% rainy days are not accounted for because there is no travel on those days. Thus, in total, 345 × (30 + 50) ≈ 27,600 km or 13,800 euros.

But we can reason out a little more finely. On an average, we will not cover exactly one-half the radius of the area. More exactly, we will go up till we reach a small circle of radius r with the same centre Lausanne, such that the nearer area to be surveyed is equal to the farther area to be surveyed. Under these conditions, we have

$$\pi R^2 = 2\pi r^2$$

or:

$$r = \frac{R}{\sqrt{2}} = \frac{25.2}{1.414} = 17.8 \text{ km, instead of 12.6 km.}$$

In other words, the method first presented gives a lower estimate of the distance to be actually travelled.

(d) *Cost of publishing the map*

The semi-circular area to be mapped has a radius of 25.2 km. Taking the scale into account, the map will be printed inside a rectangle of about 50 cm × 1 m. It is also necessary to make room for writing the legend with its printed coloured boxes. A map of one metre by one metre has to be expected. The cost will be according to this.

(e) *Statistics of the results*

A class of 34 students of l'École Polytechnique Fédérale de Lausanne did the exercise. The results are presented in Table A.2.

Since then the standards have changed slightly. The reader will perhaps not get the same results if Table 3.2 in Chapter 3 is applied.

Table A.2 Estimation of the work to be done for a soil survey around Lausanne

Estimated quantity	EPFL - GR, 1993-1994 class			
	1st decile	Mean	9th decile	Range
Area per day per team	250	415	500	From 1 to 2
Days for each team	100	142	160	From 1 to 1.6
Profiles to be opened	300	381	481	From 1 to 1.6

Also, the students calculated that the cost of salaries represented one-half of the total and the cost of analysis about 40%. They proposed, under these conditions, for reducing the expenses to have just two engineers work with the assistance of two less qualified and less paid persons. They also proposed not to have all the profiles analysed (Chap. 5, § 5.2.3). Lastly, some braves went as far as to estimate that one could be satisfied with sandwiches at midday.

A2 SOILS AND LANDSCAPES ZONING (FLAINE REGION, HAUTE-SAVOIE)

A2.1 Objective, materials needed

Execution of this exercise requires two and half hours.

The aims are to recognize the principal observable features on an aerial photograph, to make some logical deductions on functioning of the natural environment and to delimit landscape units. The materials required are
- One stereoscope per person or pair of persons.
- The three photographs numbered 268, 269 and 270 of the panchromatic aerial survey *3530 FD 74/300p Cluses* on scale 1/30,000. The following data that do not appear on the photographs should be noted: flight altitude 6340 m MSL, time (solar) 1120 hrs, date 26 July 1988. These photographs are procured from IGN-Lyon, 8 Avenue Condorcet, F69100 Villeurbanne, France; FAX 33-4-72-69-06-50.
- The topographic map *Cluses sheet on 1/50,000*.
- The geological map on 1/50,000, *Cluses sheet No. 679, 1993*.

A2.2 Photo-interpretation

Analysis of the principal features of photograph no. 269

The questions are asked very precisely, providing indications and ease in answering. Experience shows that this is useful to help the beginner observer.

(1) To identify the various surface formations that appear on the photograph. In particular:
   — alluvia (*look for the valley*);
   — hard rock (*see the top part of cliffs*);
   — scree (*look at the foot of the cliffs*);
   — alluvial cones (*look for streams that join the valley*);
   — moraines (*there are a few of them, only specialists…*);
   — and the neves!

(2) To identify the principal types of plant cover:
   — crops (*there are fields*);
   — broad-leaved trees and conifers (*the former are lighter coloured and located at lower altitude*);
   — meadows;
   — pre-forests;
   — green alder, *Aldus viridis*; (*these are slightly rounded shrubs with dense and relatively light-coloured foliage; they grow in moist areas*);
   — is the distribution of these formations unconnected to altitude?

(3) To specify the characteristics of the rocks: are the rock massifs rounded (which could lead one to think of granites or soft rocks), do they look jagged and oriented (which will correspond to schists) or are clearly seen entablatures (which could indicate limestone beds)?

(4) To examine the faults: how to explain why the plant cover is so different from one part to another of certain alignments? What other criteria can help in locating the geological faults? (in case of difficulty, consult the geological map).

Reconstruction of the functioning of the landscape

(5) Starting from the river in the west of the photograph, retrace the course of the streams in the direction of the great cliffs. Do the striae corresponding to temporary streams go up to the summit? Are streams seen in the faulted zones? Where does the water go? To sum up, are the faulted zones in soft and impermeable rock zones with surface drainage or in the hard-rock zones with water penetration through fissures and then subterranean flow?

(6) Careful examination of the faulted zones shows great irregularity in detailed relief. To what could this be due? to landslides? to fallen rocks? to karstification (non-uniform dissolution of hard limestone rocks outcropped in slabs)?

(7) To which characteristics of the environment is the distribution of vegetation finally linked to?

(8) In such a situation what criteria will you suggest for doing a pre-zoning of the observable landscapes?

Attempt at delimiting soilscapes
The term soilscape is used to denote a very small natural region (a few tens or hundreds of ha) with uniform appearance and distinct from its neighbours in its natural environmental factors (altitude, rock type, plant cover and geomorphology), having soils that are undoubtedly characteristic. These last must be studied after pre-zoning of the units in question.

The soilscapes must be precisely located on all the photographs without letting oneself go as far as distinguishing too many details whose real significance might be doubtful. It is necessary to find between 10 and 30 units. The delimitation is done on a transparent tracing sheet laid on the photograph and stuck with adhesive.

Development of the area
A ski resort, Flaine (northeast corner of the photograph), has been established. Think about the following questions:
  (9) Is it easy to open up pistes in these milieus? What solution has been implemented and do you see the trails?
  (10) Are there risks of avalanches at the resort? Do you see the riposte that has been found?
  (11) Is there risk of lack of water if too many tourists throng here in summer? Do you see the solution provided to the problem?
  (12) The area being isolated, how could the tourists be attracted to this place in summer? Do you see the efforts that have been made?
  (13) Does pastoral activity still subsist here? How are one or many mountain pastures to be recognized?

A2.3   Answers to the exercise

Responses to questions 3 to 8
  (3) Large cliffs and horizontal geological beds can be seen, leading one to suspect, quite rightly, a limestone terrain. The rocks are hard and form thick, massive beds. They are intercalated with softer rocks, which partly explains the presence, above the valley. of an irregular slope punctuated by shelves. The other reason for irregularity of the slope is the presence of faults... .
  (4) At least three faults are seen southwest of Flaine, on the other bank of the gorge: two faults are parallel and the third is at right angles to them; the alignments, strike slips, narrow depressions and different colour on either side of these faults are clearly seen. Actually, these faults separate different rocks, on either side; indeed, only certain rocks

contain, apart from calcium carbonate, insoluble residues that accumulate in place and allow soils to form and vegetation to prosper.

(5) No drainage axis is visible in the faulted region; we suspect that the water infiltrates the large (visible) or small (invisible) fissures, to come out later halfway down a large cliff in contact with a less permeable layer; about 15 speleological cavities have been seen in this zone, some of them permitting descent to 200 m below the surface.

(6) This system corresponds to a very typical karst landscape: hard, faulted limestone corroded by dissolution from the surface caused by water charged with carbon dioxide. As the limestone is pure, dissolution is complete; no residue is left behind that could form a soil; in the eastern part of photograph 270 is seen the '*Desert de Platé*', which is, as the name indicates, desolate and forms one of the most beautiful karst landscapes of France; the reader who will visit will not regret it (a cable car facilitates access).

(7) The distribution of vegetation is related to altitude (cf. Question 2), and also to rock type, as we have seen.

(8) The principal criteria for delimiting landscapes are: firstly, the altitude-climate-vegetation assemblage (cf. terracing of grasslands, forests and crop lands) and, secondly, the relief (bare slabs, scree, alluvia, etc).

Figure A.1 represents an interpretative section through these environments.

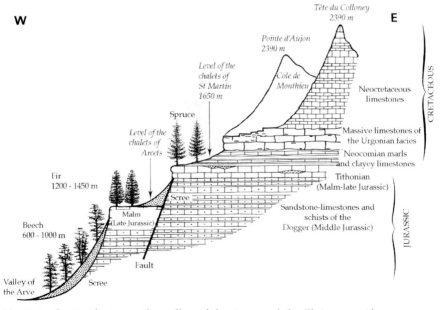

Fig. A.1 Section between the valley of the Arve and the Flaine massif.

Delimitation of the soilscapes
The reader should refer to Fig. 5.7 (Chap. 5, § 5.3.1) that represents the delimitation of the same soilscapes by 20 2-person teams of students of the EPFL (Rural Engineering, admitted in 1992-1993). The reader could then compare his/her own diagnosis to that of other observers. However, the exercise was done on photograph 36 of the 1980 flight, the path of which was a little to the east.

Answers to questions 9 to 13 pertaining to development
(9) For establishing ski trails on the karst, it was necessary to level the karst and fill the fissures, which explains all those uncoated pebbles that shine on a piste laid out in the northeast direction, towards the resort.
(10) Avalanche barriers can be seen clearly to the north of the station and below it.
(11) The small lake that can be made out in the valley southwest of the resort is used for obtaining water.
(12) There is a very large number of tennis courts here; they can be recognized by their shape, size and location close to the station.
(13) Alpine meadows can be identified if one accepts that they correspond (Chap. 3, § 3.3.4): to grass (*plant occupancy*), to gentle slopes (*relief*) and to the presence of at least one chalet with access path (man-made infrastructure). The largest of them belongs to the faulted zone.

Remarks
To complete the exercise, photographs 1002, 1003 and 1004, and also 1040, 1041, 1042 of the colour aerial survey of 1993 on scale 1/20,000 (ref. FD 74/200C-Cluses) must be procured. They pertain to the same zone and were taken at 0750 hours (first group) and at 0813 hours (second group). These photographs, taken from about 5100 m, are superior to the panchromatic images in that they, because of the larger scale, enable us to distinguish broad-leaved trees from conifers better, see some rhododendron patches in the alpine meadow, discover the new golf course of Flaine and better observe the small morainic ripples. They are of exceptional quality.

For knowing more, the following references should be consulted: Richard (1975), Party (1982), Maire (1990) and Legros et al. (1996b).

A3 PLANNING LAND USE

A3.1 Statement of the exercise

This exercise will take 3 hours to complete.

The aim is to investigate, around a Swiss village, the areas that will be suitable for extension of built-up area. Table A.3 gives the characteristics of the soilscape units to be studied.

Table A.3 Characteristics of the soilscape units to be studied

Soil-scape	Slope (%)	Land use	Soil depth (cm)	Stoni-ness (%)	Distance from the village (m)	Excess water	Texture*	Area (ha)
A	0	Recreational groves	10	20	100	No	SL	2
B	4	Crops	100	5	1000	No	LAS	3
C	10	Orchards	50	20	50	No	A	10
D	5	Commercial groves	80	30	2000	No	A	1
E	20	Meadows	60	5	2000	Strong	LAS	3
F	5	Crops	60	10	500	No	LAS	3
G	60	Protective groves	100	30	1400	No	LAS	8
H	15	Meadows	20	10	100	No	S	25
I	5	Crops	80	0	500	No	LAS	16
J	5	Crops	100	12	500	No	LAS	4
K	0	Crops	80	10	600	No	LAS	5
L	5	Crops	80	15	1500	No	A	6

* According to the French textural triangle

The first approach: complete aggregation (Chap. 8, § 8.2)
This consists of first finding the areas that must be left for agriculture in order to deduce those that can be offered for construction.
- Convert the various values appearing in the table to utility scores in relation to the objective 'area acceptable for agriculture' (Chap. 8, § 8.2.1). Use a work sheet for this organized like Table A.4. Justify your method.
- (2) Suggest a complete-aggregation method (Chap. 8, § 8.2) aimed at calculating an index of suitability for agriculture. Give the weighting of the variables with justification. Provide the score of each unit and the resulting rank.

Table A.4 Structure of the upper part of the work sheet

Soil-scape unit	Slope (%)	Land use	Soil depth (cm)	Stoni-ness (%)	Distance from the village (m)	Excess water	Texture	Area (ha)
								← Weight
							Score	Rank
A								
B								

then C, D, etc.

The second approach: partial aggregation and outclassing (Chap. 8, § 8.4)
  (3) It is accepted that soilscape unit F forms a reference standard for suitability for sustainable agriculture. Are there areas that outclass this standard for this suitability? Which are they?

The third approach: downgrading (Chap. 8, § 8.3)
  (4) Are there areas that, by all hypotheses, must not be allocated to construction? Which are they and why?

Synthesis
  (5) Do the approaches result in the same diagnosis? If not, why?
  (6) Prepare a balance sheet of the areas that can be used in priority for construction, those that are not so and the areas for which there is some doubt.
  (7) How could one remove the ambiguities?

Conclusions
  (8) Can there be conflict between the various uses of land? Does this seem generally frequent in rural areas?
  (9) What do you conclude from this exercise?

A3.2 Answers

Construction of the matrix of utility scores
Although wide latitude is allowed to the experts for making their decisions, it is necessary to take into account some obvious factors:

  Slope is unfavourable for agriculture; moreover, it increases the cost of building houses but it does not hinder construction, in fact can even be favourable for it (more unobstructed views, sunlight, etc.).

  Land use can be a constraining factor: it would appear stupid to destroy recreational groves located at the gates of the village; even a protective grove has its value for fighting erosion, wind or any other unfavourable factor.

Soil depth is favourable for agriculture and it has no effect on suitability for construction.

Stoniness is a negative factor in agriculture; it is of no consequence in construction.

Distance from the village is certainly a negative factor for building activity. It is more difficult to judge its effect on agriculture: being far is certainly a security from the disadvantages linked to the town/country fringe. But it all depends on the residence of the farmer: does he live in the village or on his lands? If he is in the village, distance can seem a negative factor.

Excess water (marsh) is an unfavourable factor in construction (wetness). Agriculture adapts to it better, especially after drainage.

Texture is to be considered in agriculture, but it is less important than slope or stoniness.

Area is neither favourable nor unfavourable. It is what it is. It will almost have no role in this exercise unless it has been ensured that the area found for building activity is adequate.

A matrix of utility scores can be set up on these principles and a system of weighting introduced to make a complete aggregation possible.

Approach by outclassing
It will be recalled that this is a matter of marking out the individuals that outclass the reference F in *all* the criteria. In any case, it does not require enumeration of the units that are rated better than F in the framework of the preceding question. If the area (which is not a quality criterion) is not considered and if the distance (the direction of whose effect is doubtful) is also forgotten, F is outclassed or equalled by B, I and K. But the case of K is worth considering. J equals F in slope, land use, distance from the village and texture, J outclasses the same F in depth. Surely its stone content is higher, but by very little (2%); we are in the zone of uncertainty (Fig. 8.7 in Chap. 8, § 8.4.1). In sum, J is worth more than F for agriculture.

Approach by downgrading
From the point of view of limiting factors, the units that are certainly not suitable for building purposes are: A (recreational grove), D (commercial grove and very far), E (too wet), G (protective grove and far) and L (distant zone).

Synthesis
The results of the method by aggregation depend on the system of notation retained; they are only given here (Table A.5) as indication.

Table A.5 Selection of areas for construction (blank), comparison of the results of the three methods

Method	A	B	C	D	E	F	G	H	I	J	K	L
Complete aggregation: elimination of areas ranked best for agriculture		■	■		■	■			■	■	■	
Outclassing: elimination of farm lands that are better than the reference		■	■		■	■			■	■	■	■
Downgrading: direct elimination of the areas having one or more limiting factors for construction		■	■	■	■	■		■			■	

Agreement is not perfect between these methods. All sorts of reasons cause the disparity, the reader could investigate them. Among them, it is seen that suitability for construction is not necessarily 'not suitable for agriculture' (cf. Fig. 8.6 in Chap. 8, § 8.2.2). Be that as it may, we can very certainly consider suitable for construction the areas that are recognized as such by these three approaches. The EPFL students (Rural Engineering major, third year, admitted in 1994-1995) proceeded in this fashion. Figure A.2 presents the synthesis of their interpretations.

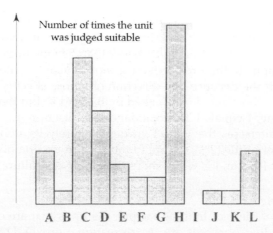

Fig. A.2 Suitability of the various units for construction, as measured by the number of students who always found this unit suitable, whatever be the reasoning used.

Many differences exist in the interpretations even though the approach method is codified in detail! But the exercise is incomplete. Of course it uses the soil characteristics (slope, stoniness, etc.) but it does not integrate their exact spatial organization. Thus use of aerial photographs and the soil map, and also a traverse of the land will enable us to see it more clearly, that is, to finally decide between the various candidate units that are suitable for construction (C and H in particular).

BIBLIOGRAPHY

Agafonoff V. 1936. *Les Sols de France au Point du Vue Pédologique*. DUNOD. 154 p. + maps.

Algret H. 1983. *Analyse de l'Aptitude au Reboisement en Cèdres sur les Bordures du Plateau de Valensole*. DAA-ENSA, Montpellier. 27 p.

Amundson R. and Yaalon D.H. 1995. History of Soil Science. E.W. Hilgard and J.W. Powell: Efforts for a joint agricultural and geological survey. *SSSAJ*, **59**: 4-13.

Anderson D. 2002. Soil survey activities and correlations by major land resource areas. *17th World Congress of Soil Science, Bangkok*. Symposium 48, paper 617.

Andrieux P., 1990. Influence de la variabilité spatiale descaractéristiques physiques des sols sur le bilan hydrologique. L'exemple d'unsystéme de sols sur barre pré'littorale (Guyane française. These, université du Montpellier II, 182 p + annexes.

André C., Dalles S. and Orth Ph. 1995. *Interpolation des Données Climatiques sur la Base de la Méthode de Thiessen Améliorée*. I and II: *Rapports Techniques* (104 pp., 80 pp.), III: *Rapports de Synthèse* (38 p.), IV: *Manuel de l'Utilisateur* (32 p.). INRA-ISIM, Montpellier.

APCA (Assemblée Permanente des Chambres d'Agriculture). 1984. *La carte des terres agricoles. Revue APCA* Supplement to No. 712. 28 p.

Arnold R.W. 1990. Fractal Dimensions of some Map Units. *Trans. 14th Int. Cong. Soil Sci.* **5**: 92-97.

Aronoff S. 1982. I, Classification Accuracy: a user approach; II, The map accuracy report: a user's view. *PERS*, **48**: 1299-1307; 1309-1312.

Aronoff S. 1985. The minimum accuracy value as an index of classification accuracy. *PERS*, **51**: 99-111.

Arrouays D. 1987. Cartographie des sols et comportements agronomiques. Comparaison de données de cartographie et d'enquêtes agronomiques en vue de la thématisation d'une carte des sols. *SdS*, **25**(1): 43-53.

Arrouays D. 1989. L'agriculteur et le pédologue cartographe, deux modes d'appréhension du milieu, une interface nécessaire. *SdS*, **27**(1): 101-104.

Arrouays D., 1995. *Analyse et Modélisation Spatiales de l'Évolution des Stocks de Carbone Organique des Sols à l'Échelle d'un Paysage Pédologique*. Thesis, ENSA de Montpellier. 155 p. + appendices.

Arrouays D., Thorette J., Daroussin J. and King D. 2001. Analyse de représentativité de différentes configurations d'un réseau de sites de surveillance des sols. *EGS*, **8**(1): 7-17.

August P., Michaud J., Labash C. and Smith C. 1994. GPS for environmental applications: Accuracy and precision of locational data. *PERS*, **60**: 41-45.

Aurousseau P. 1987. *Analyses de Terre et Banques de Données Régionales. L'exemple du Référentiel Agro-pédologique de l'Ille-et-Vilaine*. INRA, Rennes. 16 p.

Baize D. 1986. Couvertures pédologiques, cartographie et taxonomie. *SdS*, **3**(24): 227-243.

Baize D. 2000. *Guide des Analyses en Pédologie*. 2nd edition. Éditions INRA. 257 p.

Baize D. (Coordn). 1998. *A Sound Reference Base for Soils. The 'Référentiel Pédologique'*. Éditions INRA. 322 p.

Baize D. and Jabiol B. 1995. *Guide Pour la Description des Sols*. Éditions INRA. 375 p.

Baize D. and Terce M. 2002. *Les Éléments Traces Métalliques dans les Sols, Approches Fonctionnelles et Spatiales*. Éditions INRA. 565 p.

Banton O., Seguin M.K. and Cimon M.A. 1997. Mapping field-scale physical properties of soil with electrical resistivity. *SSSAJ*, **61**: 1010-1017.

Barrette J., August P. and Golet F. 2000. Accuracy assessment of wetland boundary delineation using aerial photography and digital orthophotography. *PERS*, **66**(4): 409-416.

Batjes N.H. 2000. *SOTER Summary File for Central and Eastern Europe (SOVEUR Project, version 1.0)*. FAO-ISRIC report. 12 p.

Baumgardner M.F. 1994. A World Soils and Terrain Digital Database: linkages. *Trans. 15th World Cong Soil Sci.* **6a**: 718-727.

Baveye Ph. 2002. Comment on "Modelling soil variation: past, present and future" by G.B.M. Heuvelink and R. Webster. *Geoderma*, **109**: 289-293.

Beckett P.H.T. and Burrough P.A. 1971. The relation between cost and utility in soil survey. IV: Comparison of the utilities of maps produced by different survey procedures and to different scales. *J. Soil. Sci.* **22**: 466-480.

Beckett P.H.T. and Webster R. 1971. Soil variability: a review. *Soils Fertil., Harpenden*, **34**: 1-15.

Benoit M. and Maire B. 1992. Création d'une carte d'expert. Images des zones agricoles de la Haute Marne. In (P. Buche, S. Lardon and D. King, eds). *Gestion de l'Espace Rural et SIG*. Séminaire INRA. pp. 267-274.

Bergeron J.F. Saucier J.P. Robitaille A.and robert D. 1992. Quebec Forest Ecological Classification Program. *The Forestry Chronicle*, **68**(1): 53-63.

Bertrand R. 1989. Les ressources en sols des régions soudano-sahéliennes. In *Soil, Crop and Water Management Systems for Rainfed Agriculture in the Sudano-Sahelian Zone*. Proceedings of an International Workshop. ICRISAT, Niamey. pp. 3-16.

Bertrand R. 1998a. *Du Sahel à la Forêt Tropicale. Clé de Lecture des Sols dans les Paysages Ouest-Africains*. Coll. Repères. CIRAD. 272 p.

Bertrand R. 1998b. Interpénétration des pédogenèses et des morphogenèses sur le socle granito-gneissique ouest-africain; interprétation paléoclimatique. Symposium 15, Paper 1408, *16th World Congress of Soil Science*, Montpellier.

Bertrand R. Falipou P. and Legros J.P. 1984. *Notice pour l'entrée des descriptions et analyses de sols en banque de données (STIPA)*. Doc. ACCT, Paris. 136 p.

Bierkens M.F.P and Weerts H.J.T. 1994. Application of indicator simulation to modelling the lithological properties of a complex confining layer. *Geoderma*, **62**: 265-284.

Bindzi-Tsala, 1994. Cameroon. In (J.A. Zinck, ed.) *Soil Survey: Perspectives and Strategies for the 21st Century*. ITC Publication No. 21: 68-69.

Bishop T.F.A. and McBratney A.B. 2001. A comparison of prediction methods for the creation of field-extent soil property maps. *Geoderma*, **103**: 149-160.

Bishop T.F.A., McBratney A.B. and Whelan B.M. 2001. Measuring the quality of digital soil maps using information criteria. *Geoderma*, **103:** 95-111.

Bissonnette J., Tingxian L., Ducruc J.P and Boucher F. 1995. Télédétection et modèles numériques d'altitude dans la cartographie écologique à petite échelle du Québec. In *Méthodes et Réalisations de l'Écologie du Paysage pour l'Aménagement du Territoire*. Polyscience Publications Inc., Canada. pp. 149-155.

Bock L., 1994. Analyses de sols et gestion de l'espace. *EGS,* **1**(1): 23-33.

Bojorquez-Tapia L.A., Diaz-Mondragon S. and Ezcurra E. 2001. GIS-based approach for participatory decision making and land suitability assessment. *Int. J. Geogr. Information Sci.* **15**(2): 129-151.

Bonfils P. 1989. Une évaluation du risque d'érosion dans les pays du sud de la communauté européenne (le programme CORINE). *SdS,* **27**(1):33-36.

Bonfils J. 1992. *Feuille de Lodève* Carte des sols de France au 1/100,000. (Map published, report under publication). INRA-SESCPF, Orléans.

Bonfils J. 1993. Carte des sols de France au 1/100,000 *Feuille de Lodève*. (Map published, report under publication). INRA-SESCPF, Orléans.

Bonnard L.F. 1989. Les cartes des sols en Suisse. *SdS,* **27**(1): 1-4.

Bonneric Ph. 1978. *Conception et Réalisation d'un Système Cartographique Appliqué à la Pédologie*. Mémoire d'Ingénieur du CNAM en Informatique. Montpellier. 125 p.

Bonneric P., Navarro R. and Falipou P. 1985. *Notice pour la Gestion Informatique de la Banque de Données (STIPA)*. ACCT, Paris. 83 p. + appendices.

Bonneton Ph. 1986. *Un Essai d'Évaluation de la Faisabilité du Maïs au Niveau d'une Région agricole.* Memoir Météorologie Nationale, Toulouse. 68 p.

Bornand M. 1972. *Etude Pédologique de la Moyenne Vallée du Rhône.* Notice explicative. INRA-SES, Montpellier. No. 395. 244 p. + maps.

Bornand M. 1978. *Altération des Minéraux Fluvio-glaciaires, Genèse et Évolution des Sols sur Terrasses Quaternaires dans la Moyenne Vallée du Rhône.* Doct. État. thesis, Univ. Montpellier. 329 p. + appendices.

Bornand M. and Guyon A. 1969. *Étude Pédologique dans la Vallée du Rhône. Région de Tain-Alban-St Sorlin-St Donat.* INRA-Montpellier Pub. 106. 203 p. + appendices + maps.

Bornand M. and Lehman C. 1997. *Connaissance et Suivi de la Qualité des Sols en France.* Report Ministry Agriculture, Ministry Environment and INRA. 176 p.

Bornand M. and Voltz M. 1986. Étude des conséquences d'un abaissement de nappe phréatique sur les transferts hydriques au sein d'une culture de maïs. *Soc. Hydrotech. Fr.* **IV**(7): 1-5.

Bornand M. Callot G. and Favrot J.C. 1968. La carte pédologique du Val d'Allier au 1/100,000. *Bull. AFES*, **6**: 21-29.

Bornand M., Legros J.P. and Moinereau J. 1977. *Carte Pédologique de France à 1/100,000, Feuille de PRIVAS, Notice Explicative.* SESCPF, Orléans. 256 p.

Bornand M., Legros J.P.and Rouzet C. 1994. Les banques régionales de données-sols. Exemple du Languedoc-Roussillon. *EGS*, **1**: 67-82.

Bornand M., Lagacherie Ph., and Robbez-Masson J.M. 1995. Cartographie des pédopaysages et gestion de l'espace. In *La Cartographie pour la Gestion des Espaces Naturels.* Proceedings International Colloquium, Saint-Étienne. pp. 209-220.

Bottraud J.C. 1983. *Résistivité Électrique et Étude des Sols. Application à la Cartographie et à la Caractérisation du Fonctionnement Hydrique.* Doct. Ing. thesis. Université de Montpellier USTL. 191 p.

Bouchardy J.Y. 1992. *Méthodologie pour la Spatialisation des Zones Sensibles à la Pollution Diffuse Agricole par le Phosphore, à l'Aide de la Télédétection et des SIG.* Thesis, Univ. Grenoble. 276 p.

Boulaine J. 1978. Les unités cartographiques en pédologie. Analyse de la notion de génon. *SdS*, **1**: 15-30.

Boulaine J. 1980. *Pédologie Appliquée.* Masson, 220 p.

Boulaine J. 1983. V.V. Dokoutchaiev et les débuts de la pédologie. *Rev. Hist. Sci.* **36**(3-4): 285-305.

Boulaine J. 1989. *Histoire des Pédologues et de la Science du Sol.* Éditions INRA. 285 p. + appendices.

Boulaine J. and Aubert G. 1989. Contribution de certains pédologues français à l'évolution des concepts pédologiques utilisés en cartographie. *SdS*, **27**(4): 395-411.

Boulaine J. and Girard M.C. 1970. Fiche de description de profils. *Bull. AFES*, **1**: 13-17.

Boulet R., Chauvel A., Humbel F. and Lucas Y. 1982. Analyse structurale et cartographie en pédologie. I, II, III. *Cah. ORSTOM, Série Pédologie*, **9**(4): 309-339.

Boutefoy I., Vinatier J.M. and Chafchafi A. 1999. *Schéma de Vocation des Territoires Agricoles et Forestiers, Département du Rhône*. Sol Info Rhône-Alpes. 79 p. + maps.

Bouteyre G. and Duclos G. 1994. *Carte Pédologique de France à 1/100,000, Feuille d' ARLES, Notice Explicative*. SESCPF, Orléans. 302 p.

Brabant P. 1989a. La cartographie des sols dans les régions tropicales: une procédure à 5 niveaux coordonnés. *SdS*, **27**(4): 369-395.

Brabant P. 1989b. La connaissance de l'organisation des sols dans le paysage, un préalable à la cartographie et à l'évaluation des terres. *1er Sémin. Franco-Africain de Pédologie Tropicale*. Éditions ORSTOM. pp. 65-85.

Brabant P. 1992. Pédologie et système d'information géographique. Comment introduire les cartes de sols et les autres données sur les sols dans les SIG? *Cah. ORSTOM, Série Pédologie*, **27**(2): 315-345.

Brannon G.R. and Hajek B.F. 2000. Update and recorrelation of soil surveys using GIS and statistical analysis. *SSSAJ*, **64**: 679-680.

Breiman L., Friedman J.H, Olsen R.A and Stone C.J. 1984. *Classification and Regression Trees*. Wadworth & Brooks, Wadworth Statistics/Probability Series. 358 p.

Breuning-Madsen H. and Jones R.J.A. 1998. Towards a European soil profile analytical database. In (H.J. Heineke, W. Eckelmann, A.J. Thomasson, R.J.A. Jones, L. Montanarella and B. Buckley, eds). *Land Information Systems. Developments for Planning the Sustainable Use of Land Resources*. EUR-17729-en. pp. 43-49.

Bridges E.M., Batjes N.H. and Nachtergaeles F.O. (Eds). 1998. *World Reference Base for Soil Resources. Atlas*. Acco CV, Louvain/Amersfoort, Belgium. 79 p.

Briggs D.J. and Martin D.M. 1988. CORINE. An Environmental Information System for the European Community. *Environment Review*, **2**(1): 29-34.

Brisson N., Mary B., Ripoche D., Jeuffroy M.H., Ruget F., Nicoullaud B., Gate P., Devienne F., Antonioletti R., Dürr C., Richard G., Beaudoin N., Recous S., Tayot X., Plenet D., Cellier P., Machet J.M., Meynard J.M. and Delecolle R. 1998. STICS, a generic model for the simulation of crops and their water and nitrogen balance. I. Theory and parametrization applied to wheat and corn. *Agronomie*, **18**: 163-175.

Bruckert S. 1987. Clé dichotomique de classement des sols agricoles Francs-Comtois situés en climat tempéré semi-continental. *Annales scientifiques de l'Université de Franche-Comté, Besançon*. 4ème Série, **7**: 17-29.

Bruckert S. 1989a. Déterminisme de la répartition des sols et de leur variabilité spatiale sur les plateaux calcaires (exemple du Jura). *C.R. Acad. Sci. Paris*, Vol. **309**, Série II: 1615-1622.

Bruckert S. 1989b. Désignation et classement des sols agricoles d'après des critères de situation et d'organisation: application aux terres franc-Comtoises. *Agronomie*, **9**: 353-361.

Brunet R., 1987. *La Carte, Mode d'Emploi*. Fayard/Reclus. 269 p.

Buchter B., Hinz C. and Fluhler H. 1994. Sample size for determination of coarse fragment content in a stony soil. *Geoderma*, **63**: 265-275.

Bui E.N. and Moran C.J. 2001. Disaggregation of polygons of surficial geology and soil using spatial modelling and legacy data. *Geoderma*, **103**: 79-94.

Burrough P.A. 1993. Soil variability: a late 20th century view. *Soils Fertil., Harpenden*, **56**(5): 529-562.

Burrough P.A. 1994. The technologic paradox in soil survey: New methods and techniques of data capture and handling. In *Soil Survey, Perspectives and Strategies for the 21st Century*. FAO-ISRIC publication 21. pp. 15-22.

Burrough P.A. and Franck A.U. 1995. Concepts and paradigms in spatial information: are current geographical information systems truly generic? *Int. J. Geogr. Inf. Systems*, **9**(2): 101-116.

Burrough P.A. and McDonnell R.A. 1998. *Principles of Geographical Information Systems*. Oxford University Press. 333 p.

Burrough P.A., van Gaans P.F.M. and Hootsmans R. 1997. Continuous classification in soil survey: spatial correlation, confusion and boundaries. *Geoderma*, **77**: 115-135.

C.P.C.S (compilation). 1967. *Classification Française des Sols*. Mimeographed document, Grignon. 96 p.

Caloz R. 1992. *Télédétection Satellitaire*. Course notes, Rural Engineering, École Polytechnique Fédérale, Lausanne. 136 p.

Canada, 1981. *A Soil Mapping System for Canada: Revised*. 53 pp. Available on Internet on the Cansis site.

Cazemier D. 1999. *Utilisation de l'Information Incertaine Dérivée d'un Base de Données Sols*. Thesis, ENSAM, Montpellier. 171 p. + appendices.

CEC, 1985. *Soil Map of the European Communities, 1:1 million*. CEC-DG-VI, Luxembourg. 124 p.

Chamayou H. and Legros J.P. 1989. *Les Bases Physiques, Chimiques et Minéralogiques de la Science du Sol*. Presses Universitaires de France. 595 p.

Chang H.K.C., Liu S.H. and Tso C.K. 1997. Two-dimensional template-based encoding for linear quadtree representation. *PEERS*, **63**(11): 1275-1282.

Chaplot V., Walter C., Curmi P. and Hollier-Larousse A. 2001. Mapping field-scale hydromorphic horizons using Radio-MT electrical resistivity. *Geoderma*, **102**: 61-74.

Chaplot V. and Walter C. 2002. The suitability of quantitative soil-landscape models for predicting soil properties at a regional level. Symposium 48, paper 2331, *17th World Soil Science Congress, Bangkok*. 13 p.

Cheng T., Molenaar M. and Lin H. 2001. Formalizing fuzzy objects from incertain classification results. *Int. J. Geogr. Inf. Sci.* **15**(1): 27-42.

Chery Ph. 1990. *Modélisation Spatiale de la Sensibilité au Ruissellement et à l'Érosion. Recherche sur la Combinaison de Données Cartographiques du Milieu (Région Nord-Pas-de-Calais)*. DEA Fédéral de Pédologie. 30 p. + maps + appendices.

Chery Ph., Le Bissonnais Y., King D. and Daroussin J. 1991. Définition et délimitation des Unités Spatiales de Fonctionnement (USF) du ruissellement et de l'érosion. In (P. Buche, S. Lardon and D. King, eds.). *Gestion de l'Espace Rural et SIG*. Séminaire INRA. pp. 134-147.

Chrétien J. 2000. *Régions Naturelles, Pédopaysages et Sols de la Côte d'Or*. INRA-Orléans. 194 p.

Chrisman N.R. 1982. A theory of cartographic error and its measurement in digital data bases. *Proc. AUTO-CARTO 5*, pp. 159-168.

Chrisman N.R. and Lester M.K. 1991. A diagnostic test for error in categorial maps. *Proc. AUTO-CARTO 10*, **6**: 330-348.

Collet C., 1986. GIS management and analysis. In: *Geographical Information Technology in the Field of Environment*. Swiss Federal Institute of Technology, EPF-Lausanne. pp. 270-369.

Collet C., 1992. *Systèmes d'Information Géographique en Mode Image*. Presses Polytechniques et Universitaires Romandes, Lausanne. 186 p.

Collet L.W. 1955. *Carte Géologique de la Suisse. Genève-Lausanne. Notice*. Kummerly & Frey, Berne.

Collins M.E. and Doolittle J.A. 1987. Using ground-penetrating radar to study soil microvariability. *SSSAJ*, **51**: 491-493.

Collins M.E, Doolittle J.A. and Rourke R.V. 1989. Mapping depth to bedrock on a glaciated landscape with ground-penetrating radar. *SSSAJ*, **53**: 1806-1812.

Congalton R.G. 1991. A review of assessing the accuracy of classifications of remotely-sensed data. *Remote Sens. Environ*. **37**: 35-46.

Congalton R.G. 1997. Exploring and evaluating the consequences of vector-to-raster and raster-to-vector conversion. *PEERS*, **63**(4):425-434.

Cook S.E., Corner R.J., Grealish G., Gessler P.E. and Chartres C.J. 1996. A rule-based system to map soil properties. *SSSAJ*, **60**: 1893-1900.

CORINE, 1992. *Corine Soil Erosion Risk and Important Land Resources in the Southern Regions of the European Community*. EUR 13233-EN, 97 p.

Coulson M.R.C. and Ellehoj E.A. 1993. Soil map legends as tools of cartographic communication. *Catena*, **20**: 257-271.

CPCS. 1967. *Classification des Sols*. Mimeographed document, Grignon. 96 p.

d'Or D. and Bogaert P. 2001. Fine scale soil texture. Estimation using soil maps and profile descriptions. In *Geostatistics for Environmental Applications*. Kluwer Academic Publishers, Netherlands. pp. 453-462.

Dabas M., Duval O., Bruand A. and Verbeque B. 1995. Cartographie électrique en continu: apport à la connaissance d'une couverture de sol développée sur matériaux deltaïques. *EGS*, **2**(4): 257-268.

Danneels V. and Schwartz C. 1993. *Étude Méthodologique sur l'Exploitation d'une Banque Régionale de Données d'Analyses de Terre. Application à la Région Nord-Pas-de-Calais.* DERF-ISA, Lille. 36 p. + appendices.

Davis T.J. and Keller C.P. 1997. Modelling uncertainty in natural resource analysis using fuzzy sets and Monte Carlo simulation: slope stability prediction. *Int. J.Geogr. Inf. Sci.* **11**(5):409-434.

Day J.M. and McMenamin J. 1983. *Manuel de Description des Sols sur le Terrain.* Revision 1982, IRT No. 82-52. Direction Générale de la Recherche Agriculture Canada, Ottawa, Ontario. 109 p. + appendices.

DECADE (compilation). 1984. *Cartographie et Développement.* Coordination: Y. Poncet and A. Lalau-Keraly. Ministère des Relations Extérieures, Coopération, Développement, Paris. 181 p.

Deckers J.A., Nachtergaele F.O. and Spaargaren O.C. (eds.). 1998. *World Reference Base for Soil Resources. Introduction.* ACCO, Louvain/Amersfoort, Belgium. 165 p.

Deffontaines J.P. and Lardon S. 1993. *Itinéraires Cartographiques et Développement.* Éditions INRA. 286 p.

Delcros Ph., 1993. *Écologie du Paysage et Dynamique Végétale en Zone de Montagne.* Doctoral thesis. Université Grenoble I. 263 p.

Delecolle R. 1989. Indices et modèles, outils du zonage agrométéorologique. In *Le Zonage Agro-pédo-climatique.* Séminaire INRA. pp. 17-38.

Delecour F. and Kindermans M. 1977. *Manuel de Description des Sols.* Service de la Science du Sol. Faculté des Sciences Agronomiques, Gembloux., 111 p.

Delecour F., Legros J.P. and Rousseaux G. 1978. *Étude de Faisabilité pour la Création de Banques de Données Pédologiques en Afrique.* ACCT, Paris. 43 p.

Delvaux-Galernaux N. 1987. *Saisie Informatique de Terrain pour la Description du Sol. Programme SONDO.* SESCPF-INRA 16 p.

Demolon A. 1949. *La Génétique des Sols.* Que sais-je ?, PUF. 133 p.

Dent D. and Young A. 1993. *Soil Survey and Land Evaluation.* George Allen & Unwin, London. 278 p.

Dicks S.E. and Thomas H.C.L. 1990. Evaluation of thematic map accuracy in a land-use and land-cover mapping program. *PERS*, **56**(9): 1247-1252.

Djili K. 2000. *Contribution à la Connaissance des Sols du Nord de l'Algérie. Création d'une Banque de Données Informatisées et Utilisation d'un Système d'Information Géographique pour la Spatialisation et la Valorisation des Données Pédologiques.* Thesis, Institut National Agronomique d'Alger.

Dobson J.E. 1993. Commentary: A conceptual framework for integrating remote sensing, GIS and geography. *PERS*, **59**(10): 1491-1496.

Dokoutchaiev V. 1900. *Zones Verticales des Sols, Zones Agricoles, Sols du Caucase.* Ministry of Finance, St. Petersburg. 56 p.

Domburg P., de Gruijter J.J. and van Beek P. 1997. Designing efficient soil survey schemes with a knowledge-based system using dynamic programming. *Geoderma*, **75**: 183-201.

Doolittle J.A. and Collins M.E. 1998. A comparison of EM induction and GPR methods in areas of karst. *Geoderma*, **85**: 83-102.

Douay F.1994. Approche morpho-pédo-paysagère de la Flandre Intérieure. Résumé. *4e Journées Nationales d'Études des Sols*. AFES, Lille. 47 p.

Duchaufour Ph. 1956. *Pédologie. Applications Forestières et Agricoles*. École National des Eaux et Forêts, Nancy. 310 p.

Duchaufour Ph. 1983. *Pédologie. Pédogenèse et Classification*. Masson. 491 p.

Duchaufour H. and Party J.P. 1986. *Étude Pédologique Préalable au Drainage. Secteur de Référence du Pays de Hanau*. INRA-ONIC-CEMAGREF. 240 p. + appendices + maps.

Duclos G. 1980. Appréciation de l'aptitude à la mise en valeur forestière agricole des zones accidentées de Provence. *Eau et Aménagement de la Région Provençale*, **24**: 1-15.

Duclos G. 1985. Les cartes départementales des terres agricoles. Proposition d'une méthode pour la région provençale. *SdS*, **3**: 139-150.

Ducruc J.P. 1991a. *La Carte Écologique: son Contenu et ses Utilisations*. Pub. 42, Ministère de l'Environnement, Quebec. 18 p.

Ducruc J.P. 1991b. *Planification Écologique. Le Cadre Écologique de Référence: Les Concepts et les Variables de la Classification et de la Cartographie Écologique*. Ministère de l'Environnement, Quebec. 40 p.

Dudal R., Bregt A.K. and Finke P.A. 1995. Feasibility of the creation of a Soil Map of Europe at a scale of 1/250,000. In: *European Land Information Systems for Agro-Environmental Monitoring*. EUR-16232-en. pp. 207-220.

Dumanski J. 1994. Strategies and opportunities for soil survey information and research. In *Soil Survey, Perspectives and Strategies for the 21st Century*. Pub. FAO-ITC 21, pp. 36-41.

Dunand-Divol F. 1988. *Interface de Vérification de la Qualité des Descriptions et Analyses de Sols (STIPA)*. Status report. INRA-ISIM, Montpellier. 78 p.

Dupis A., Vincent B., Gauthier F. and BOSSUET G. 1991. Application de la radio-magnéto-tellurique aux études préalables au drainage agricole. *SdS*, **29**, 231-243.

Dupuis J. 1960. Carte pédologique schématique au 1/million du nord-est de la France. *Annales de l'Institut National Agronomique*, **46**: 3-12.

Durand R. and Legros J.P. 1981. Cartographie automatique de l'énergie solaire en fonction du relief. *Annales Agronomiques*, **1**(1): 31-39.

Eastman J.R. 1993. *IDRISI. Update Manual*. Clark University, Worcester MA, USA. 210 p.

ECSS (Expert Committee on Soil Survey). 1987. *Soil Survey Handbook*. Vol. 1. G.M. Coen (ed.) Land Resource Research Centre Technical Bulletin 1987-9E, Research Branch, Agriculture Canada, Ottawa.

EEA (European Environment Agency). 2000. *Down to Earth: Soil Degradation and Sustainable Development in Europe*. UNEP, Environment Issue Series No. 16. 32 p.

Engels P., Bock L., Neven C. and Mathieu L. 1993. Exemples d'utilisation de la carte des sols de Belgique comme base d'un système d'information géographique pour la gestion des terroirs. *Bull. Rech. Agron. Gembloux*, **28**(2-3): 181-195.

Engels P., Bock L., Neven C., Laroche J. and Mathieu L. 1994. *La Carte des Sols de Belgique comme Base d'un Système Géographique de Référence et de Gestion de l'Espace à Différents Niveaux de Perception (Résumé)*. Quatrièmes Journées Nationales. AFES, Lille. 19 p.

ESWG (Ecological Stratification Working Group). 1995. *A National Ecological Framework for Canada. Report and National Map at 1/7,500,000 scale*. Agriculture and Agri-Food Canada, Research Branch, Centre for Land and Biological Resources Research and Environment Canada, State of the Environment Directorate, Ecozone Analysis Branch, Ottawa/Hull. 27 p.

EUR-17729-en. 1998. *Land Information Systems. Developments for Planning the Sustainable Use of Land Resources*. (H.J. Heineke, W. Eckelmann, A.J. Thomasson, R.J.A. Jones, L. Montanarella and B. Buckley, eds.). 546 p.

European Union. 2001. *Base de Données Géoréférencées des Sols pour l'Europe. Manuel de Procédures, Version 1.1*. EUR-18092-fr. 174 p.

Falipou P. and Legros J.P. 2002. Le système STIPA-2000 d'entrée et édition des données pour la base nationale de sols DONESOL.II. *EGS*, **9**: 55-70.

Falipou P., Legros J.P. and Bertrand R. 1981. Un algorithme général pour le calcul des textures à partir de l'analyse granulométrique. *Sols*, **5**: 29-40.

Falque M.C. 1994. *Evaluation des Paysages. Une Approche Méthodologique, une Typologie, un Outil Adapté*. Mastère en Systèmes d'Informations Localisées, ENGREF-ENSAM-INAPG. 60 p.

FAO. 1966. *Soil Map of Europe, Scale 1/2,500,000*. Explanatory text by R. Dudal, R. Tavernier and D. Osmond' European Commission on Agriculture. 120 p.

FAO. 1995. *Global and National Soil and Terrain Digital Databases (SOTER). Procedures Manual*. World Soil Resources Report 74, Rev-1. 122 p.

FAO-ISRIC-UNESCO. 1990. *Guidelines for Soil Profile Description*. 3rd Edition. 70 p.

FAO-UNESCO. 1988. *Revised Legend, Soil Map of the World*. World Soil Resources Report 60. 119 p.

Favrot J.C. 1981. Pour une approche raisonnée du drainage agricole en France. La méthode des Secteurs de référence. *C.R. Acad. Agric. Fr.* **67**: 716-729.

Favrot J.C. 1987. Études et recommandations préalables au drainage: la méthode des Secteurs de référence. Enseignements et prolongements de l'Opération drainage ONIC-Ministère de l'Agriculture (1980-1985). *C.R. Acad. Agric. Fr.* **73**(4): 23-32.

Favrot J.C. 1989. Une stratégie d'inventaire cartographique à grande échelle: la méthode des Secteurs de référence. *SdS*, **27**: 351-368.

Favrot J.C. 1990. *Cahier des Clauses Techniques Générales pour la Réalisation de l'Étude Pédologique d'un Secteur de Référence*. Vol. 5 of Programme Inventaire, Gestion et Conservation des Sols de France. Ministère de l'Agriculture, Paris.

Favrot J.C. 1994. *Note de Synthèse sur le Volet Pédologique et sur le Répertoire Informatisé REFERSOLS*. INRA-Montpellier. 34 p.

Favrot J.C. and Bouzigues R. 1975. *Études Préliminaires en Vue du Drainage des Terres Agricoles du Sud-Ouest du Département de l'Eure*. Cartes et Légende Générale. 9 vols. INRA, Ministère Agriculture. 1000 p.

Favrot J.C. and Lagacherie Ph. 1993. La cartographie automatisée des sols: une aide à la gestion écologique des paysages ruraux. *C.R. Acad. Agric. Fr.* **79**: 61-76.

Fisher P.F and Lindenberg R.E. 1989. On distinction among cartography, remote sensing and Geographic Information Systems. *PERS*, **55**: 1431-1434.

Foody G.M. 1992. On the compensation for chance agreement in image classification accuracy assessment. *PERS*, **58**:1459-1460.

Forman R.T.T. and Godron M. 1986. *Landscape Ecology*. John Wiley & sons, New-York, 640 p.

FRIDLAND V.M. 1976. *Pattern of the Soil Cover* (translated from Russian). Israel Program for Scientific Translations, Jerusalem. 291 p.

Gaddas F. 2001. *Proposition d'une Méthode de Cartographie des Pédopaysages. Application à la moyenne Vallée du Rhône*. Thesis. Institut National Agronomique, Paris-Grignon.

Gao J. 2001. No-differential GPS as alternative source of planimetric control for rectifying satellite imagery. *PERS*, **67**: 49-55.

Gardiner M.J., 1989. *Compilation of Soil Analytical Data for the EC. Computerization of Land Use Data*. Pisa 1987. EUR 11151. pp. 61-65.

Gascuel-Odoux Ch., Grimaldi M. and Veillon L. 1991. Apport de la géostatistique à l'analyse morphologique du sol. *SdS*, **29**: 189-209.

Gascuel-Odoux Ch., Walter C. and Voltz M. 1993. Intérêt du couplage des méthodes statistiques et de cartographie des sols pour l'estimation spatiale. *SdS*, **31**: 193-213.

Gaucher G. 1977. Milieu naturel, pédogenèse et prospection du terrain en pédologie agricole. *L'Agronomie Tropicale* **32**(1): 11-30.

Gaultier J.P. 1990. *Projet DONESOL. Étude Détaillée*. 4 vols. INRA, Versailles. 200 p.

Gaultier J.P., Daroussin J. and Yart G. 1992a. L'interface ARC/INFO-ORACLE pour la gestion des données pédologiques. In (P. Buche, S. Lardon and D. King, eds.). *Gestion de l'Espace Rural et SIG*. Séminaire INRA. pp. 321-328.

Gaultier J.P., Legros J.P., Bornand M., King D., Favrot J.C. and Hardy R. 1992b. L'organisation et la gestion des données pédologiques spatialisées: le projet DONESOL. *Revue de Géomatique*, **3**(3): 235-253.

Gegout J.C. 1995. *Étude des Relations entre les Ressources Minérales du Sol et la Végétation Forestière dans les Vosges*. Thesis, Univ. Nancy I. Vol. 1, 214 pp.; Vol. 2 (appendices), 115 p.

Geoderma. 1994. Special issue, volume **62**: PEDOMETRICS 92. *Developments in Spatial Statistics for Soil Science*. Edited by J.J. de Gruijter, R. Webster and D.E. Myers. Elsevier. 326 p.

Girard C.M. and Girard M.C. 1975. *Application de la Télédétection à l'Étude de la Biosphère*. Masson. 175 p.

Girard C.M. and Girard M.C. 1999. *Traitement des Données de Télédétection*. Coll. Technique et Ingénierie, Dunod. 560 p.

Girard M.C. 1983. *Recherche d'une Modélisation en Vue d'une Représentation Spatiale de la Couverture Pédologique*. Thèse d'État, Université de Paris 7. 295 p. + appendices.

Girard M. 1989. La cartographie en horizons, *SdS*, **27**: 41-44.

Girard M.C. 1995. Apports de l'interprétation visuelle des images satellitaires pour l'analyse spatiale des sols. *EGS*, **2**: 7-24.

Girard M.C. and King D. 1988. Un algorithme interactif pour la classification des horizons de la couverture pédologique: DIMITRI. *SdS*, **26**: 81-102.

Girard M.C., Mougenot B. and ranaivoson A. 1991. Présentation d'un modèle d'organisation et d'analyse de la structure des informations spatialisées: OASIS. In *Caractérisation et Suivi des Milieux Terrestres en Région Arides et Tropicales*. ORSTOM. pp. 341-350.

Girard M.C, Yongchalermchai C. and Girard C.M. 1992. Analyse d'un espace par la prise en compte du voisinage. In (P. Buche, S. Lardon and D. King, eds.). *Gestion de l'Espace Rural et SIG*. Séminaire INRA. pp. 349-359.

Gomez A., Juste C., Desenfants C., Brunt T. and Lopez C. 1986. *Rapport de Synthèse sur l'Échantillonnage des Sols en Vue du Suivi de l'Évolution des teneurs en Métaux Lourds*. Ministère de l'Environnement. 25 p.

Goodchild M.F. and Gopal S. (eds.). 1989. *Accuracy of Spatial Databases*. Taylor & Francis. 290 p.

Goodchild M.F., Guoquing S. and Shiren Y. 1992. Development and Test of an Error Model for Categorial Data. *Int. J. Geogr. Inf. Systems*, **6**: 87-104.

Goovaerts P. 1999. Geostatistics in soil science: state-of-the-art and perspectives. *Geoderma*, **89**: 1-45.

Gras R. 1994. *Sols Caillouteux et Production Végétale*. Éditions INRA. 175 p.

Gras F., Hesse A., Tillier C., Tessier D. and Zimmer D. 1997. La prospection électrique : une méthode adaptée à la cartographie et à la connaissance de l'état hydrique des sols. Cas des sols de Lorraine. *EGS*, **4**: 161-174.

Gratier M. 1995. *Système de Cartographie des Sols Employé dans le Canton de Vaud*. Doc Service de l'Aménagement du Territoire, Canton de Vaud. 6 p. + appendices.

Gratier M. and Kissling A. 1994. *Informatisation de la Cartographie Pédologique*. Doc Service de l'Aménagement du Territoire, Canton de Vaud, 6 p.

de Gruijter J.J and ter BRAAK C.J.F. 1990. Model-free estimation from spatial samples: a reappraisal of classical sampling theory. *Mathematical Geology*, **22**: 407-415.

de Gruijter J.J., Walvoort D.J.J. and van Gaans P.F.M. 1997. Continuous soil maps—a fuzzy set approach to bridge the gap between aggregation levels of process and distribution models. *Geoderma*, **77**: 169-195.

Grzebyk M. and Dubrucq D. 1994. Quantitative analysis of distribution of soil types: existence of an evolutionary sequence in Amazonia. *Geoderma*, **62**: 285-298.

Guyot Ph. and Bornand M. 1987. Cartes départementales des terres agricoles; intégration des données sols et des données économiques. *SdS*, **25**: 1-16.

GWDSM. 2004. *Proc. Global Wkshp Digital Soil Mapping.* Montpellier (France), 14-17 September 2004. Workshop Agro, AFES, INRA, IUSS. Compact disc (Communications + summaries). Order from lagacherie@ensam.inra.fr

Hadzilacos T. 1996. On layer-based systems for undetermined boundaries. (P.A. Burrough and A.U. Frank, eds.). In *Geographic objects with Indeterminate Boundaries*. GISDATA 2. Taylor & Francis. pp. 237-255.

Harding J.S. and Winterbourn M.J. 1997. An Ecoregion Classification of the South Island, New Zealand. *J. Environ. Management*, **51**: 275-287.

Hartung S.L., Scheinost S.A and Ahrens R.J. 1991. Scientific Methodology of the National Cooperative Soil Survey. In *Spatial Variabilities of Soils and Landforms*. SSSA Special Publication No. 28. Madison WI, USA. pp. 39-48.

Helms D., Effland A.B.W. and Durana P.J. (eds.). 2002. *Profiles in the History of the U.S. Soil Survey*. Iowa State University Press. 331 p.

Henric J.F. 1995. *Impact Agricole d'un Changement Climatique Simulé en Languedoc. Calibration, Validation et Exploitation Spatiale du Modèle de Bilan Hydrique ACCESS3D*. DAA-ENSA, Montpellier. 56 p. + appendices.

Heuvelink G.B.M. and Webster R. 2001. Modelling soil variation: past, present, and future. *Geoderma*, **100**: 269-301.

Heuvelink G.B.M. and Webster R. 2002. Reply to comment on 'Modelling soil variation: past, present and future' by Philippe Baveye. *Geoderma*, **109**: 295-297.

Hewitt A.E, 1993. Predictive modelling in soil survey. *Soils Fertil., Harpenden*, **56**: 305-314.

Hodgson J.M. (ed.). 1976. *Soil Survey Field Handbook: Describing and Sampling Soil Profiles*. Soil Survey Technical Monograph No. 5, Harpenden, England. 99 p.

Hodgson J.M. 1978. *Soil Sampling and Soil Description*. Clarendon Press, Oxford. 241 p.

Hord R.M. and Brooner W. 1976. Land-use map accuracy criteria. *PERS*, **42**: 671-677.

Hubrechts L., Vander Poorten K., Vanclooster M. and Deckers J. 1998. From soil survey to quantitative land evaluation in Belgium. In (H.J. Heineke, W. Eckelmann, A.J. Thomasson, R.J.A. Jones, L. Montanarella and B. Buckley, eds.). *Land Information Systems. Developments for Planning the Sustainable Use of Land Resources*. EUR-17729-en. pp. 91-100.

Hudson W.D. 1987. Correct formulation of the Kappa coefficient of agreement. *PERS*, **53**: 421-422.

Hudson B.H. 1992. The soil survey as paradigm-based science. *SSSAJ*, **56**: 836-841.

IGCS. 1993. *Cahier des Charges de Déontologie*. Programme inventaire, gestion et conservation des sols. Ministère de l'Agriculture—INRA. 11 p.

Ibáñez J.J. and de Alba S., 2000. Pedodiversity and scaling laws: sharing Martín and Rey's opinion on the role of the Shannon index as a measure of diversity. *Geoderma*, **98**: 5-9.

Inman D.J., Freeland R.S., Ammons J.T. and Yoder R.E. 2002. Soil investigations using electromagnetic induction and ground-penetrating radar in south Tennessee. *SSSAJ*, **66**: 206-211.

ISSS-ISRIC-FAO. 1998. *World Reference Base for Soil Resources*. Word Soil Resources Report No. 84. 88 p.

Jacquet-LaGrèze E. and Siskos Y. 1983. *Méthode de Décision Multicritère*. Hommes et Techniques, Boulogne-Billancourt. 167 p.

Jamagne M. 1967. *Bases et Techniques d'une Cartographie des Sols*. Ann. Agron. Hors Serie, **18**. Éditions INRA. 142 p. + maps.

Jamagne M., 1993. Évolution dans les conceptions de la cartographie des sols. *Pedologie, Ghent*, **43**: 59-115.

Jamagne M., Bornand M. and Hardy R. 1989. La cartographie des sols en France à moyenne échelle. Programmes en cours et évolution des démarches. *SdS*, **27**: 301-318.

Jamagne M., King D., Daroussin J. and Le Bas C. 1993. Évolution et état actuel des programmes européens de connaissance et de gestion des sols. *Bull. Rech. Agron. Gembloux*, **28**: 135-163.

Jamagne M., King D., Le Bas C., Daroussin J., Burrill A. and Vossen P. 1994. Creation of an European Soil Geographic Database. *Trans. 15th World Soil Sci. Cong*. (Acapulco), **6a**: 728-742.

Jamagne M., Hardy R., King D. and Bornand M. 1995a. La base de données géographiques des sols de France. *EGS*, **2**: 153-172.

Jamagne M., Le Bas C., Berland M. and Eckelmann W. 1995b. Extension of the E.U. database for the soils of Central and Eastern Europe. In *European Land Information Systems for Agro-Environmental Monitoring*. EUR-16232-en. pp. 85-99.

Jankowski P. 1995. Integrating geographical information systems and multiple criteria decision-making methods. *Int. J. Geogr. Inf. Systems*, **9**: 251-273.

Jankowski P., Andrienko N. and Andrienko G. 2001. Map-centred exploratory approach to multiple criteria spatial decision making. *Int. J. Geogr. Inf. Sci.* **15**: 101-127.

Janssen L.L.F and van der Wel F.J.M. 1994. Accuracy assessment of satellite derived land-cover data: a review. *PERS*, **60**: 419-425.

Jenny H. 1941. *Factors of Soil Formation. A System of Quantitative Pedology.* McGraw-Hill, New-York.

Joerin F. 1995. Méthodes multicritères d'aide à la décision et SIG pour la recherche d'un site. *Revue Int. Géomatique*, **5**: 37-51.

Joerin F., Thériault M. and Musy A. 2001. Using GIS and outranking multicriteria analysis for land-use suitability assessment. *Int. J. Geogr. Inf. Sci.* **15**: 153-174.

Jones C.A. and Kiniry J.R. 1986. *CERES-Maize: A Simulation Model of Maize Growth and Development.* Texas A & M University Press, College Station TX, USA.

Jones R.J.A., Buckley B. and Jarvis M.G. 1998. European Soil Database: information access and data distribution procedures. In (H.J. Heineke, W. Eckelmann, A.J. Thomasson, R.J.A. Jones, L. Montanarella and B. Buckley, eds.). *Land Information Systems. Developments for Planning the Sustainable Use of Land Resources.* EUR-17729-en. pp. 19-31.

Journel A.G. and Huijbregts Ch. J. 1989. *Mining Geostatistics.* Academic Press, 600 p.

Jung D. 1994. *Spatialisation de Données Météorologiques et Propagation de l'Erreur d'Estimation au Travers d'un Modèle de Bilan Hydrique Régional.* DAA-ENSAM, Montpellier. 52 p. + appendices.

Kaufmann A. 1990. *Test Tactile Pour la Classification des Principaux Types de Sols.* French adaptation and translation by J.A. Neyroux and P.F. Lavanchy. Mimeographed document. 11 p.

Kilian J. 1974. Étude du Milieu Physique en Vue de son Aménagement. *L'Agronomie Tropicale,* **29**: 141-153.

Kindermans M. 1976. *Contribution à l'Élaboration d'une Banque de Données Pédologiques.* Travail de fin d'Études, Faculté des Sciences Agronomiques, Gembloux. 134 p.

King D. 1986. *Modélisation Cartographique du Comportement des Sols Basée sur l'Étude de la Mise en Valeur du Marais de Rochefort.* Thèse de Docteur-Ingénieur, INA, Paris-Grignon. 243 p.

King D. and Duval O. 1988. *LOGOS. Logiciels pour l'Étude de la Géographie des Sols (Version 3.1).* INRA-SESCPF. 112 p.

King D. and Girard M.C 1988. Réflexions sur la classification des profils de la couverture pédologique. Proposition d'un algorithme: VLADIMIR. *SdS*, **26**: 239-254.

King D. and Montanarella L. 2002. Inventaire et surveillance des sols en Europe. *EGS*, **9**: 137-148.

King D. and Saby N. 2001. Analyse de la représentativité des cartes pédologiques au 1/100,000 pour la connaissance des sols du territoire français. *EGS*, **8**: 247-267.

King D., Le Bissonnais Y., Hardy R., Eimberck M. and King C. 1991. Combinaison spatiale d'informations pour l'évaluation des risques de ruissellement à l'échelle régionale. In (P. Buche, S. Lardon and D. King, eds.). *Gestion de l'Espace Rural et SIG*. Séminaire INRA. pp. 149-166.

King D., Jamagne M., Chrétien J. and Hardy R. 1994. Soil-space oganization model and soil functioning units in Geographical Information Systems. *Trans. 15th World Soil Sci. Cong.* (Acapulco), **6a**: 743-757.

King D., Burrill A., Daroussin J., Le Bas C., Tavernier R. and van Ranst E. 1995. The E.U. Soil Geographic Database. In *European Land Information Systems for Agro-Environmental Monitoring*. EUR-16232-en. pp. 43-59.

King D., Jamagne M., Arrouays D., Bornand M., Favrot J.C., Hardy R., Le Bas C. and Stengel P. 1999. Inventaere cartographique et surveillance des sols on France *EGS*, **6**: 215-228.

Koukoulas S. and Blackburn G.A. 2001. Introducing new indices for accuracy evaluation of classified images representing semi-natural woodland environments. *PERS*, **67**: 400-510.

Krebs P. 1995. Dommage que les travaux soient déjà terminés. *VIA, le Magazine du Rail, SBB-CFF-FFS*, **2/95**: 6-9.

Laaribi A. 2000. *SIG et Analyse Multicritère*. Éditions Hermès. 190 p.

Laflamme G. and Goyette N., 1994. Banque d'informations référentielles sur les sols québécois: étapes de réalisation. *Agrosol*. Québec, **7**(2): 8-11.

Laflamme G. and Rompre M. 1993. Programme de recherches et développement en classification et cartographie des sols. *Agrosol*, Québec, **6**(1): 14-19.

Laflamme G., Rompre M., Carrier D. and Ouellet L. 1989. *Étude Pédologique du Comté de Mégantic*. Service de Recherche en Sols, Québec. 89-0050. 160 p. + maps.

Lagacherie Ph. 1984. *Étude Pédologique Préalable au Drainage; Secteur de Référence de la Bresse Jurassienne*. Doc. INRA-ONIC-Ministère de l'Agriculture-SES. **563**. 148 p. + appendices + map.

Lagacherie Ph. 1992. *Formalisation des Lois de Distribution des Sols pour Automatiser la Cartographie Pédologique à Partir d'un Secteur Pris Comme Référence. Cas de la Petite Région Naturelle 'Moyenne Vallée de l'Hérault'*. Thesis, Université de Montpellier II. 175 p. + appendices.

Lagacherie Ph. and Holmes S. 1997. Addressing geographical data errors in a classification tree for soil unit prediction. *Int. J. Geogr. Inf. Sci.* **11**: 183-198.

Lagacherie Ph. and Voltz M. 2000. Predicting soil properties over a region using sample information from a mapped reference area and digital elevation data: a conditional probability approach. *Geoderma*, **97**: 187-208.

Lagacherie Ph., Legros J.P. and Burrough P.A. 1995. A soil survey procedure using the knowledge of soil pattern established on a previously mapped reference area. *Geoderma*, **65**: 283-301.

Lagacherie Ph., Andrieux P. and Bouzigues R. 1996. Fuzziness and uncertainty of soil boundaries: from field reality to coding in GIS. In (P.A. Burrough and A.U. Frank, eds.). *Geographic Objects with Indeterminate Boundaries*. GISDATA 2. Taylor & Francis. pp. 275-286.

Lagacherie Ph., Cazemier D.R., van Gaans P.F.M. and Burrough P.A. 1997. Fuzzy k-means clustering of fields in an elementary catchment and extrapolation to a larger area. *Geoderma*, **77**: 197-216.

Lagacherie Ph., Robbez-Masson J.M., Nguyen-The N. and Barthes J.P. 2001. Mapping of reference area representativity using a mathematical soilscape distance. *Geoderma*, **101**: 105-118.

Lahmar R., Aurousseau P. and Bresson L.M. 1989.Analyse de contenu d'une carte pédologique en horizons: les formules du sol. *SdS*, **17**: 45-48.

Lammers D.A and Johnson M.G. 1991. Soil mapping concepts for environmental studies. In *Spatial Variabilities of Soils and Landforms*. SSSA Special Publication **28**: 149-160.

Lark R.M. 2000. A geostatistical extension of the sectioning procedure for disaggregating soil information to the scale of functional models of soil processes. *Geoderma*, **95**: 89-112.

Lark R.M. and Bolam H.C. 1997. Uncertainty in prediction and interpretation of spatially variable data on soils. *Geoderma*, **77**: 263-282.

Ledreux Ch. 1992. *Le Projet SAPRISTI, Système d'Aide à la Prédiction Intelligente des Sols par Traitement Informatique*. Mémoire d'Ingénieur du CNAM, Montpellier., 93 p. + appendices.

Leeds-Harrison P.B. and Rounsevell M.D.A. 1993. The impact of dry years on crop water requirements in Eastern England. *J. Inst. Wat. Environ. Management*, **7**: 497-505.

Leenhardt D. 1991. *Spatialisation du Bilan Hydrique. Propagation des Erreurs d'Estimation des Caractéristiques du sol au Travers des Modèles de Bilan Hydrique*. Doctoral thesis, ENSA-Montpellier. 129 p. + appendices.

Legros J.P. 1973. Précision des cartes pédologiques, la notion de finesse de caractérisation. *SdS*, **2**: 115-128.

Legros J.P. 1978. Recherche et contrôle numérique de la précision en cartographie pédologique: (1) Précision dans la délimitation des sols. *Ann. Agron.* **29**: 499-519; (2) Précision dans la caractérisation des unités de sols. *Ann. Agron.* **29**: 583-601.

Legros J.P. 1986. Cartographie des paysages pédologiques dans les Alpes humides. Agrométéorologie des régions de moyenne montagne. *Colloque INRA,* **39**: 119-127.

Legros J.P. 1990. Francophonie et banques de données de sols. *INRA Mensuel*, **49**: 4-5.

Legros J.P. and Argeles J. 1973. *Enquête sur la Précision des Descriptions de Profils*. Mimeographed document. SES No. 211, INRA, Montpellier. 15 p.

Legros J.P. and Bonneric Ph. 1979. Modélisation informatique de la répartition des sols dans le Parc Naturel Régional du Pilat. *Annales de l'Université de Savoie, Sci. Nat.* **4**: 63-68.

Legros J.P. and Bornand M. 1985. *Étude Pédologique de la France à 1/100,000. Feuille de St-Étienne*. SESCPF-INRA Orléans.

Legros J.P. and Bornand M. 1989. Systèmes d'information géographique et zonage agro-pédo-climatique. In *Séminaire INRA: Le zonage agro-pédo-climatique*. Commission d'agrométéorologie de l'INRA. pp. 101-115.

Legros J.P. and Bornand M., 1992. Cartographie numérique des sols. Premier bilan. In *Journées d'Études: SIG et Gestion des sols*. IATE-EPFL et Société Suisse de Pédologie, Lausanne. pp. 47-57.

Legros J.P. and Nortcliff S. 1990. Conception d'un vocabulaire pour la description du milieu naturel et des sols. *Pedologie, Ghent*, **40**(2): 195-213.

Legros J.P., Antonioletti R. and Genre-Grandpierre G. 1986. Evaluation de l'énergie solaire en zone de montagne. Agrométéorologie des régions de moyenne montagne. *Coll. INRA, Paris*, **39**: 87-93.

Legros J.P., Dorioz J.M. and Party J.P. 1987. Répartition des milieux calcaires, calciques et acidifiés en haute montagne calcaire humide. Conséquences agronomiques et écologiques. *Doc. Carto. Éco. Grenoble*, **30**: 137-157.

Legros J.P., Falipou P. and Dunand-Divol F. 1992. Vérification de la qualité de l'information dans les bases de données de sols. *SdS,.* **30**: 117-131.

Legros J.P., Baldy C., Fromin N. and Bellivier D. 1994. Crop models: principle and adaptation to the problem of climate change. In: *Soil Responses to Climate Change*. NATO ASI Series: Global Environmental Change No. 23. Springer-Verlag. pp. 72-98.

Legros J.P., Leenhardt D., Voltz M. and Cabelguenne M. 1996a. Scientific basis of Euro-ACCESS-II. In: *Euro-ACCESS*, publication of DG-XII of the European Community. pp. 19-34.

Legros J.P., Kölbl O. and Falipou P., 1996b. Délimitation d'unités de paysage sur des photographies aériennes. Eléments de réflexion pour la définition d'une méthode de tracé. *EGS*, **3**: 113-124.

Legros J.P., Emery C. and Falipou P. 1997. Exploitation simultanée de plusieurs missions photographiques en cartographie des sols. Analyse, modélisation et application. *EGS*, **4**: 265-277.

Legros J.P., Bornand M., Jung D. and Mayr T. 1999. Map overlay of soil and climatic data, the Languedoc-Roussillon example. *6th International Meeting, Soils with Mediterranean Type of Climate,* Barcelona, 4-9 July. Extended abstract. pp. 448-451.

Legros J.P., Martin S., Baize D., Riviere J.M. and Lepretre A. 2001. Accumulation de cuivre et de zinc dans les sols recevant du lisier de porc. Étude d'un site de l'observatoire de la qualité des sols en Bretagne. In: (D. Baize and M. Tercé, coordinators). *Eléments Traces Métalliques dans les Sols, Approches Fondamentales et Spatiales*. Éditions INRA. 570 p.

Leleux A., Aurousseau P. and Roudaut A. 1988. Synthèse cartographique régionale à partir de données d'analyses de terre. *SdS*, **26**: 29-40.

Leparoux P. 1988. *Recherche des Relations entre les Contraintes Agronomiques et les Caractéristiques Pédologiques du Sol*. Doc. ENSA-Rennes, Science du Sol, SDS 399. 68 p. + appendices.

LIAT (compilation). 1985. *Logiciel d'Interprétation Automatique des Analyses de Terre. Cahier des Clauses Techniques Particulières.* Ministère de l'Agriculture-DERF-INRA. 2 vols.

Lo C.P. and Watson L.J. 1998. The influence of geographic sampling methods on vegetation map accuracy evaluation in a swampy environment. *PEERS*, **64**: 1189-1200.

Louchard X. 1994. *Étude des Stratégies de Prédiction des Propriétés du sol à Partir de la Connaissance d'un Secteur de Référence.* DAA-ENSA, Montpellier. 69 p.

Loveland P.J., Legros J.P., Rounsevell M.D.A., de la Rosa D. and Armstrong A. 1994. A spatially distributed soil agroclimatic and soil hydrological model to predict the effects of climate change within the European Community. *Trans. 15th World Soil Sci. Cong.* (Acapulco), **6a**: 83-100.

Lowell K.E. 1994. Probabilistic temporal GIS modelling involving more than two map classes. *Int. J. Geogr. Inf. Systems*, **8**: 13-41.

Lucot E. and Gaiffe M. 1994. *Cartographie des Massifs Forestiers Témoins sur Substrats Calcaires du Nord-Est de la France.* Laboratoire de Pédologie, Université de Franche-Comté. 12 p. + appendices.

Lucot E. and Gaiffe M. 1995. Méthode pratique de description des sols forestiers caillouteux sur substrat calcaire. *EGS*, **2**: 91-104.

Mabbutt J.A. 1968. Review of concepts of land classification. In (G.A. Stewart, ed.). *Land Evaluation.* Macmillan of Australia. pp. 11-28.

Maclean A., D'Avello T.P. and Shetron S.G. 1993. The use of variability diagrams to improve the interpretation of digital soil maps in a GIS. *PERS*, **59**: 223-228.

Madsen H.B., 1989. Elaboration of a soil profile and analytical database connected to the EC-soil map and climatic data. *Proceedings of a Workshop in the Community Programme for Coordination of Agricultural Research*, Wageningen. EUR-12039-en. pp 119-132.

Madsen H.B. and Jones R.J.A. 1995. The establishment of a soil profile analytical database for the European Union. In *European Land Information Systems for Agro-Environmental Monitoring.* EUR-16232-en. pp. 61-69.

Maire R. 1990. *La haute montagne calcaire.* Chapitre 1: les hauts karsts de Platé et du Haut-Giffre. *Karstologia Memoires*, **3**: 19-65.

Malczewski J. 1996. A GIS-based approach to multiple criteria group decision-making. *Int. J. Geogr. Inf. Sci.* **10**: 955-971.

Maling D.H. 1989. *Measurements from Maps. Principles and Methods of Cartometry.* Pergamon Press. 577 p.

Margulis H. 1954. *Aux Sources de la Pédologie.* Publications ENSAT, Toulouse. 85 p.

Martín M.A. and Rey J.M. 2000. On the role of Shannon's entropy as a measure of heterogeneity. *Geoderma*, **98**: 1-3.

Martin S., Baize D., Bonneau M., Chaussod R., Cieselski H., Gaultier P., Lavelle P., Legros J.P., Lepretre A. and Steckeman T. 1999. Le suivi de la qualité des sols en France: la contribution de l'Observatoire de la qualité des sols. *EGS*, **6**: 215-230.

Matheron G. 1970. *La Théorie des Variables Régionalisées et ses Applications*. Les cahiers du Centre de Morphologie Mathématique de Fontainebleau. Vol. 5. 212 p.

Mathieu C., Gay M. and Sanchez E. 1993. Évaluation de la fertilité des sols en Midi-Pyrénées: une approche cartographique pour mieux valoriser les analyses de terre. *Bull. Rech. Agron. Gembloux*, **28**: 359-375.

Maystre L.Y and Bollinger D. 1999. *Aide à la Négociation Multicritère. Pratique et Conseils*. Presses Polytechniques et Universitaires Romandes. 192 pp.

Maystre L.Y., Pictet J. and Simos J. 1994. *Méthodes Multicritères ELECTRE*. Presses Polytechniques et Universitaires Romandes. 323 p.

McBratney A.B., de Guijter J.J and Brus D.J. 1992. Spatial prediction and mapping of continuous soil classes. *Geoderma*, **54**: 39-64.

McBratney A.B., Minasny B., Cattle S.R. and Vervoort R.W. 2002. From pedotransfer functions to soil inference systems. *Geoderma*, **109**: 41-73.

McCracken R.J. 1990. North Carolina's roots in soil survey, pedology and soil conservation. *Newsletter ISSS*. Working Group on History, Philosophy and Sociology of Soil Science. pp. 5-6.

McHaffie P. 2000. Surfaces: tacit knowledge, formal language, and metaphor at the Harvard Lab for Computer Graphics and Spatial Analysis. *Int. J. Geogr. Inf. Sci.* **14**: 755-773.

McKeague J.A. 1995. Soil survey and genesis and classification research in Canada. *Can. J. Soil Sci.* **75**: 3-9.

McKeague J.A. and Stobbe P.C. 1978. *History of Soil Survey in Canada 1914-1975*. Canada Department of Agriculture, Historical Series No. 11. 30 p.

McKenzie N.J. and Ryan P.J. 1999. Spatial prediction of soil properties using environmental correlation. *Geoderma*, **89**: 67-94.

McRae S.G. 1988. *Practical Pedology*. John Wiley & Sons, New York. 253 pp.

McSweeney K., Gessler P.E., Slater B.K., Hammer R.D., Bell J.C. and Petersen G.W. 1994. Towards a new framework for modelling the soil-landscape continuum. In *Factors of Soil Formation—a Fiftieth Anniversary Retrospective*. SSSA Special Publication No. **33**: 127-145.

Metral R. 1997. Conception et organisation d'une base de données des petites régions naturelles françaises. DAA-ENSA, Montpellier. 73 p. + appendices.

Meunier F. and Hardy B. 1986. La cartographie thématique à l'ORSTOM. *ORSTOM Actualités*, **13**: I-IV.

Meyer M. 1984. *Application de la Sismique Réfraction à la Prospection Pédologique des Sols et des Formations Superficielles*. Docteur-Ingénieur thesis. ENSA-Montpellier. 166 p.

Meyer-Roux J. 1987. *The Ten Years Research and Development Plan for the Application of Remote Sensing in Agriculture Statistics*. CEC-DGVI, JRC-Ispra. 23 p.

Meyer-Roux J. and Montanarella L. 1998. The European Soil Bureau. In (H.J. Heineke, W. Eckelmann, A.J. Thomasson, R.J.A. Jones, L. Montanarella and B. Buckley, eds.). *Land Information Systems. Developments for Planning the Sustainable Use of Land Resources*. EUR-17729-en. pp. 3-10.

Meynen E. (ed.) 1973. *Multilingual Dictionary of Technical Terms in Cartography*. International Cartographic Association, Commision II. Franz Steiner Verlag, Wiesbaden. 573 p.

Morton L.S. and Evans C.V. 1996. Soil radioactivity and the soil survey: field data collection for series interpretation. *SSSAJ*, **60**: 531-536.

Musy A. 1991. GIS applications in water management. In *Geographical Information Technology in the Field of Environment*. Swiss Federal Institute of Technology (EPF-Lausanne). pp. 367-398.

Neven C. and Engels P. 1993. *Développement d'un Outil pour la Gestion Intégrée des Exploitations et des Terroirs du Sud-Est de la Belgique*. G.R.E.O.A. (non-profit-making organization) Programme de Développement Intégré (PDI). Projet No. 8924BL0290. Project report. 74 p. + appendices.

Nicoullaud B. 1988. Un exemple d'aide à la décision en culture légumière de plein champs: choix des parcelles et des calendriers culturaux pour la petite carotte. *BTI*, 426-427.

Nolin M.C. and Lamontagne L. 1990. *Étude Pédologique du Comté de Richelieu (Québec). Vol. 1. Description et Interprétation des Unités Cartographiques*, 287 pp.; *Vol. 2. Description et Classification des Séries de Sol*, 115 p. + maps. Agriculture Canada, Direction Générale de la Recherche.

Nolin M.C. and Lamontagne L. 1991. Fiabilité d'une étude pédologique détaillée réalisée en terrain plat. *Can. J. Soil Sci.* **71**: 339-353.

Nordt L.C., Jacob J.S. and Wilding L.P. 1991. Quantifying map unit composition for quality control in soil survey. In *Spatial Variabilities of Soils and Landforms*. SSSA Special Publication No. 28. Madison WI, USA. pp. 183-197.

Norris J.M. 1971. The application of multivariate analysis to soil studies. I. Grouping of soils using different properties. *J. Soil Sci.* **22**: 69-80.

O.Q.S. 1988. *Manuel de l'Observatoire de la Qualité des Sols*. Projet, Ministère de l'Environnement, Paris. 100 p.

Odeh I.O.A., McBratney A.B. and Chittleborough D.J. 1992. Soil pattern recognition with fuzzy-c-means: application to classification and soil-landform interrelationships. *SSSAJ*, **56**: 505-516.

OFEFP, OFR & SVI, 1993. *Informations Concernant l'Étude de l'Impact sur l'Environnement (EIE), EIE et Infrastructure Routières, Guide pour l'Établissement de Rapports d'Impacts*. Prepared by l'Office Fédéral de l'Environnement, des Forêts et du Paysage, l'Office Fédéral des Routes and l'Association Suisse des Ingénieurs en Transports. 150 p.

Oldeman L.R. and van Engelen V.W.P. 1993. A World Soils and Terrain Digital Database (SOTER). An Improved Assessment of Land Resources. *Geoderma*, **60**: 309-325.

Oldeman L.R., Hakkeling R.T.A and Sombroeck W.G., 1990. *World Map of the Status of Human-Induced Soil Degradation*. ISRIC-UNEP, Wageningen, Netherlands. 27 p.

Olivier M.A and Webster R. 1989. Geostatistically constrained multivariate classification. *Geostatistics*, **1**: 383-395.

Olson C.G. and Doolittle J.A. 1985. Geophysical techniques for reconnaissance investigations of soils and surficial deposits in mountainous terrains. *SSSAJ*, **49**: 1490-1498.

ORSTOM. 1985. *Répertoire des Cartes 1946-1984*. Éditions de l'ORSTOM. Paris. 89 p.

ORSTOM. 1995. *Répertoire des Cartes 1985-1995*. Éditions de l'ORSTOM. Paris. 33 p.

Ototzky P. 1900. *Guide Scientifique Sommaire de la Section Pédologique Russe à l'Exposition Universelle de Paris* Ministère de l'Agriculture et des Domaines, St-Petersburg. 23 p.

Ouattara S. 1980. *Étude Critique d'un Banque de Données (STIPA). Adaptation aux Conditions Africaines*. DEA Option Pédologie, USTL-ENSAM. 45 p. + appendices.

Ozenda P. 1986. *La Cartographie Écologique et ses Applications*. Masson. 160 p.

Party J.P. 1982. *Interrelations Sol-Faune en Milieux à Fortes Contraintes. Observations, Revue Bibliographique, Perspectives*. DEA ENSA-INRA-Montpellier. 76 p. + appendices.

Party J.P., Revol P., Duchaufour H. and Thaler A. 1986. *Les Sols et les Contraintes Agricoles du Canton de St Symphorien de Lay (Loire)*. Contrat de Pays, Région Rhône-Alpes. 78 pp. + map on 1/25,000 + appendices.

Pavat J.L. 1986. *Contribution à l'Étude de la Ressemblance entre Types de Sols. Application aux Secteurs de Référence*. DEA, ENSAM-USTL. 75 p. + appendices.

Pedro G. 1984a. La pédologie cent ans après la parution du 'Tchernozem russe' de B.B. Dokoutchaiev (1883-1983). *SdS*, **2**: 81-92.

Pedro G. 1984b. L'association française pour l'étude du sol. Son rôle dans le développement de la Science du Sol en France (1934-1984). In *Livre Jubilaire du Cinquantenaire de l'AFES*. pp. 19-40.

Pedro G. 1989a. L'approche spatiale en pédologie. Fondements de la connaissance des sols dans le milieu naturel. Réflexions liminaires. *SdS*, **27**: 287-300.

Pedro G. 1989b. Présentation d'ouvrage: Carte de France de l'hydromorphie à l'échelle des petites régions naturelles par Ph. Lagacherie. *C.R. Acad. Agric. Fr.* **75**: 81-83.

Pellerin J. and de Queiroz-Neto J.P. 1992. Relations entre la distribution des sols, les formes et l'évolution géomorphologique du relief dans la haute vallée du Rio do Peixe (État de Sao-Paulo, Brésil), *SdS*, **30**(3): 133-147.

Pereira J.M.C. and Duckstein L. 1993. A multiple criteria decision-making approach to GIS-based land suitability evaluation. *Int.J. Geogr. Inf. Systems*, **7**: 407-424.

Phillips J.D. 1998. On the relations between complex systems and the factorial model of soil formation (with discussion). *Geoderma*, **86**: 1-42.

Pillet G. and Longet R. 1989. *Les Sols Faciles à Perdre et Difficiles à Regagner*. Georg Eshel, Geneva. 136 p.

Platou S.W., Norr A.H. and Madsen H.B. 1989. Digitization of the EC Soil Map. In *Proc. Int. Workshop Computerization Land Use Data*, Pisa 1987. CEC-DGVI, EUR 11151-en. pp. 12-24.

Poncet Y. and Lalau-Keraly A. 1984. *Cartographie et Développement*. Groupe de 'DECADE', Ministère de la Coopération, France. 181 p.

Pontius R.G. Jr. 2000. Quantification error versus location error in comparison of categorial maps. *PERS*, **66**: 1011-1016.

Pourgaton M. 1977. *Introduction à l'Étude Statistique de la Notion de Province Pédologique*. DEA ENSA Montpellier. 42 p.

Prélaz-Droux R. 1995. *Système d'Information et Gestion du Territoire*. Presses Polytechniques et Universitaires Romandes. 156 p. + appendices.

Prélaz-Droux R. and Musy A. 1991. *Revalorisation des Matériaux d'Excavation de Tunnel: le Cas du Rehaussement des Terreni Carcale, au Tessin*. Société Suisse de Mécanique des Sols et des Roches, Doc. 124. pp. 3-8.

Prélaz-Droux R. and Musy A. 1992. Aménagement rural et développement régional: le cas du rehaussement des Terreni Carcale, en Suisse. *C.R. Acad. Agric. Fr.*, **78**(4): 57-73.

Proctor M.E., Siddons P.A., Jones R.J.A., Bellamy P.H. and Keay C.A. 1998. LandIS—a land information system for the UK. In (H.J. Heineke, W. Eckelmann, A.J. Thomasson, R.J.A. Jones, L. Montanarella and B. Buckley, eds.). *Land Information Systems. Developments for Planning the Sustainable Use of Land Resources*. EUR-17729-en. pp. 219-233.

Purnell M.F. 1994. Soil survey information supply and demand: international policies and stimulation programmes. In (J.A. Zinck, ed.) *Soil Survey: Perspectives and Strategies for the 21st Century*, ITC publication No. 21. pp. 30-35.

Pury Ph. (de) and Dupasquier P. 1993. *Étude Détaillée des Sols sur l'Emprise de la N16-Secteur 06*. Doc. Int. Ponts et Chaussées, Canton du Jura.

Pury Ph. (de) and Dupasquier P. 1995. Sols. In *Projet Définitif, Étude d'Impact sur l'Environnement*. Routes Nationales Suisses, N16, Section 3, Doc. Int. Ponts et Chaussées. pp. 145-158.

Puterski R., Carter J.A., Hewitt M.J., Stone H.F, Fisher L.T. and Slonecker E.T. 1990. GPS Technology and its Application in Environmental programs. GIS Technical Memorandum 3. In *Remote Sensing Thematic Accuracy Assessment: a Compendium*. Am. Soc. Photogrammetry and Remote Sensing. pp. 172-234.

Ragg J.M. 1977. The recording and organization of soil filed data for computer areal mapping. *Geoderma*, **19**: 81-89.

Raunet M. 1989. Les terroirs rizicoles des Hautes-terres de Madagascar: environnements physiques et aménagements. *L'Agronomie Tropicale*, **44**(2): 69-91.
Rhoades J.D. 1993. Electrical conductivity methods for measuring and mapping soil salinity. *Adv. Agron.* **49**: 201-251.
Richard L. 1975. Carte écologique des Alpes au 1/50,000. Feuilles de Cluses et Chamonix. *Documents Cartographie Écologique*, **16**: 65-96.
Risler E. 1898. *Géologie Agricole*. Vols. 1 to 4. Berger-Levrault et Cie, Paris.
Ritchie, J.T. and Otter S. 1984. Description and performance of CERES-Wheat: A user-oriented wheat yield model. In *ARS Wheat Yield Project. ARS-38*. Natl. Tech. Inf. Serv., Springfield, VA. pp. 159-175.
Robbez-Masson J.M. 1994. *Reconnaissance et Délimitation de Motifs d'Organisation Spatiale. Application à la Cartographie des Pédopaysages.* Thesis, ENSA Montpellier. 157 p. + appendices.
Robbez-Masson J.M., Barthes J.P., Bornand M., Falipou P. and Legros J.P. 2000. Bases de données pédologiques et systèmes d'informations géographiques. L'exemple de la Région Languedoc-Roussillon. *Forêt Méditerranéenne*, **21**(1): 88-98.
Robitaille A. 1989. *Cartographie des Districts Écologiques. Normes et Techniques.* Ministère des Forêts, Quebec. 111 p.
Robitaille A. 1992. *Cartographie des Districts Écologiques. Concepts, Objectifs, Méthodes et Documents Générés.* Document ER-3145, Ministère des Forêts, Service des Inventaires Forestiers, Quebec. 10 p.
Robitaille A. and Grondin P. 1992. *Guide sur l'Utilisation des Produits de la Cartographie des Districts Écologiques en Vue de l'Élaboration du Plan Général d'Aménagement Forestier.* Document Ministère des Forêts, Quebec. 51 p. + appendices.
Rogowski A.S. and Wolf J.K. 1994. Incorporating variability into soil map unit delineations. *SSSAJ*, **58**: 163-174.
Rompre M. 1994. Observatoire de la qualité des sols agricoles du Québec. *AGROSOL*, **7**(2): 3-7.
Roques T. 1990. *Estimation de la Granulométrie d'un Échantillon à Partir de l'Appréciation de la Texture et de la Granulométrie.* DEA USTL-ENSA Montpellier. 26 p. + appendices.
Rosenfield G.H. and Fitzpatrick K. 1986. A coefficient of agreement as a measure of thematic classification accuracy. *PERS*, **52**: 223-227.
Rounsevell M.D.A. 1993. A review of soil workability models and their limitations in temperate regions. *Soil Use and Management*, **9**(1): 15-21.
Rounsevell M.D.A., 1999. *Spatial Modelling of the Response and Adaptation of Soil and Land Use Systems to Climate Change—An Integrated Model to Predict European Land Use* (IMPEL). Final Report, Commission of the European Communities, DG-XII. 210 p.
Rounsevell M.D.A. and JONES R.J.A. 1993. A soil and agroclimatic model for estimating machinery work-days: the basic model and climatic sensitivity. *Soil Tillage Res.* **26**: 179-191.

Roy B. 1985. *Méthodologie Multicritère d'Aide à la Décision*. Economica, Paris. 423 p.

RP. 1995. *Référentiel Pédologique*. Éditions INRA. 332 p.

Ruellan A. and Dosso M. 1993. *Regards sur le Sol*. Foucher-AUPELF. 192 p.

Ruellan A., Dosso M. and FRITSH F., 1989. L'analyse structurale de la couverture pédologique. *SdS*, **27**: 319-334.

Schär U., Ryniker K., Schmid K., Haefeli Ch. and Rutsch R.F. 1971. *Atlas Géographique de la Suisse à 1/25,000. Feuille 1145, Bieler See*.

Schärlig A. 1985. *Décider sur Plusieurs Critères*. Presses Polytechniques et Universitaires Romandes. 304 p.

Schellentrager G.W. and DOOLITTLE J.A. 1991. Using systematic sampling to study regional variation of a soil map unit. In *Spatial Variabilities of Soils and Landforms*. SSSA Special Publication No. 28. Madison WI, USA. pp. 199-212.

Scholes R.J., Skole D. and Ingram J.S. 1994. *A Global Database of Soil Properties. Proposal for Implementation*. Report of the Global Soil Task Group, IGBP-DIS. 30 p.

Schowengerdt, R.A. 1983. *Techniques for Image Processing and Classification in Remote-Sensing*. Academic Press, New York. 249 p.

Schulze D.G., Nagel J.L., van Scoyoc G.E., Henderson T.L., Baumgardner M.F. and Stott D.E. 1993. Significance of organic matter in determining soil colors. In *Soil Color*. SSSA Special Publication No. 31. 159 p.

Schwartz C. and Douay F. 1992. Établissement d'une clé de détermination des types de sols (exemple de la Flandre intérieure). Abstract. *Troisièmes Journées Nat. AFES*. Lausanne.

Schwartz C., Walter C., Claudot B., Bouédo Th. and Aurousseau P. 1997. Synthèse nationale des analyses de terre réalisées entre 1990 et 1994. I. Constitution d'une banque de données cantonale. *EGS*, **4**(3): 191-204.

Sede M.H. (de), Prélaz-Droux R., Claramunt C. and Vidale L. 1991. Un système d'information environnementale à référence spatiale (SIERS) pour la gestion globale de l'environnement. In (P. Buche, S. Lardon and D. King, eds.). *Gestion de l'Espace Rural et SIG*. Séminaire INRA. pp. 53-65.

SESCPF. 1991. *Evaluation Spatiale de la Sensibilité à l'Érosion Hydrique des Terres Agricoles de la Région Nord-Pas-de-Calais*. Doc. INRA-Conseil Régional. 207 p.

Sharpley A.N. and Williams J.R. (eds.). 1990. *EPIC Erosion/Productivity Impact Calculator. I. Model Documentation*. USDA Technical Bulletin No. 1768. 235 p.

Shea K.S. and McMaster R.B. 1989. Cartographic generalization in a digital environment: when and how to generalize. *Proc. AUTO-CARTO 9*, 56-67.

Simonneaux V. 1987. *Mesure de la ressemblance entre des groupes de sondages à la tarière et des profils de référence*. DEA INAPG-INRA. 61 p. + appendices.

Simos J. 1990. *Evaluer l'Impact sur l'Environnement*. Presses Polytechniques et Universitaires Romandes. 261 p.

Skidmore A.K. 1989. A comparison of techniques for calculating gradient and aspect from a gridded digital elevation model. *Int. J. Geogr. Inf. Systems*, **3**: 323-334.

Smyth A.J. and Dumanski J. 1993. *FESLM. An International Framework for Evaluating Sustainable Land Management*. FAO World Soil Resource Report No. 73. 74 p.

Smyth A.J. and Dumanski J. 1994. Progress towards an international framework for evaluating sustainable land management. *Trans. 15th World Soil Sci. Cong.* (Acapulco), **6a**: 373-378.

Soto P. and Bouche M.B. 1993. ECORDRE: base de données relationnelle pour toutes données écologiques y compris les activités humaines. In *Les Systèmes d'Information Environnementale*. ICAPE, Le Bourget du Lac. pp. 69-93.

SSEW (Soil Survey of England and Wales). 1965. *Soil Map of England and Wales*. Rothamsted Experimental Station, Harpenden, Herts.

Stehman S.V. 1992. Comparison of systematic and random sampling for estimating the accuracy of maps generated from remotely sensed data. *PERS*, **58**: 1343-1350.

Stolt M.H., Baker J.C. and Simpson T.W. 1991.Bucket auger modification for obtaining undisturbed samples of deep saprolite. *Soil Sci.* **151**: 179-182.

Story M. 1986. Accuracy assessment: a user's perspective. *PERS*, **52**: 397-399.

Tabi M. and Carrier D. 1993. Le Service des Sols, 50 ans au service de l'agriculture québécoise. *Agrosol*, **6**(1): 3-13.

Tandarich J.P., Darmody R.G., Follmer L.R. and Johnson D.L. 2002. Historical development of soil and weathering profile concepts from Europe to the United States of America. *SSSAJ*, **66**: 335-346.

Tardieu H., Rochfeld A. and Colletti R. 1987. *La Méthode Merise. Principes et Outils*. Vol. 1. Les Éditions d'Organisation.

Tardy Y. 1993. *Pétrologie des Latérites et des Sols Tropicaux*. Masson. 459 p.

Tavernier R. 1950. La cartographie des sols en Belgique. Comptes rendus de recherches. Travaux du comité pour l'établissement de la carte des sols et de la végétation de la Belgique. *IRSIA*, **4**: 25-35.

Teachman G.E., Benham E.C., Ditzler C.A and Ahrens R.J., 1995. Automated Pedon Description. *Agron. Abstr.* Annual Meetings, St Louis, Missouri. p 60.

Theocharopoulos S.P., Petrakis P.V. and Trikasoula A. 1997. Multivariate analysis of soil grid data as a soil classification and mapping tool: the case study of a homogeneous plain in Vagia Viota, Greece. *Geoderma*, **77**: 63-79.

Thomas A.L., King D., Dambrine E., Couturier A. and Roque J. 1999. Predicting soil classes with parameters derived from relief and geologic materials in a sandstone region of the Vosges mountains (Northeastern France). *Geoderma,* **90**: 291 - 305.

Thomasson A.J. and Jones R.J.A. 1989. Computer mapping of soil trafficability in the U.K. In *Computerization of Land Use Data.* Commission of the European Communities. EUR-11151-en. pp. 97-109.

Tomlinson R.F. 1987. Current and potential uses of geographical information systems. The North American Experience. *Int. J. Geogr. Inf. Systems,* **1**: 203-218.

Triantafilis J. and McBratney A.B. 1993. *Application of continuous Methods of Soil Classification and Land Suitability Assessment in the Lower Namoi Valley.* Divisional Report No. 121. CSIRO Division of Soils. 172 p.

Urbano G. 1994. *Le Programme Inventaire Gestion et Conservation des Sols. Bilan Avril 1990 – Avril 1994.* Ministère de l'Agriculture-DERF. 17 pp.

USDA. 1998. *Keys to Soil Taxonomy,* 8th edition. Natural Resources Conservation Service. 327 p.

Utset A., Lopez T. and Diaz M. 2000. A comparison of soil maps, kriging and a combined method for spatially predicting bulk density and field capacity of ferralsols in the Havana-Matanzas Plain. *Geoderma,* **96**: 199-213.

Van den Driessche R. and Garcia-Gomez A.M. 1972. Distances non paramétriques entre profils. *Rev. Écol. Bio. Sol,* **9**(4): 617-628.

Van den Driessche R, Garcia-Gomez A., Giey A. and Aubry A.M. 1975. POSEIDON, procédure opérationnelle en statistique et informatique pour données en langage naturel. *Cah. ORSTOM, Série Pédol.,* **13**(3/4): 507-510.

Vandendriessche H., Hendrickx G. and Bries J. 1993. Soil fertility and adjusted fertilizer recommendations for arable land and grassland in Belgium: a review for the period 1989-1991. *Bull. Rech. Agron. Gembloux,* **28**(2-3): 377-391.

Van Koninckxloo M. and Huart J. 1994. Valorisation de données pédologiques par la confection de cartes de richesse. Summary. *Quatrièmes Journées Nationales AFES,* Lille. 52 p.

Van Niel T.G. and McVicar T.R. 2002. Experimental evaluation of positional accuracy estimates from a linear network using point- and line-based methods. *Int. J. Geogr. Inf. Sci.* **16**: 455-473.

Van Orshoven J., Maes J., Vereecken H., Feyen J. and Dudal R. 1988. A structured database of Belgian soil profile data. *Pedologie, Ghent,* **38**(2): 191-206.

Van Orshoven J., Vandenbroucke D., Cammaer R. and Feyen J. 1991. Soil mapping in Belgium. In *Soil and Groundwater Report I. Soil Survey—a Basis for European Soil Protection.* Commission of the European Communities. EUR-13340-en. pp. 1-5.

Van Waveren E.J. and Bos A.B. 1988. *ISIS: ISRIC Soil Information System. User Manual.* Technical Paper No. 15, ISRIC, Wageningen. 55 p.

Vaudour E. 2001. *Les Terroirs Viticoles. Analyse Spatiale et Relation avec la Qualité du Raisin.* Thesis, INA-PG. 343 p.
Vedy J.C. 1990. *Gestion et Conservation des Sols à Risques.* Course Notes. EPFL-IATE-Pédologie. 92 p.
Veregin H. 1989. Error modelling for the map overlay operation. In *Accuracy of Spatial Databases.* Taylor & Francis. pp. 3-18.
Vigouroux A., Berger J.F. and Bussi C. 1987. La sensibilité du pêcher au dépérissement bactérien en France: relations avec la nutrition. *Agronomie*, **7**: 483-495.
Villamayor F.P. and Huddleston J.H. 1991. New method for characterizing soil contrast. *SSSAJ*, **55**: 767-772.
Vink A.P.A., 1975. *Land Use in Advancing Agriculture.* Springer-Verlag, 394 p.
Voltz M. 1986. *Variabilité Spatiale des Propriétés Physiques du Sol en Milieu Alluvial. Essai de Cartographie Quantitative des Paramètres Hydrodynamiques.* Docteur-Ingénieur thesis, ENSA-Montpellier. 198 p.
Voltz M. and Webster R. 1990. A comparison of kriging, cubic splines and classification for predicting soil properties from sample information. *J. Soil Sci.* **41**: 473-490.
Voltz M., Lagacherie P. and Louchard X. 1997. Predicting soil properties over a region using sample information from a mapped reference area, *European J. Soil Sci.* **48**: 19-30.
Vossen P. 1992. Forecasting national crop yields of EC countries: the approach developed by the Agriculture Project. In *The Application of Remote Sensing to Agricultural Statistics.* Proceedings of Conference, Villa Carlotta, Belgirate, Italy, 1991. EUR-14262-en, JRC, Ispra. pp. 159-176.
Vossen P. and Meyer-Roux J. 1995. Crop monitoring and yield forecasting activities of the Mars Project. In *European Land Information Systems for Agro-Environmental Monitoring.* EUR-16232-en. pp. 11-29.
Wagner G., Mohr M.E., Sprengart J., Desaules A., Theocharopoulos S., Muntau H., Rehnert A., Lischer P. and Quevauviller Ph. 2000. *Comparative Evaluation of European Methods for Sampling and Sample Preparation of Soil.* Contract SMT4-CT96-2085, European Commission. EUR-19701-en. 207 p.
Walter C. 1990. *Estimation de Propriétés du Sol et Quantification de Leur Variabilité à Moyenne Échelle: Cartographie Pédologique et Géostatistique dans le Sud de l'Ille et Vilaine (France).* Thesis, Université de Paris VI. SDS 436. 172 p.
Walter C., Schwartz C., Claudot B., Bouédo Th. and Aurousseau P. 1997. Synthèse nationale des analyses de terre réalisées entre 1990 et 1994. II. Description statistique et cartographique de la variabilité des horizons de surface des sols cultivés. *EGS*, **4**(3): 205-220.
Warren S.D., Johnson M.O., Goran W.D and Diersing V.E. 1990. An automated, objective procedure for selecting representative field sample sites. *PERS*, **26**: 333-335.

Webster R. 2000. Is soil variation random? *Geoderma*, **97**: 149-163.
Webster R. and Olivier M.A. 1990. *Statistical Methods in Soil and Land Resource Survey*. Oxford University Press. 316 p.
Wielemaker W.G., de Bruin S., Epema G.F. and Veldkamp A. 2001. Significance and application of the multi-hierarchical land system in soil mapping. *Catena*, **43**: 15-34.
Wilding L.P. and Drees L.R. 1983. Spatial Variability in Pedology. In (L.P. Wilding, N.E. Smeck and G.F. Hall, eds.). *Pedogenesis and Soil Taxonomy. Vol. 1. Concepts and Interactions*. Elsevier, Wageningen. pp. 83-116.
Williot B., 1995. Variabilité spatiale et risques d'erreurs dans l'analyse des horizons holorganiques forestiers. *EGS*, **2**: 73-83.
Wischmeier W.M. and Smith D.D. 1978. *Predicting Rainfall Erosion Losses. A Guide to Conservation Planning*. Agric. Handbk No. 537, US Dept. Agric., Washington DC. 58 p.
Woodcock C.E. and Gopal S. 2000. Fuzzy set theory and thematic maps: accuracy assessment and area estimation. *Int. J. Geogr. Inf. Sci.* **14**: 153-172.
Yaalon D.H. 1989. The earliest soil maps and their logic. *Newsletter, ISSS Working Group History, Philosophy Sociology Soil Sci.* pp. 2-7.
Young F.J., Maatta J.M. and Hammer R.D. 1991. Confidence interval for soil properties within map units. In *Spatial Variabilities of Soils and Landforms*. SSSA Special Publication No. 28: 213-229.
Young F.J., Hammer R.D. and Williams F. 1997. Estimation of map unit composition from transect data. *SSSAJ*, **61**: 854-861.
Young F.J., Hammer R.D. and Williams F. 1998. Evaluating central tendency and variance of soil properties within map units. *SSSAJ*, **62**: 1640-1646.
Zhang J. and Kirby R.P. 1999. Alternative criteria for defining fuzzy boundaries based on fuzzy classification of aerial photographs and satellite images. *PEERS*, **65**: 1379-1387.
Zhu A.X. 1997. A similarity model for representing soil spatial information. *Geoderma*, **77**: 217-242.
Zhu A.X., Band L., Vertessy R. and Dutton B. 1997. Derivation of soil properties using a Soil Land Inference Model (SoLIM). *SSSAJ*, **61**: 523-533.
Zhu A.X., Hudson B., Burt J., Lubich K. and Simonson D. 2001. Soil mapping using expert knowledge and fuzzy logic. *SSSAJ*, **65**: 1463-1472.
Zinck J.A. 1994. Introduction. In *Soil Survey: Perspectives and Strategies for the 21st Century*. FAO Publication (ITC) No. 21: 2-6.

INDEX

absolute positional error   181
accuracy : boundaries   184
accuracy : raster map   186
accuracy : vector map   190
Aerial Photographs   78
aerial photographs : ordering   80
Aerial Photographs : types   78
Agricultural geology   56
Agricultural planning   12
agronomic data   310
aligned systematic sampling   39
analyses : ordering   176
analysis : display   310
'analytical' display   62
anamorphosis   58
Applications of Mapping   8
Arbitrary sampling   166
Arc   203
Areal Error   187
attribute ambiguity   46
Auger holes   161, 166
Augers   88
azonal soils   21

Best Possible Map   186
bias   51
Block diagrams   8
Bottom-Up approach   294
Boundaries : accuracy   184
Boundaries : modelling   36
boundaries : simplification   268
Boundaries : validation   182
Box   3
Buffer   59, 210

Calculation of weights   285
CanSIS   340
capability map   274
cardinality   216
cartogram   57
categorial mapping   49
categorical map   2
Catenas   29
chloropleth map   2
choice function   284
chorological study   9
Chronosequences   30
Classification Success Index   190
Climatic Data   325
Climatic risks   302
Climosequences   30
closed legend   66
Coarse fragments   114
codification   125
Colour   116
Complete Aggregation   281, 284
Components   218
composite samples   157
Computer validation   144
concept soil   23
conceptual data model   217
Confusion matrix   186
Connectivity analysis   213
Consistency index   292
consociation   27
constraints   283
consumer risk   188
content : prediction   238
contiguity analysis : raster model   212

contiguity analysis : vector model   210
Contingency Table   186
contract for   73
convolution   212
Coordinates   6
CORINE   337
CORINE project   337
correct percentage   187
corridor of transition   48
Credibility   292
crisp limits   37, 48, 50
crop models   299

decision matrix   281
defuzzification   46
density of auger holes   167
description of map units   169
Detection of the boundary   163
Digital Elevation Model (DEM)   297
digitization   203
dissimilar soils   27, 193
dissimilarity   258
distance   258
documents   97
Dokoutchaiev   17
'DONESOL' system   217
Downgrading   288
draft map : construction   172
draft map : scale   67
draft maps   172
Drawbacks of thematic mapping   303

Ecodistricts   322
'ecological mapping'   320
Ecoprovinces   322
Ecoregions   322
Ecozones   322
edge filters   213
Effervescence   122
Electric sounding   93
Electromagnetic induction (EMI)   93
Elementary soil areal   2
elementary soil areas   209
elementary soil delineation   2
entity-relation model   216
Entry of auger-hole descriptions   168
epsilon band   184
Error Matrix   186

Error of Commission   187
Error of Omission   187
Estimation of cost   75, 78, 80
Estimation of Times   75
European Soil Bureau   336
Evaluation matrices   281
Example of a legend   177

Field Book For Describing And Sampling Soils   102
field hand books   100
field sheets   131
field texture   108
fineness of characterization   197
fractals   209
Free survey   49
functional models   277
Functioning of the landscape   83
fuzzy approach   46
Fuzzy k-mean   264
fuzzy logic   264
fuzzy set theory   38
fuzzy vector   46

generalization : procedure   244, 248
Geostatistical mapping   42
geostatistics   55
Global Pedon Data Base   334
Global Positioning System   91
glonous   100
glossaries   129
graphic information   3
grid mapping   39
Ground Penetrating Radar   92
Guidelines for Soil Profile Descriptions   101

heterogeneity of a map delineation   193
high-pass filters   213
Horizon boundaries   105
horizons   25
Hydrochloric acid   89

Identification of horizons   123
image soil   23
impurity   27
inaccuracy   270
inclusions   27

indeterminacy   270
inexactness of classification   46
Inset maps   7
intergrade soil   64
intergrade   64
Internet?   xiv
intrazonal soil   21
ISO standards   101
isolinear maps   2
isopleth maps   2

juxtaposition of soils   29

Kappa coefficient   189
Knife   90
kriging   44

land evaluation   273
Land Suitability Analysis   284
land system approach   30
Large scale   4
layout of auger holes   166
legend   175
legend : construction   175
limit : drawing   162
limit : treatment of the map   174
Limit : prediction   236
local variation   193
Locational error   181
Low-pass filters   213

Machinery work-days   302
macroscope'   xiii
Major Land Resource Areas (MLRA)   230, 325
map : countries of Europe   339
map : of Canada   339
map : of France   348
map : of Africa   357
map : of Australia   359
map : of Belgium   343
map : of great Britain   347
map : of Russia   361
map : of China   361
map : Switzerland   356
map legibility   174
map of : Europe   335
map of USA   342

Map overlay   205
Map unit   2
Map unit: definition   2
mapping criteria   183
mapping efficiency   69
Mapping errors   181
mapping teams   172
maps of the world   332
MARS project   338
Mechanistic models (process models)   277
mental model   165
metadata   228
'minute'   50
Misclassification Matrix   186
model : deterministic   277
model : diffuse   275, 281
model : statistical   276
monofactorial map   275
monoliths   22
Morphopedology   51
Morton numbers   211
Mottles   119
Moving window   212
Multicriteria Analysis   274
multiple attribute utility theory   282
Munsell Soil Color Charts   116

named soils   193
Nodes   203
Non-compensatory models   288
nugget variance   43
Number of profiles required   196

object soil   23
Observation density   75
open legend   66
Opening of profile pits   152
Organic matter   120
organizational scheme   148
Outclassing   290
outranking method   290
overall accuracy   186

pairwise comparison matrix   285
parametric system   284
partial aggregation   290
Particle-size distribution   107

pedon   26
pedotransfer functions   276
pedotransfer rules   276
Perkal's epsilon error band   184
photo-interpretation   82, 84
physiographic mapping   30
Phytoecology   53
Pickaxe   89
pixels   202
polygon attribute table   204
Polygon   203
Potential error   190
pre-zoning   181
probability vectors   47
process model   148
producer risk   188
producer's accuracy   187
profile   26
Profile : representativeness   197
profile description   100, 103, 152
Profile drawing   146
profile model   25
Profiles : management of the list   173
project   337
projection system   6
purity   192

quadtree   211
Quality Checks   142, 181
quality checks : on description   142
quality checks : on drawings   181
quantification error   187

Radiomagnetotelluric determinations   94
Random sampling   166
range   43
raster mode   211
raster model   202
rating system   284
raw score   282
reference area (RA)   34
reference profile   180
relational data model   217
Relational Database Management System (RDBMS)   215

relative positional error   181
remote sensing   54, 86
resistivity meter   93
Risks of erosion   302
run-length codes   211

sample : size   187
sample areas   254
Sample collection   155
sample size   187, 195
Scale   4
Scales   67
score : raw   282
score : standardized   282
'sectons'   39
segmentation algorithm   248
Segmentation of Space   241, 279
Seismic refraction   94
semantic information   3
semi-variance   43
series mapping   26
sill variance   43
similar soils   193
similarity between soil : calculation   258
similarity between soil : validation   266
Similarity between soil types   33
Similarity Pattern Index (SPI)   191
similarity   258
sketch   50
sliver polygons   206
small natural regions of soils   30
small scale   4
Soil association   29
Soil complex   29
Soil data base : Canada   232
Soil data base : Netherland   232
Soil data base : Europe   336
Soil data base : France   222
soil data base : world   334
Soil data bases : USA   229
Soil description   100
soil factor equation   18
Soil Landscape Hierarchical Model   32
Soil Landscape   322
soil map of Europe   22
Soil Survey Manual   102
Soil unit   3
soil-landscape laws   248

soil-landscape paradigm   18
soil-landscape unit   30
'soilscape' models   30
Solar-energy map   297
SOTER   232, 334
Spatial analysis   210
Spatial representation   57
Standard mapping symbols   65
standardized score   282
state factor model   18
statistical mapping   59
Statistical models   276
STIPA-2000 system   137, 223
stochastic models   277
Strata   218
Stratified random sampling   149
Stratified sampling   148, 166
structural analysis   27
suitability   11
suitability map   275
synthetic display   62
Systematic sampling   166

tabular data   3
target-property model   276
taxadjuncts   63
teams   96
texture   107
texture class   108
The profile model   25
thematic, attribute or descriptive error   190

thematic mapping   273
thematic mapping : advantages   306
Tomograms   61
Top-Down approach   294
Topological model   204
toposequences   30
trafficability   302
transects   166
transitional zone   48
triangular texture diagram   108
type profile   180
Types of boundary   162

Undifferentiated soil group   29
undisturbed samples   156
user's accuracy   187
'utility functions'   282

variograms   43
vector data model   202
vector system   203
Vergière cube method   156
Vertices   204
veto threshold   293
virtual scale   68, 69
vulnerability map   274

Weight of sample   155
workability   302

Zonal soil   20